Comprehensive Analytical Chemistry
Volume 70

Monitoring of Air Pollutants: Sampling, Sample Preparation and Analytical Techniques

Advisory Board

Comprehensive Analytical Chemistry

Volume 70

Monitoring of Air Pollutants: Sampling, Sample Preparation and Analytical Techniques

Edited by

Patricia B.C. Forbes
University of Pretoria, Pretoria

ELSEVIER

AMSTERDAM • BOSTON • HEIDELBERG • LONDON
NEW YORK • OXFORD • PARIS • SAN DIEGO
SAN FRANCISCO • SINGAPORE • SYDNEY • TOKYO

Elsevier
Radarweg 29, PO Box 211, 1000 AE Amsterdam, Netherlands
The Boulevard, Langford Lane, Kidlington, Oxford OX5 1GB, UK
225 Wyman Street, Waltham, MA 02451, USA

ISBN: 978-0-444-63553-2
ISSN: 0166-526X

Contents

Contributors to Volume 70 xiii
Series Editor's Preface xv
Preface xvii

Section I
Introduction

1. **Perspectives on the Monitoring of Air Pollutants**

Patricia B.C. Forbes

1. **Background to Air Pollution** 3
2. **Potential Effects of Air Pollution** 4
3. **The Importance of Quality Control in Environmental Analyses** 5
4. **Statistical Tools for Environmental Monitoring** 7
5. **The Scope of the Book** 8

References 8

Section II
Sampling Air Pollutants with Associated Sample Preparation Techniques

2. **Passive Diffusion Sampling Devices for Monitoring Ambient Air Concentrations**

J.J. Pienaar, Johan Paul Beukes, Pieter Gideon Van Zyl, C.M.B. Lehmann, J. Aherne

1. **Introduction** 14
2. **Basic Principles** 15
   2.1 Types of Passive Sampling Devices 15
   2.2 Governing Equations 20
3. **Passive Sampler Network Design** 25
   3.1 Selection and Quantity of Sampling Locations 27
   3.2 Sampling Frequency and Duration 28
   3.3 Field Siting Criteria 28
   3.4 Collocation with Reference Measurements 30

4. **Methodology** 33
4.1 Network Operations 34
4.2 Laboratory Methods for Sampling of Inorganic Gases 35
4.3 Laboratory Methods for Sampling of Organic Gases 39
5. **Quality Assurance and Quality Control** 41
5.1 Laboratory QA and QC 41
5.2 Field QA and QC 42
5.3 Comparison with Reference Methods 42
5.4 Data Validation and Reporting 42
6. **Case Studies** 43
6.1 Mapping Regional Pollutant Concentrations 43
6.2 Determining Long-Term Ammonia Trends: The NADP/AMoN 46
7. **Conclusions** 49
References 50

3. **Biomonitors**

Patricia B.C. Forbes, Leandri van der Wat, Eve M. Kroukamp

1. **Introduction** 54
2. **Lichens** 56
2.1 Introduction 56
2.2 Lichens as Biomonitors of Inorganic Air Pollutants 58
2.3 Lichens as Biomonitors of Organic Air Pollutants 59
3. **Mosses** 65
3.1 Introduction 65
3.2 Mosses as Biomonitors of Air Pollutants 65
4. **Conifer Needles** 67
4.1 Introduction 67
4.2 Conifer Needles as Biomonitors of Organic Air Pollutants 73
4.3 Conifer Needles as Biomonitors of Inorganic Air Pollutants 74
5. **Tree Bark** 76
5.1 Introduction 76
5.2 Tree Bark as a Biomonitor of Air Pollutants 76
6. **Other Plants** 78
7. **Sampling Considerations** 78
7.1 Sampling of Biomonitors for Organic Analytes 83
7.2 Sampling of Biomonitors for Inorganic Analytes 84
8. **Analyte Extraction Methods** 85
8.1 Soxhlet Extraction 85
8.2 Microwave-Assisted Extraction 87
8.3 Ultrasound-Assisted Solvent Extraction 88
8.4 QuEChERS 93
8.5 Sample Cleanup 95
9. **Analytical Techniques** 98
9.1 Organic Analytes 98
9.2 Inorganic Analytes 100

10. Conclusions, Challenges and Future Outlook 100
References 102

4. Ambient Air Sampling with Whole-Air, In-Field Concentration and Particulate Matter (PM) Methodologies

Jason S. Herrington

1. Introduction 109
2. Very Volatile Organic Compounds/Volatile Organic Compounds 112
   2.1 Whole-Air Sampling 112
3. Volatile Organic Compounds/Semi-volatile Organic Compounds 132
   3.1 In-field Concentration 132
4. Semi-volatile Organic Compounds/Particulate Matter 143
   4.1 Methodology 144
   4.2 Preparation 144
   4.3 Fractionators 144
   4.4 Applications 147
   4.5 SVOCs/Artifacts 147
   4.6 Storage 148
5. High Volume Sampling 148
6. Conclusions 150
References 151

5. Denuders

Patricia B.C. Forbes, Egmont R. Rohwer

1. Introduction 156
   1.1 Definitions 156
   1.2 The Need for Denuder-Based Sampling 156
   1.3 Reasons for Determining Gas and Particle-Associated Analyte Concentrations Separately 159
   1.4 Disadvantages of Conventional Filter/Adsorbent-Based Sampling Systems 159
2. Types of Denuders 162
3. Theory of Denudation 165
   3.1 Air Flow Considerations 165
   3.2 Particle Transmission Efficiency 166
   3.3 Gas Phase Collection Efficiency 168
4. Denuder Sampling Applications 171
   4.1 Denuders Used for the Sampling of Inorganic Analytes 171
   4.2 Denuders Used for the Sampling of Organic Analytes 173
5. Denuder Sampling Considerations 175

6. **Analyte Extraction Methods and Analytical Techniques Employed** 179
6.1 Extraction of Analytes, Pre-concentration and Use of Internal Standards 179
6.2 Analytical Techniques Generally Employed 179
7. **Challenges and Future Outlook with Respect to Denuder Sampling of Air Pollutants** 179
References 180

6. Automated Continuous Air Monitoring

Johan Paul Beukes, Andrew Derick Venter, Miroslav Josipovic, Pieter Gideon Van Zyl, Ville Vakkari, Kerneels Jaars, Marcell Dunn, Lauri Laakso

1. **Introduction** 183
2. **Measurement Motives and the Consequences Thereof** 184
2.1 Air Quality versus Scientific Research Interest 184
2.2 Selection of Monitoring Devices 190
2.3 Measurement Accreditation 193
2.4 Site Selection 195
2.5 Data Logging Requirements 197
3. **Data Handling, Quality and Processing** 200
3.1 Working with Large Data Sets 200
3.2 Data Quality 201
3.3 Data Cleaning 203
4. **Conclusions** 204
References 206

7. Novel and Emerging Air-Sampling Devices

Jonathan Grandy, Saba Asl-Hariri, Janusz Pawliszyn

1. **Introduction to Emerging Microextraction Technologies** 209
2. **SPME Devices and Techniques** 210
2.1 Basic Fundamentals of Equilibrium-Based Microextraction Methods 211
2.2 Selection and Characteristics of SPME Sorbent Coatings for Air Sampling 212
2.3 Description of SPME Fibre Geometry and Basic Operation 215
3. **Calibration of SPME for On-site Sampling** 216
3.1 External Calibration of Gaseous Samples Using Portable Standard Gas Generating Vials 216
3.2 Interface Model Based on Diffusive Laws 218
3.3 Kinetic Calibration Techniques 219
4. **Needle Trap Devices** 221
4.1 Description of NTD Geometry and Extraction Mechanism 222

4.2 Active Sampling 227
4.3 Passive TWA-Based Sampling 228
4.4 Characterisation of Particulate Matter Using Needle Trap Sampling 229
4.5 Application of NT Sampling 231
5. Concluding Remarks 233
References 234

Section III
Analysis of Air Pollutants

8. Spectroscopic and Chromatographic Techniques and Methodologies for the Determination of Metals, Metalloids and Ions in Atmospheric Aerosols

Patricia Smichowski, Darío Gómez

1. Introduction 240
2. Physical Aspects of Atmospheric Particles 241
2.1 Size Distribution 242
2.2 Morphology 243
3. Chemical Profile of Aerosols 244
3.1 Chemical Composition 244
3.2 Chemical Markers of Emission Sources 245
4. Analytical Techniques 254
4.1 Spectroscopic Techniques for Elemental Analysis 255
4.2 Ion Chromatography for Anions and Cations 261
5. Concluding Remarks 263
References 264

9. Chromatographic Techniques for Organic Analytes

Pérola de Castro Vasconcellos, Gisele Olímpio da Rocha, Elina Bastos Caramão, Maria Elisabete Machado, Laiza Canielas Krause

1. Introduction to Chromatography 268
1.1 General Fundamental Concepts 268
1.2 Optimizing Resolution and Separation 272
1.3 Main Types of Chromatography 272
2. Sample Introduction 277
2.1 Split versus Splitless Liquid Injections 279
2.2 Liner Types 280
2.3 Thermal Desorption Injection 282
3. Stationary Phases Commonly Employed in the Chromatographic Analysis of Air Pollutants 283
3.1 Polar versus Nonpolar Stationary Phases 285
3.2 Analyte-Specific Stationary Phases 287

3.3 Orthogonality for GC × GC Separations 288
3.4 Developments in LC Stationary Phases 288
4. Chromatography Detectors Commonly Employed in Air Pollutant Analysis 289
4.1 General Requirements of Detectors for this Application 289
4.2 Flame Ionisation Detector 290
4.3 Electron Capture Detector 291
4.4 Nitrogen Phosphorus Detector 292
4.5 Photoionisation Detector 293
4.6 Mass Spectrometer (MS) Detector 294
5. Selected Applications of Chromatography in the Analysis of Air Pollutants 295
5.1 Persistent Organic Pollutants 297
5.2 Polycyclic Aromatic Hydrocarbons and Their Derivatives 298
5.3 Volatile Organic Compounds 301
6. Challenges and Future Outlook 304
References 304

10. Mass Spectrometry and Ion Mobility Spectrometry

Fatkhulla K. Tadjimukhamedov, Patricia B.C. Forbes

1. Introduction 311
2. Ionisation Methods 313
2.1 Electron Ionisation 313
2.2 Secondary Ionisation Methods 314
2.3 Laser Desorption/Ionisation 315
3. Methods of Sample Introduction 316
3.1 Thermal Desorption 316
3.2 Membrane Introduction Mass Spectrometry 317
4. Instrumentation 318
4.1 Mass Analysers 318
4.2 Gas Chromatography—Mass Spectrometry 319
4.3 Portable/Miniature Mass Spectrometers 320
4.4 Aerosol MS 321
4.5 Ion Mobility Spectrometry 323
4.6 Proton Transfer Reaction—Mass Spectrometry 325
5. Conclusions 326
References 326

11. Microscopic Single-Particle Analytical Methods for Aerosol Characterisation

Hong Geng, Dhrubajyoti Gupta, Chul-Un Ro

1. Introduction 332
2. Energy-Dispersive Electron Probe X-Ray Microanalysis 333
2.1 Fundamentals 333
2.2 Applications 334

3. **Scanning Transmission X-Ray Microscopy** 338
3.1 Fundamentals 338
3.2 Applications 338
4. **Molecular Vibrational Imaging Spectroscopies** 339
4.1 Fundamentals 339
4.2 Applications of ATR-FTIR Imaging for Characterisation of Individual Atmospheric Particles 340
4.3 Applications of RMS for Characterisation of Individual Atmospheric Particles 341
5. **Miscellaneous Methods** 342
5.1 Time-of-Flight Secondary Ion Mass Spectrometry 342
5.2 Atomic Force Microscopy 342
5.3 Optical Microscopy 343
6. **Application of Multiple Techniques for Characterisation of the Same Single Particles** 344
6.1 Combined Application of ED-EPMA and ATR-FTIR Imaging 345
6.2 Combined Application of ED-EPMA and RMS 355
6.3 Combined Application of ED-EPMA, ATR-FTIR Imaging and RMS 358
7. **Conclusions** 363
References 364

Section IV
Concluding Comments

12. **Concluding Comments and Future Outlook**

Patricia B.C. Forbes

References 371

Index 373

Contributors to Volume 70

J. Aherne, Environmental and Resource Studies, Trent University, Peterborough, ON, Canada

Saba Asl-Hariri, Department of Chemistry, University of Waterloo, Waterloo, ON, Canada

Johan Paul Beukes, Unit for Environmental Sciences and Management, Faculty of Natural Sciences, North-West University, Potchefstroom Campus, Potchefstroom, South Africa

Elina Bastos Caramão, Nucleo de Estudos Coloidais, Universidade Tiradentes – UNIT, Aracaju, Sergipe, Brazil; Institute of Chemistry, Federal University of Rio Grande do Sul – UFRGS, Porto Alegre, Rio Grande do Sul, Brazil

Gisele Olímpio da Rocha, Institute of Chemistry, Federal University of Bahia – UFBA, Salvador, Bahia, Brazil

Pérola de Castro Vasconcellos, Institute of Chemistry, University of São Paulo – USP, São Paulo, State of São Paulo, Brazil

Marcell Dunn, Unit for Environmental Sciences and Management, North-West University, Potchefstroom Campus, Potchefstroom, South Africa

Patricia B.C. Forbes, Department of Chemistry, Faculty of Natural and Agricultural Sciences, University of Pretoria, Pretoria, South Africa

Hong Geng, Institute of Environmental Science, Shanxi University, Taiyuan, China

Darío Gómez, Comisión Nacional de Energía Atómica, Gerencia Química, Buenos Aires, Argentina

Jonathan Grandy, Department of Chemistry, University of Waterloo, Waterloo, ON, Canada

Dhrubajyoti Gupta, Department of Chemistry, Inha University, Incheon, Korea

Jason S. Herrington, Innovations Group Restek Corporation, Bellefonte, PA, USA

Kerneels Jaars, Unit for Environmental Sciences and Management, North-West University, Potchefstroom Campus, Potchefstroom, South Africa

Miroslav Josipovic, Unit for Environmental Sciences and Management, North-West University, Potchefstroom Campus, Potchefstroom, South Africa

Laiza Canielas Krause, Nucleo de Estudos Coloidais, Universidade Tiradentes – UNIT, Aracaju, Sergipe, Brazil

Eve M. Kroukamp, Department of Chemistry, Faculty of Natural and Agricultural Sciences, University of Pretoria, Pretoria, South Africa; Spectrum Analytical Facility, University of Johannesburg, Johannesburg, South Africa

Lauri Laakso, Unit for Environmental Sciences and Management, North-West University, Potchefstroom Campus, Potchefstroom, South Africa; Finnish Meteorological Institute, Helsinki, Finland

C.M.B. Lehmann, National Atmospheric Deposition Program, Illinois State Water Survey, Prairie Research Institute, University of Illinois at Urbana-Champaign, Champaign, IL, USA

Maria Elisabete Machado, Institute of Chemistry, Federal University of Rio Grande do Sul – UFRGS, Porto Alegre, Rio Grande do Sul, Brazil

Janusz Pawliszyn, Department of Chemistry, University of Waterloo, Waterloo, ON, Canada

J.J. Pienaar, Unit for Environmental Research and Management, Faculty of Natural Sciences, North-West-University, Potchefstroom, South Africa

Chul-Un Ro, Department of Chemistry, Inha University, Incheon, Korea

Egmont R. Rohwer, Department of Chemistry, Faculty of Natural and Agricultural Sciences, University of Pretoria, Pretoria, South Africa

Patricia Smichowski, Comisión Nacional de Energía Atómica, Gerencia Química, Buenos Aires, Argentina; Consejo Nacional de Investigaciones Científicas y Técnicas (CONICET), Buenos Aires, Argentina

Fatkhulla K. Tadjimukhamedov, US Pharmacopeial Convention, Rockville, MD, USA

Ville Vakkari, Finnish Meteorological Institute, Helsinki, Finland

Leandri van der Wat, Department of Chemistry, Faculty of Natural and Agricultural Sciences, University of Pretoria, Pretoria, South Africa

Pieter Gideon Van Zyl, Unit for Environmental Sciences and Management, Faculty of Natural Sciences, North-West University, Potchefstroom Campus, Potchefstroom, South Africa

Andrew Derick Venter, Unit for Environmental Sciences and Management, North-West University, Potchefstroom Campus, Potchefstroom, South Africa

Series Editor's Preface

I am delighted to introduce this new CAC title edited by Patricia Forbes on *Monitoring of Air Pollutants*: Sampling, sample preparation and analytical techniques. This book is an excellent addition to the CAC series and it reinforces our interest to expand analytical chemistry to other fields, like air quality monitoring. Since the mid-1990s we had CAC books on environmental analytical chemistry applied to water, sediments and/or biological samples, but no one was focused on the challenges of analysing air pollutants.

The volume that you have now in your hands covers the main pillars of air pollutants monitoring in 12 chapters. Six chapters address sampling techniques. It is well recognised that sampling air pollutants requires specific instrumentation as compared to aquatic, solid or biological samples. Different air sampling devices, including biomonitoring with lichens or moss, or high volume and continuous monitoring systems are described here. The second part of the book has four chapters on spectroscopic, chromatographic, mass spectrometric as well as microscopic and surface analytical methodologies. Patricia, as editor, wrote the first and final chapters of this book reporting perspectives as well as the future outlook in this field.

The book offers a comprehensive view of the analytical methodologies for air pollutants. It is a useful tool for newcomers and practitioners that want to be introduced in this new field of research. But also can serve other disciplines, like environmental epidemiology needing air monitoring data to establish human population risks. Several knowledge's spheres like analytical chemistry, environmental chemistry and ecological risk assessment with human risk assessment are integrated within this book. The final message is a better understanding of the air pollution problems at global scale. In this respect, air pollution brings the issue to the public's attention, as the impacts on human health can create media interest and pressure for government response. As the frequency and intensity of high air pollution levels increases around the world, i.e., heavy traffic episodes at megacities, researchers should focus on creating reliable data that help governments reduce the burden of air pollution.

Finally I would like to thank especially Patricia for the amount of work, time and expertise devoted as editor of the book. I would like to acknowledge as well the various well-known authors for their contributions in compiling such a world-class and timely book for the CAC series.

D. Barceló
IDAEA-CSIC, Barcelona and ICRA, Girona, 2 September 2015
Editor in Chief of the Comprehensive Analytical Chemistry Series

Preface

The importance of monitoring pollutants in ambient air arises from the linkages which have been found between these air pollutants and both human health and environmental effects. In order to correctly elucidate such exposure and effect relationships and to understand temporal and spatial concentration fluctuations, the concentrations of the air pollutants need to be accurately determined. It is therefore vital that both sampling and analysis of these compounds are performed methodically, with due consideration to best practice and analytical quality control procedures.

In terms of monitoring ambient air pollutants, two options are possible: either the target analyte is identified prior to sampling, or the potential components in the air sample which are relevant to the outcome of the monitoring campaign are not known beforehand. In each case (targeted vs nontargeted monitoring, respectively), a different approach in terms of sampling and analysis will be appropriate in order to ensure that scientifically sound conclusions can be drawn from the results obtained.

The aim of the monitoring campaign will also impact on the choice of analytical technique, as the requirements for monitoring ambient air for legislative compliance reporting are typically well defined, for example, whilst research studies may allow for more flexibility in the choice of methodologies and for explorative analytical method development. In addition, pollutants may be emitted into the air from a range of natural and anthropogenic sources. Source apportionment is needed to understand the relative contribution of each source to the ambient air concentrations, and this may be facilitated by the use of sophisticated analytical protocols for atmospheric samples.

Analytical chemistry is a rapidly evolving field, with new technologies and improvements in existing techniques driving the science forward. This results in the potential to detect a wider range of compounds at ever decreasing levels, with more accuracy and selectivity. Consequently, the ambit of environmental pollutants of potential concern with respect to human health and the environment is fluid. In addition, the production and use of new compounds, as well as the development of new industrial processes and emission abatement processes for existing technologies, leads to changing focus in terms of which ambient air pollutants receive the most attention from the scientific community and regulators alike.

Analytical chemists working in this exciting field therefore need to assess which sampling, sample preparation and analytical methods are most appropriate for a particular application. This is no simple task, as there are numerous options available for each air pollutant. Besides the factors already mentioned, equipment availability and cost may also be important considerations in making such decisions. This book aims to assist readers in determining which option(s) are most appropriate by providing an in-depth resource on the monitoring of ambient air pollutants, covering the basic principles, recent developments and important applications in this field. Discussion of the pros and cons of the methods described, along with the inclusion of local, regional and global ambient air-monitoring case studies, will further assist readers in their choice of appropriate analytical tools.

This book is targeted at practitioners in the field of air quality monitoring, specifically environmental scientists (including analytical and environmental chemists), consultants, researchers and postgraduate students. We trust it will appeal to those in both academia and industry due to the inclusion of both fundamental and practical information.

Part I provides a brief introduction to the monitoring of air pollutants, whilst Part II (Chapters 2 to 7 inclusive) focuses on the sampling of air pollutants with associated sample preparation techniques. Chapters 2 and 3 describe passive sampling methods for ambient air pollutants, namely passive (or diffusive) sampling devices based on a sorbent medium; and biomonitors (including lichens, moss and conifer needles), respectively. These sampling approaches are cost-effective and do not require extensive monitoring equipment, therefore they find application in wide-scale monitoring campaigns and in the monitoring of remote areas, for example. Chapter 4 provides an overview of whole air sampling and in-field concentration of a range of ambient air pollutants, from very volatile organic compounds to semivolatile organic compounds and particulate matter. Many of these compounds are regulated therefore important references to standard methods are included. Chapter 5 focuses on the simultaneous sampling of gas and particle phase analytes, such as semivolatile organic air pollutants, using denuder technology. Such sampling is of importance as different phases of the same analyte lead to differences in environmental fate and human health effects. Chapter 6 discusses practical aspects relating to automated continuous ambient air monitoring, including site selection and data handling, whilst Chapter 7 describes novel and emerging air sampling devices, including solid phase microextraction and silicone-based sampling methods.

Part III of the book focuses on the analytical techniques employed in air pollutant monitoring. Chapter 8 deals with the analysis of inorganic analytes (metals, metalloids and ions) present in atmospheric aerosols using ion chromatography and a range of spectroscopic techniques. Chemical markers of emission sources are discussed with reference to geological minerals, marine aerosols, traffic emissions and secondary aerosols. Chapter 9 relates to gas and liquid chromatographic analyses of organic air pollutants, including sample introduction and stationary phase and detector options which are commonly utilised in air pollutant analysis. Specific reference is given to the analysis of persistent organic pollutants, polycyclic aromatic hydrocarbons and volatile organic compounds. Chapter 10 discusses mass spectrometric methods in terms of ionisation methods, sample introduction and different mass spectrometer technologies employed in air pollutant monitoring. Important developments in portable systems are also noted. Chapter 11 focuses on microscopic characterisation of aerosols and related recently developed single-particle analytical methods, where microscopy can provide information on physicochemical properties and compositional information of aerosols, useful in source apportionment studies. Case studies related to mineral, marine and urban particulate matter are included.

Finally, Chapter 12 (Part IV) of the book provides concluding comments and future outlook with respect to ambient air pollution monitoring.

I am indebted to those who contributed to this book; your expertise and enthusiasm has allowed for the creation of a comprehensive and useful text, and it was a pleasure working with you. I would like to thank the series editor, Damià Barceló Cullerés, for putting his trust in me to complete this volume and the publishing team from Elsevier for their assistance. I would also like to thank my family, Andrew and Juliet, for their support and patience during the months of preparation of this book.

Patricia Forbes

August 2015

Section I

Introduction

Chapter 1

Perspectives on the Monitoring of Air Pollutants

Patricia B.C. Forbes
Department of Chemistry, Faculty of Natural and Agricultural Sciences, University of Pretoria, Pretoria, South Africa
E-mail: patricia.forbes@up.ac.za

Chapter Outline

1. Background to Air Pollution 3
2. Potential Effects of Air Pollution 4
3. The Importance of Quality Control in Environmental Analyses 5
4. Statistical Tools for Environmental Monitoring 7
5. The Scope of the Book 8
References 8

1. BACKGROUND TO AIR POLLUTION

The atmosphere consists of a thin layer of gases surrounding the surface of the earth. The atmosphere is divided into several layers on the basis of temperature, where the troposphere extends from the earth's surface to an altitude of approximately 11 km, whilst the stratosphere is from 11 to approximately 50 km [1]. Atmospheric gases are primarily inorganic (nitrogen, oxygen, argon, carbon dioxide, water vapour, etc.); however, there are also organic species present (such as methane and other hydrocarbons). Air pollutants may be regarded as substances present in the atmosphere at levels which can potentially have a negative impact on human health or the environment. These pollutants may arise from natural sources, such as volcanoes, forest fires or sandstorms, or from anthropogenic activities. Many air pollutants are known to undergo long-range transport in the atmosphere; therefore, they may be detected far from their source.

Until well into the nineteenth century, air pollution was mainly considered to arise from emissions of particulate matter (soot or smoke) and sulphur dioxide, primarily from domestic combustion and industrial processes [2]. Additional air pollutants received attention following the popularisation of

Comprehensive Analytical Chemistry, Vol. 70. http://dx.doi.org/10.1016/bs.coac.2015.09.001

motor vehicles, due to exhaust emissions of nitrogen oxides, carbon monoxide, hydrocarbons and lead [2]. Atmospheric science has evolved as new industrial processes and chemicals are introduced, which may result in elevated levels of compounds in the atmosphere. The scope of compounds considered as air pollutants is therefore increasing, which is also due to developments in analytical techniques allowing for the detection of compounds at ever lower levels. The results of ongoing studies on the effects of chemical species on man and the environment also inform the choice of compounds targeted in air-monitoring campaigns, as do regulations which impose limits on the concentrations of certain air pollutants.

This book focuses on the monitoring of air pollutants in the troposphere and therefore describes analytical methods for target analytes which are relevant in this context. It is important to note that pollutants cycle in the environment; therefore an air pollutant may be transformed in the atmosphere into other species via photochemical reactions, for example, or be removed from the atmosphere via wet or dry deposition processes. As such, air pollutants may become contaminants of water systems or soil. Another important consideration is the impact of meteorology on air quality: many atmospheric processes and reactions may be affected by ambient temperature, wind speed, humidity, and solar radiation, for example. Local inversion conditions which form under stable atmospheric conditions may also lead to elevated concentrations of air pollutants. These meteorological parameters, as well as wind direction, are therefore often critical both in the choice of sampling location and in the interpretation of air pollutant monitoring results. Detailed means of measuring meteorological parameters are outside the scope of this book; however, typical instrumentation employed in this regard at continuous air monitoring stations is briefly covered in Chapter 6.

2. POTENTIAL EFFECTS OF AIR POLLUTION

Air pollutants may have a range of effects on humans and the environment, from diminished visibility to human health effects. A number of these are briefly introduced in this section.

Health effects may arise from acute or chronic exposure to an air pollutant or a mixture of pollutants, and the effects include respiratory illness, heart disease and carcinogenicity. Diesel engine exhaust emissions, for example, were declared carcinogenic to humans in 2012 due to evidence supporting that exposure to these emissions was associated with an increased risk for lung cancer [3]. Exposure to particulate matter has been associated with numerous health effects including heart disease [4,5] and respiratory symptoms [5]. These effects have been shown to be presented more frequently in vulnerable individuals, such as children [6] and those with asthma and diabetes [5]. It is largely the human health effects of air pollutants that has led to the promulgation and implementation of air quality legislation.

In addition, air pollution can cause a number of environmental effects, such as photochemical smog arising from secondary air pollutant formation upon emission of nitrogen oxides and hydrocarbons under stagnant air conditions with high solar radiation. Besides the human health impacts associated with smog, it also reduces visibility and damages materials and plants. The emission of acid-forming air pollutants, such as sulphur dioxide and nitrogen oxides, may lead to the formation of acid rain, which may result in acidification of water bodies, damage to crops and forests, and enhanced weathering of buildings, for example.

The global environmental effect of ozone depletion in the stratosphere arose from (largely the previous) emission of chlorofluorocarbon refrigerants, whilst global warming is an environmental topic that continues to receive attention, due to concerns regarding the numerous potential direct and indirect effects of elevated concentrations of greenhouse gases (particularly carbon dioxide) in the atmosphere. These global environmental issues have led to international conventions and treaties, such as the Montreal Protocol, the Rotterdam Convention and the Stockholm Convention, in order for collective efforts to be made to minimise emissions and potential impacts.

As previously mentioned, the cycling of air pollutants results in these compounds being distributed throughout the environment; therefore, air pollution can lead to contamination of water and land resources. Due to these numerous potential effects, pollution of the atmosphere incurs significant direct and indirect costs to society. The health and environmental effects of air pollutants are difficult to quantify due to the vast range of parameters which may have an impact thereon; thus these effects are the subject of debate, controversy and extensive study. Effective monitoring of air pollutants has a vital role to play in the study and alleviation of air pollution via the determination of ambient air pollutant levels, as well as geographical and temporal trends in these levels, pollutant sources, cause and effect mechanisms and so on.

3. THE IMPORTANCE OF QUALITY CONTROL IN ENVIRONMENTAL ANALYSES

In order to ensure that the results obtained in environmental analyses, including ambient air monitoring, are meaningful and fit for purpose, it is important that appropriate quality control procedures are implemented.

Commercial analytical laboratories which process environmental samples are usually required to be accredited to perform such analyses, which is especially important when the results thereof are for legislative compliance purposes. In research laboratories, accreditation is typically not appropriate; however, quality control measures should still be implemented, and new sampling and analytical techniques which these laboratories may develop should be compared to traditional or standard methods, as far as possible. The analysis of certified reference materials (CRMs), which have defined analyte concentrations, is considered best practice in assessing the performance of an analytical method; however, suitable CRMs may not be available for some ambient air sample matrices and analytes. CRMs are commercially available for particulate matter (dust) and biomonitors (such as lichens), for example. Inter-laboratory proficiency testing schemes provide a means of comparison between laboratories in terms of the results obtained for the same environmental sample analyses at different facilities.

It is important to regularly calibrate instrumentation prior to analysis, and this includes in-field instruments in addition to laboratory-based equipment. For real-time gas phase analysers, such as those described in Chapter 6, calibrant gas mixtures in gas cylinders are generally required. These can be purchased from suppliers at user-defined concentrations and compositions.

Method validation is a critical component of good laboratory practice and allows for the assessment of analytical method performance in terms of whether it is fit for purpose. Method validation involves the determination of a number of key method performance parameters: precision, accuracy, bias, selectivity, limit of detection, limit of quantitation, linearity and ruggedness. It should be noted that not all of these are relevant to every analytical method, as precision and limits of quantitation are not important in terms of qualitative analyses, for example. For a detailed description of analytical method validation and the design of experiments to determine these method performance parameters, the reader is referred to specialist texts on this topic, such as Ref. [7]. Here the performance parameters are briefly defined, as they are mentioned in a number of chapters of this book.

Precision relates to the closeness of or agreement between independent results obtained under defined conditions, and it is often expressed as relative standard deviation. Method precision determinations should include sample preparation steps in addition to instrumental analysis and should represent the range of samples expected for the particular application in terms of concentration range and sample type [7]. Accuracy provides an indication of how close the analytical result obtained is to the true value (such as that provided on a CRM certificate), whilst bias indicates whether the results are systematically higher or lower than the true value, which is the result of systematic error(s).

Selectivity is an important consideration when deciding on an appropriate method for a particular ambient air-monitoring application; therefore, it needs to be given sufficient attention during method development. Selectivity relates to how well a particular analytical method or detection technique can discriminate between analytes. Interferences can arise when analytes are structurally similar (such as the simultaneous determination of airborne planar polychlorinated biphenyls and polychlorinated dibenzo-*p*-dioxins by gas chromatography–mass spectrometry) or have physicochemical similarities. Interferences may also arise from the sample matrix or the sampling material (such as a sorbent) which has been employed. Potential interferences therefore need to be well defined to allow for the most appropriate method to be determined and for any potential uncertainties in the results thus obtained to be clearly understood.

It is necessary for a chosen analytical method to be able to detect (and quantify in the case of quantitative analyses) the analyte of interest down to a level which is required for the application. This may be legislative limit values or typical ambient levels expected from previous studies reported in the literature, for example. The limit of detection (LOD) is the minimum concentration of the target analyte which can be detected by the analytical method with a specified level of confidence, whilst the limit of quantitation (LOQ) is the lowest concentration of the analyte that can be determined with an acceptable level of uncertainty [7]. The LOD and LOQ are usually obtained from the standard deviation (s) of results obtained from replicate analysis of a blank sample (which does not contain the analyte of interest). A value of $3s$ is typically used for the LOD, whilst the LOQ is higher, usually $10s$ [8]. Before analytical results are reported, it must be ensured that the concentrations fall within the range over which the method has been demonstrated to be fit for purpose and typically within the linear range of the method.

Potential sources of error (gross, random and systematic errors) which may affect analytical results should be well understood and minimised where possible. The ruggedness of the method in terms of its ability to withstand small changes in conditions such as pH, temperature or flow rate also needs to be determined as appropriate, to prevent unintended variations in results.

Quality control is underpinned by statistics, which is briefly discussed in the following section.

4. STATISTICAL TOOLS FOR ENVIRONMENTAL MONITORING

The evaluation and interpretation of environmental monitoring data is often a complex and time-consuming task; however, it is imperative that it is done correctly to ensure that appropriate conclusions are drawn from the monitoring campaign or study. Often the results can have serious direct and indirect consequences for people (such as human health-related outcomes) and the environment (for example, an indication of degradation and negative impacts). But the consequences can be even more widespread, potentially impacting on public perception; company profitability; legislative compliance and thereby litigation with associated fines and jail sentences. It is usually very costly to try and reverse the impacts of incorrect environmental reporting, if this is even possible, as incorrect perceptions once formed may be difficult to change.

Environmental and analytical scientists therefore have a significant responsibility to report correctly and to present results objectively, ensuring the uncertainties and limitations in the results are clearly defined.

Statistics (including chemometrics) provides useful tools for the evaluation, interpretation and presentation of environmental data sets in a concise and understandable manner, as these data sets can be very large (refer to Chapter 6, which deals with continuous automated air monitoring, for example). Principal component analysis (PCA) is a useful multivariate data analysis tool for environmental monitoring, which can be used to extract dominant patterns from data matrices [9]. It is important that the user of these statistical tools determines the correct statistical procedure and understands what the results mean including the context in which they can be reported. Statistics should be considered right from the planning stage of a sampling campaign or project, to correctly determine a suitable sample size and representative sampling strategy. The reader is referred to two texts, which deal with the topic of statistics for analytical chemistry in detail, Refs [7,10] and to Ref. [11] regarding statistics related to environmental sampling and the analysis of environmental data.

5. THE SCOPE OF THE BOOK

This book provides a comprehensive reference text on air pollutant monitoring, in that it addresses experimental approaches to sampling and sample preparation (Part II), as well as analytical technologies (instrumental methods) (Part III), which are applicable to a wide variety of air pollutants. The book provides an in-depth resource on the monitoring of ambient air pollutants and covers basic principles, recent developments and important applications in this field.

Ambient air is described by the United States Environmental Protection Agency (US EPA) as 'the portion of the atmosphere external to buildings and breathed by the general public' [12]. The book therefore does not focus on stack sampling (industrial or emissions monitoring) of air pollutants. Occupational hygiene methods relating to air pollution and the assessment of indoor air quality are also outside the scope of this book. In addition, odour monitoring; radioactivity measurements of airborne nuclides; and remote sensing of atmospheric species are not covered. A number of these topics are mentioned in various chapters of the book, where the reader is often referred to comprehensive texts dealing with them. There are also a number of companion texts in the *Comprehensive Analytical Chemistry* series, which link to topics covered in this book which readers are referred to [13−16].

REFERENCES

[1] S.E. Manahan, Environmental Chemistry, seventh ed., Lewis Publishers, USA, 2000.
[2] D.M. Elsom, Atmospheric Pollution, a Global Problem, second ed., Blackwell, UK, 1992.
[3] IARC (International Agency for Research on Cancer), Diesel Engine Exhaust Carcinogenic, Press Release No. 213, 2012.
[4] A. Peters, Toxicology and Applied Pharmacology 207 (2005) S477−S482.
[5] A.D. Kappos, P. Bruckmann, T. Eikmann, N. Englert, U. Heinrich, P. Höppe, E. Koch, G.H.M. Krause, W.G. Kreyling, K. Rauchfuss, P. Rombout, V. Schultz-Klemp, W.R. Thiel,

H.-E. Wichmann, International Journal of Hygiene and Environmental Health 207 (2004) 399–407.
[6] M. Kampa, E. Castanas, Environmental Pollution 151 (2008) 362–367.
[7] S.L.R. Ellison, V.J. Barwick, T.J.D. Farrant, Practical Statistics for the Analytical Scientist, A Bench Guide, second ed., RSC Publishing, UK, 2009.
[8] V. Thomsen, D. Schatzlein, D. Mercuro, Spectroscopy 18 (12) (2003) 112–114.
[9] S. Wold, K. Esbensen, P. Geladi, Chemometrics and Intelligent Laboratory Systems 2 (1987) 37–52.
[10] J.N. Miller, J.C. Miller, Statistics and Chemometrics for Analytical Chemistry, sixth ed., Pearson, UK, 2010.
[11] W.W. Piegorsch, A.J. Bailer, Analyzing Environmental Data, Wiley, England, 2006.
[12] US EPA, http://www.epa.gov/airquality/airdata/ad_glossary.html, (accessed 22.08.15).
[13] L. Ramos (Ed.), Comprehensive Two Dimensional Gas Chromatography, Elsevier, Netherlands, 2009.
[14] R. Greenwood, G. Mills, B. Vrana (Eds.), Passive Sampling Techniques in Environmental Monitoring, Elsevier, Netherlands, 2007.
[15] J. Pawliszyn, Sampling and Sample Preparation in Field and Laboratory – Fundamentals and New Directions in Sample Preparation, Elsevier, Netherlands, 2002.
[16] D. Pérez-Bendito, S. Rubio, Environmental Analytical Chemistry, Elsevier, Netherlands, 1998.

Section II

Sampling Air Pollutants with Associated Sample Preparation Techniques

Chapter 2

Passive Diffusion Sampling Devices for Monitoring Ambient Air Concentrations

J.J. Pienaar,[1,*] Johan Paul Beukes,[1] Pieter Gideon Van Zyl,[1] C.M.B. Lehmann[2] and J. Aherne[3]

[1]*Unit for Environmental Research and Management, Faculty of Natural Sciences, North-West-University, Potchefstroom, South Africa;* [2]*National Atmospheric Deposition Program, Illinois State Water Survey, Prairie Research Institute, University of Illinois at Urbana-Champaign, Champaign, IL, USA;* [3]*Environmental and Resource Studies, Trent University, Peterborough, ON, Canada*

**Corresponding author: E-mail: kobus.pienaar@nwu.ac.za*

Chapter Outline

1. Introduction 14
2. Basic Principles 15
2.1 Types of Passive Sampling Devices 15
2.1.1 Samplers for Inorganic Gases 17
2.1.2 Samplers for Organic Gases 20
2.2 Governing Equations 20
3. Passive Sampler Network Design 25
3.1 Selection and Quantity of Sampling Locations 27
3.2 Sampling Frequency and Duration 28
3.3 Field Siting Criteria 28
3.4 Collocation with Reference Measurements 30
4. Methodology 33
4.1 Network Operations 34
4.2 Laboratory Methods for Sampling of Inorganic Gases 35
4.2.1 Sampler Preparation 35
4.2.2 Sampler Extraction and Analytical Preparation 36
4.2.3 Sample Analysis 37
4.2.4 Precautions and Known Interferences 37
4.3 Laboratory Methods for Sampling of Organic Gases 39
4.3.1 Sampler Preparation 39
4.3.2 Sampler Deployment 39

Comprehensive Analytical Chemistry, Vol. 70. http://dx.doi.org/10.1016/bs.coac.2015.09.002

4.3.3 Sampler Desorption and Analysis 39
4.3.4 Precautions and Known Interferences 41
5. Quality Assurance and Quality Control 41
5.1 Laboratory QA and QC 41
5.2 Field QA and QC 42
5.3 Comparison with Reference Methods 42
5.4 Data Validation and Reporting 42
6. Case Studies 43
6.1 Mapping Regional Pollutant Concentrations 43
6.1.1 Determining Exceedance of Critical Loads in South Africa 43
6.1.2 The DEBITS Network 44
6.2 Determining Long-Term Ammonia Trends: The NADP/AMoN 46
7. Conclusions 49
References 50

1. INTRODUCTION

Passive, also termed 'diffusive', samplers rely on passive diffusion of air pollutants to a sorbent medium without mechanical convection of air. They provide a cost-efficient method to measure gaseous air concentrations in regions where it is impractical to use active methods (e.g. filter packs or continuous monitors) owing to capital cost or infrastructure concerns. They are particularly useful in regional air quality studies that assess 'hotspots' and spatial patterns, which can ultimately guide placement of continuous monitoring instruments. This chapter focuses on sampling of ambient air using passive samplers. A number of key considerations for air quality monitoring are identified, including siting, sampling design and analytical methods.

The site operational requirements for passive air sampling are very basic, although their use necessitates an analytical laboratory facility for sampler preparation and analysis. The use of this type of sampler in ambient air quality studies is popular, owing to their simplicity and convenience in terms of deployment. They are small in size, of low mass, operate silently and do not require electricity or field calibration. As such, they can be deployed in remote or high-risk environments as they are less susceptible to damage or theft [1]. They can also be easily deployed by nonspecialists, which makes spatial and temporal collection of ambient air data possible [2,3].

The development of passive samplers can be traced back to 1853 when filter papers were impregnated with potassium iodide to measure ground-level ozone (O_3) [4]. The first reported attempt to quantitatively detect airborne pollutants using a passive sampler was patented by Gordon and Lowe in 1927 to capture carbon monoxide with palladium chloride salts [5], a method that was at most semi-quantitative. The first serious effort to apply quantitative diffusive sampling was made by Palmes and Gunnison (1973) whereby a mercuric chloride solution was used as a sorbent for capturing sulphur dioxide (SO_2), with analytical determination following the colourimetric method described by West and Gaeke [6,7].

The use of passive samplers for ambient air quality monitoring of inorganic gases was developed into a viable and practical option in the early 1970s with the 'Palmes Tube' diffusion passive sampler [6]. The commercial use of these samplers was further developed by Martin Ferm at the Swedish Environmental Institute [8]. This research showed that passive samplers could fulfil a much needed role in monitoring of selected air pollutants in remote or high-risk environments when supported by a well-equipped analytical laboratory and good quality control procedures.

The Palmes Tube was later developed for nitrogen dioxide (NO_2) monitoring; it consisted of a simple closed-end acrylic tube ~8 cm long with an internal diameter of 1 cm [9]. Nitrogen dioxide gas migrated into the open end by molecular diffusion and was collected at the closed end on a stainless-steel grid coated with triethanolamine. Gaseous NO_2 was trapped as nitrate (NO_3^-) and reduced to nitrite (NO_2^-) before being measured spectrophotometrically using sulphanilamide and naphthylethylenediamine as reactants.

The earliest ammonia (NH_3) sampler used an acid-impregnated glass fibre pad as the collection medium [10]; owing to high interference from volatile amines, the glass fibre diffusion screen at the front of the sampler was replaced with a charcoal-impregnated glass fibre filter pre-treated with alcoholic potassium hydroxide (KOH) containing 0.1% surfactant. The charcoal served to adsorb amines as they diffused into the sampler, while the KOH, aided by a wetting agent, prevented the charcoal from irreversibly adsorbing any NH_3 as this would result in an underestimation of ambient concentrations.

The development of passive O_3 monitors did not take place until the 1990s when Monn and Hangartner used a Palmes-type tube with a 1,2-di(4-pyridyl) ethylene adsorption medium which produced an ozonide upon reaction with atmospheric O_3 [11]. The aldehyde resulting from the hydrolysis of this compound was then measured spectrophotometrically. A detection limit of 3 $\mu g\ m^{-3}$ for a 1 week sampling period was reported, and measurements correlated well with continuous UV monitors.

The examples provided above are selected examples from a wide variety of available sampler technologies. Further information on the development and usage of diffusive samplers is provided in review articles [4,12,13].

2. BASIC PRINCIPLES

2.1 Types of Passive Sampling Devices

Passive air sampling is based on the laminar diffusion of a pollutant and its subsequent trapping by absorption or chemical reaction with a sorbent medium.

The sampling rate of diffusion passive samplers is governed by the geometry of the sampler. A large variety of passive sampler configurations are commercially available (Figure 1). Each configuration has advantages and disadvantages, and the selection of a certain type of passive sampler strongly depends on the purpose or goal of the study.

Irrespective of shape and size, all passive samplers exhibit the same basic physical components. Each sampler consists of an impregnated filter or molecular sieve medium that adsorbs the pollutant gas, and a diffusion barrier (either a gas-permeable membrane or diffusion path) with known physical characteristics (pore diameter and effective diffusion distance), which define an effective sampling rate. Assuming that the sorption to the medium is complete and irreversible, the sampling rate can be calculated from the cross-sectional area perpendicular to the transport direction, and the distance over which the gas has to diffuse (see Section 2.2). The sampling rate is directly proportional to its cross-sectional area and inversely proportional to its diffusion path length. It is essential that the transport occurs only by molecular diffusion and that no gas is lost to the walls of the sampler.

Quantitative information on atmospheric concentrations can only be obtained if the sampler is operated in a so-called linear uptake range (see Figure 2), where uptake of the pollutant on the sorbing medium is well below the point of saturation. This sorption curve is dependent on the geometry and configuration of the sampler, the air pollutant sampled and the sorbent type. These parameters must be determined in field and laboratory studies, and they are typically provided by the sampler manufacturers.

There are several differences between passive and active samplers that should be understood with respect to their suitability in environmental monitoring and use in air quality studies (Table 1). As compared with continuous monitors, passive samplers are more suited to chronic exposure assessments over wider regions.

A significant difference between active and passive samplers is in the response time. Active samplers obtain readings over a short timescale (seconds) with the average of a number of data points taken over a longer timescale (minutes to hours). In contrast, passive samplers provide a time-integrated

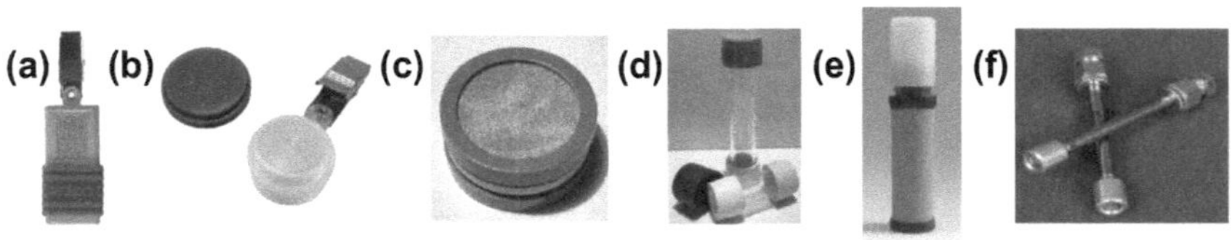

FIGURE 1 Examples of commercially available diffusion passive samplers: (a) SKC ULTRA passive sampler for organic vapours [14], (b) SKC passive sampler for nitrous oxide [14], *Images (a) and (b) provided with permission from SKC Inc.* (c) Ferm badge sampler for ammonia [15], (d) Gradko tube for nitrogen dioxide [16], (e) Radiello sampler for various gases [17] and (f) PerkinElmer sorbent tube for organic gases [18].

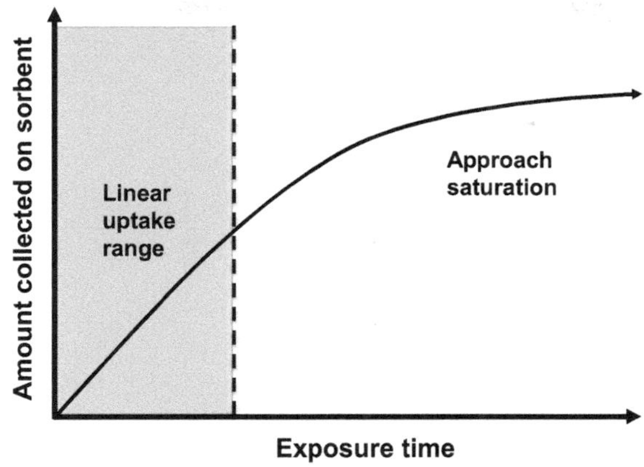

FIGURE 2 Illustration of an uptake curve for a passive diffusion sampler.

pollutant concentration over the entire exposure period. Obvious differences in the outputs of active and passive samplers are clear (Figure 3); while detailed information on pollution peaks cannot be obtained when using passive samplers, there is also no loss of data during the time period when the active sampler was not operational, as indicated by the interrupted signal. The long-term averages of the two methods are comparable (e.g. typically within $\pm 20\%$) if adequate calibration processes for both monitors have been followed.

2.1.1 Samplers for Inorganic Gases

The international Deposition of Biogeochemically Important Trace Species (DEBITS) programme of the International Global Atmospheric Chemistry (IGAC) project together with the support of the Global Atmospheric Watch (GAW) of the World Meteorology Organization (WMO) adopted passive diffusion samplers for monitoring inorganic gaseous species in tropical regions

TABLE 1 Differences between Active and Passive Air Monitors

Parameter	Active Air Monitor	Passive air monitor
Sampling timescale	5 min–1 h average	1 day–1 month average
Infrastructure requirements	Major: electric power, shelter, security	Minimal: no power, minimal field infrastructure
Human health assessment	Acute and chronic exposure; human health and welfare	Chronic exposure; human welfare
Network suitability	Individual monitoring site	Regional air quality survey

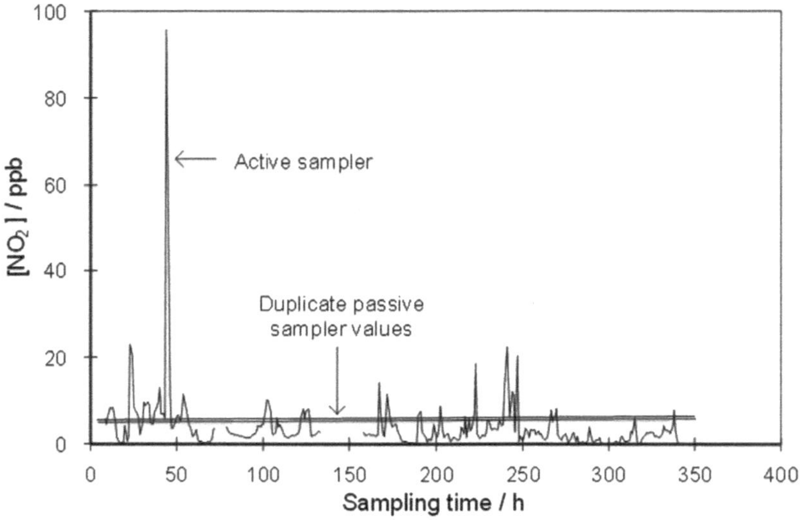

FIGURE 3 Comparison of active and passive sampler responses [28].

around the world [2,3,19–22]. The samplers and analytical procedures used in the DEBITS programme are described here to illustrate the principles and procedures of diffusion passive sampling. Similar methods are in use in other continental-scale air quality studies, including the National Atmospheric Deposition Program's Ammonia Monitoring Network (NADP/AMoN) [23].

In general, four types of passive sampler geometries are used for inorganic gases: badge [24], tube [8,9], radial [25] and cartridge [26,27]. The sampler sorption rate, concentration range and saturation capacity are important when selecting a particular sampler type. Radial and badge samplers have relatively short diffusion path lengths; as such they have high uptake rates and are more suited for short-term (e.g. daily to weekly mean) monitoring studies or sampling in regions with low ambient air concentrations. Tube samplers have longer diffusion path lengths and are more often used for long-term (weekly to monthly mean) studies or sampling in regions with high ambient air concentrations. The choice of a particular sampler is also influenced by prevailing environmental factors, including temperature and relative humidity.

The Ferm-type sampler is used extensively in ambient air quality studies (including DEBITS) owing to its simple design and analytical laboratory requirements. The theory and function of these samplers is described below. Other sampler configurations have similar physical characteristics and principles of operation. The Ferm-type sampler (Figure 4) consists of a polypropylene ring with a snap on cap on both sides, with the inlet cap removed during sampling. The impregnated filter is attached to one side of the sampler, on top of a snap-on cap. The inlet of the sampler has a 1 μm pore size PTFE (polytetrafluoroethylene or Teflon) filter to prevent convection by airflow and

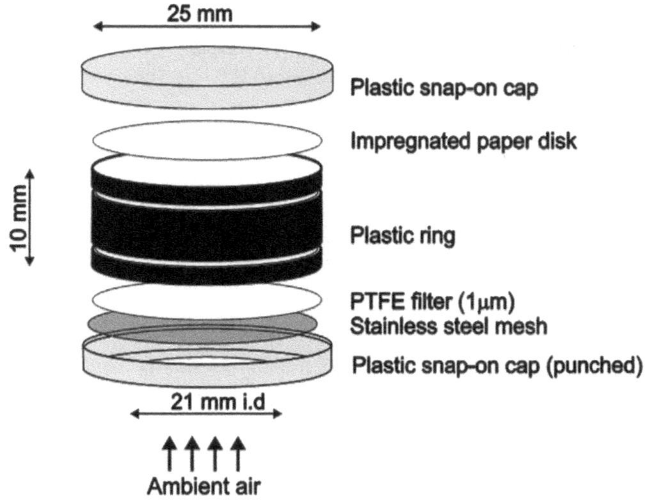

FIGURE 4 Schematic of the Ferm-type passive diffusive sampler used in the international Deposition of Biogeochemically Important Trace Species (DEBITS) network [8,28].

to prevent aerosols from impacting on the impregnated filter, behind a steel mesh to limit air turbulence. At the rear end of the sampler, the gases are trapped by a sorbent filter (typically a filter paper disk) that has been impregnated with a chemical specific to the pollutant of interest [8].

For inorganic gases, the sorbent filter reacts with the pollutant to form an ionic species (Table 2); basic solutions are used to trap acidic gases (SO_2 or NO_2), and acidic solutions are used to trap alkaline gases (e.g. NH_3). The capture of NO_2, SO_2 and NH_3 is achieved by the use of potassium iodide (KI), potassium carbonate (K_2CO_3) and $C_6H_8O_7{\cdot}H_2O$ (citric acid) absorbing solutions, respectively [8]. Ozone is sampled with the use of a sodium nitrite

TABLE 2 Pollutant Gas, Coating Solution, Analytical Reproducibility and Typical Detection Limits of Passive Samplers [3,29–31]

Pollutant Gas	Coating Solution in 50 ml Methanol	Reproducibility	Detection Limit
NO_2	KI or 0.5 g NaOH + 4.0 g NaI	16%	$0.10 \pm 0.05\ \mu g\ m^{-3}$
SO_2	K_2CO_3 or 0.5 g NaOH	11%	$0.30 \pm 0.10\ \mu g\ m^{-3}$
NH_3	1.0 g citric acid	10%	$0.10 \pm 0.06\ \mu g\ m^{-3}$
HNO_2	0.5 g NaOH	25%	$0.15 \pm 0.07\ \mu g\ m^{-3}$
O_3	$NaNO_2$		

($NaNO_2$) solution [29]. Subsequently, Ferm has reported using sodium iodide (NaI) and sodium hydroxide (NaOH) absorbing solutions for NO_2 and SO_2, respectively [30]. The samplers are extracted in the laboratory, and the amount of pollutant sorbed is quantitatively measured using standard analytical methods.

2.1.2 Samplers for Organic Gases

The physical principles of passive samplers for organic gases differ from those of inorganic gases in that pollutant capture is typically through physical adsorption rather than reaction. It is only in a very few applications where chemical reactions for capturing a pollutant of interest are desirable. In typical applications, tube samplers filled with an adsorption medium are used (Figure 5).

Monitoring of organic gases for air quality typically focuses on benzene, toluene and xylene (BTX) or benzene, toluene, ethyl benzene and xylene (BTEX) compounds owing to the very large number of organic compounds present in the atmosphere, which are mostly at rather low ambient concentrations. These compounds originate mostly from anthropogenic activities and are known to have adverse effects on human health.

The targeted compounds dictate the sorbent material, varying from polar to nonpolar compounds (see Table 3). The U.S. Environmental Protection Agency (EPA) and other organisations (e.g. the Health and Safety Laboratory, Health and Safety Executive, Sheffield, United Kingdom) have developed several methods (e.g. TO-1, TO-2, TO-14 and TO-17) for the determination of toxic organic compounds in ambient air. These methods describe thermal desorption systems as well as solid sorbent types that are available commercially. The sorbents can be used singly or in multisorbent packings. Multisorbent packings allow for increasing sorbent strength to facilitate a wider volatility range. The higher molecular weight compounds are retained on the least retentive sorbent, normally packed at the beginning of the tube, and the more volatile compounds are adsorbed further into the tube on stronger adsorbent materials. The higher molecular weight compounds never encounter the stronger adsorbents, thereby improving the efficiency of the thermal desorption process during analyses [33].

2.2 Governing Equations

The effective sampling rate of passive samplers is governed by Fick's first law of diffusion [34]. This law states that mass transport occurs from regions of high

FIGURE 5 Example of a commercially available diffusion passive sampler for adsorption of organic gases [32].

TABLE 3 Characteristics of Commercially Available Sorbents for Volatile and Semivolatile Organic Compounds

Sorbent	Volatility Range (bp = Boiling Point)	Maximum Desorption Temperature (°C)	Surface Area ($m^2 g^{-1}$)	Examples of Analytes
CarbotrapC®/ CarbopackC®	n-C_8 to n-C_{20}, alkyl benzene and aliphatic compounds	400	12	Alkyl benzenes, polychlorinated biphenyls (PCBs)
Tenax® TA	n-C_7 to n-C_{30}	350	35	Aromatics except benzene, nonpolar compounds (bp > 100 °C), such as aliphatic hydrocarbons from n-C_7 and less volatile polar compounds (bp > 150 °C)
Tenax® GR	bp 100 °C–350 °C n-C_7 to n-C_{30}	350	35	Alkyl benzenes, PCBs, and the same as for Tenax TA
Carbotrap®/ CarbopackB®	n-C_5 to n-C_{14}	400	100	Volatile organic compounds (VOCs), including ketones, aldehydes and all polar compounds within the volatility range specified
Chromosorb® 102	bp 50 °C–200 °C	250	350	VOCs including oxygenated compounds and haloforms less volatile than methylenechloride
Chromosorb® 106	bp 50 °C–200 °C	250	750	VOCs including hydrocarbons from n-C_5 to n-C_{12}; volatile oxygenated compounds

From EPA [33].

concentration to regions of low concentration, at a flux that is proportional to the concentration gradient (Eqn. (1)). The proportionality constant D is the diffusion coefficient of the target gas (see Table 4). This can be written as:

$$J = -D\frac{\mathrm{d}C}{\mathrm{d}L} \tag{1}$$

where

$J =$ diffusion flux of the gas in the direction of concentration gradient (μg m^{-2} s^{-1})
$D =$ diffusion coefficient of the gas (m^2 s^{-1}; see Table 4)
$C =$ concentration of the pollutant (μg m^{-3})
$L =$ effective diffusion path length (m)
$-\frac{\mathrm{d}C}{\mathrm{d}L} =$ instantaneous concentration gradient of pollutant in direction of airflow

Since:

$$\frac{\mathrm{d}X}{\mathrm{d}t} = J \cdot A \tag{2}$$

where

$X =$ amount of pollutant trapped on paper disk corrected for the blank (μg)
$A =$ cross-sectional area of diffusion path (m^2)
$t =$ sampling time (s)

By combining (1) with (2) to eliminate J:

$$\frac{\mathrm{d}X}{\mathrm{d}t} = -A \cdot D\frac{\mathrm{d}C}{\mathrm{d}L} \tag{3}$$

Integrating and rearranging (3) yields:

$$C_{\mathrm{Avg}} = \frac{X}{D \cdot t} \cdot \frac{L}{A} \tag{4}$$

TABLE 4 Physical Constants and Parameters for Selected Gases

Gas	Molar Mass (g mol^{-1})	Diffusion Coefficient (10^{-10} m^2 s^{-1})	Approximate Collection Rate at Ambient Levels (pg s^{-1})
SO_2	64.06	1.30	2.0
NO_2	46.01	1.52	1.7
NH_3	17.03	2.51	1.5
O_3	48.00	1.48	18.0

Adapted from Ferm [8].

The term L/A is the sum of all L/A contributions from the sampler components; i.e. $L/A = [(L_R/A_R) + (L_F/A_F) + (L_N/A_N) + (L_S/A_R)]$ where L_R and A_R are the thickness and area of the plastic ring, respectively, L_F is the thickness of the diffusion filter (typically PTFE, as illustrated in Figure 4) and A_F is the total area of the pores through which diffusion takes place, L_N and A_N are the same parameters for the steel mesh and L_S is the length of the static air layer. It must be noted that diffusion may not take place over the full diameter of the PTFE filter and steel mesh (i.e. 21 mm vs 25 mm in the example provided in Figure 4). For non-indoor sampling locations, the average thickness of the static air layer has been found to be 1.5 mm [8]. It can be measured by exposing two samplers together; the first a normally assembled sampler and a second sampler with the sorbent filter placed directly after the PTFE filter. Under these circumstances, the diffusion path length of the second sampler would be $L_F + L_N + L_S$, and the amount of pollutant accumulated by the sorbent would differ accordingly. As such, L_S is then easily calculated by comparison of the two results.

The sampler geometry can be modified to change the pollutant uptake rate (Eqn. (4)). The sampler body can be lengthened into a tube, which is a configuration used prior to the development of badge samplers. Tube-type samplers can be used in any location where the wind speed is 0.1 cm s^{-1} or more [35], i.e. where the thickness of the static air layer would not cause a large relative error in the effective diffusion path length. The opposite is true for badge-type samplers, which have a high area to length ratio, as such, the effects caused by a static air layer can be significant. The dimensions and configuration of the DEBITS passive sampler (Figure 4) produce $L/A = 35.0\ m^{-1}$. Variation in the size of the static air layer affects this term at ~7% per mm thickness.

In the Ferm-type sampler (Figure 4), the thickness of the 25 mm diameter steel mesh is 160 μm with a porosity of 40%, while the 25 mm diameter PTFE filter is 175 μm thick with a porosity of 85%. If other materials are used, their physical characteristics have to be taken into account when calculating the ambient concentration of the gases measured.

The diffusion coefficient is inversely proportional to ambient pressure and directly proportional to absolute temperature (to the power of 1.5). Ambient concentrations are thus preferably expressed as mixing ratios [36] or ppb units (i.e. mm^3 pollutant gas per m^3 moist air under sampling conditions) as the pressure dependence is then eliminated. Under these circumstances, the concentration is inversely proportional to the square root of absolute temperature, fluctuating by less than 0.2% per °C at ambient temperatures. The ideal gas law can be used in this case to calculate the mixing ratio as follows:

$$C_{\text{Avg}}(\text{ppb}) = \frac{1000X \cdot R \cdot T}{Mr \cdot D \cdot t} \cdot \frac{L}{A} \tag{5}$$

where T is the absolute temperature during the sampling period, M_r is the molar mass of the gas in question and R (gas constant) = 8.31 J K^{-1} mol^{-1}. The amount of pollutant trapped on the filter (X) can be evaluated by multiplying the concentration of the analyte of interest in the sample (μg dm^{-3}) by the total volume of the sample (dm^3).

For organic gases, the uptake rate can either be theoretically derived from Fick's Law (Eqn. (1)) or determined experimentally. The ideal theoretical uptake rate (U) is given by:

$$U = \frac{D \cdot A}{L} \tag{6}$$

where

A = cross-sectional area of the diffusive path, or the equivalent sorption surface (cm^2)
L = length of static layer of sample (cm)
D = diffusion coefficient (cm^2 min^{-1})
U = ideal uptake rate (ml min^{-1})

Knowing the ideal uptake rate, the atmospheric concentrations of the compound can then be calculated:

$$C = \frac{X}{U_p \cdot t} \tag{7}$$

where

C = atmospheric concentration (cm^3 min^{-1})
U_p = ideal uptake rate (cm^3 min^{-1})
X = amount of the gas adsorbed on the tube (ng)
t = time of exposure (min)

Uptake rates are usually given in units of ng ppm^{-1} min^{-1} or pg ppb^{-1} min^{-1} to give atmospheric concentrations in ppm or ppb. Note: U_p is the uptake rate at ambient conditions. The conversion of U (cm^3 min^{-1}) to U_p (ng ppm^{-1} min^{-1}) is given by:

$$U_p = \frac{U \cdot M \cdot 293 \cdot P}{24.5 \cdot T \cdot 101.1} \tag{8}$$

where

U = ideal diffusive uptake rate (ng ppm^{-1} min^{-1})
M = molecular mass of gas (g mol^{-1})
P = atmospheric pressure during sampling (kPa)
T = temperature during sampling (K)
24.5 dm^3 mol^{-1} = molar gas volume at 760 mm Hg and 298 K

The theoretical uptake rate is only an approximation of the actual uptake rate for each component, for a specific tube and packing material, but it does not take into account factors that can influence the real uptake rate. The theoretical and experimental uptake rates for BTX compounds differ greatly (Table 5).

3. PASSIVE SAMPLER NETWORK DESIGN

In general, passive samplers are well suited to mapping pollutant concentrations in cities and residential areas, such as along roads and highways, and across large regions, including continents, owing to the fact that they capture time-integrated observations (varying from days to months). Sampler network design, including the density and distribution of sampling sites, is ultimately determined by the project goals. For example, sampling sites for regional mapping should consider both sites representing background pollution levels as well as localized pollution sources. This better represents regional pollution patterns for interpolation and modelling purposes. Additional measurements may be desired in the close vicinity of strong sources (hot spots) that are not necessarily representative of the larger area. It is recommended to deploy duplicate or triplicate samplers to assess the reproducibility of the measurements and fulfil quality assurance and quality control (QA/QC) requirements. Further details are provided in Section 5.

Passive sampling is suitable for the following purposes:

- Providing general assessment of air quality in regions with little to no prior ambient air quality data to determine general exposure of human health and the environment to air pollutants. Data may provide impetus for additional studies using continuous monitors.
- Determining optimal locations for continuous, long-term monitors by measuring spatial trends and anomalies in a region, including areas with elevated or improved air quality owing to local topography or climate.
- Providing information on the 'area of influence' or representivity of existing continuous monitors by measuring air pollution gradients in regions (especially those in proximity to emissions sources), changes in topography or localized climate. Such data can guide the suitability of existing continuous monitors for modelling studies and air quality control implementation plans.
- Extending existing continuous stack monitoring from pollutant sources and mandated 'fence-line' monitoring at pollution sources to determine the wider impact thereof on affected communities and regions. Such data can be correlated with time and quantity of emissions to identify the impact of specific emitters through source apportionment.
- Performing long-term trend studies across regions, especially when a direct data comparison of local conditions is desired.
- Personal dosimeters for human exposure studies.

TABLE 5 Theoretical and Experimental Uptake Rates for BTX Compounds

Component	Diffusion Coefficient ($cm^2\ s^{-1}$)	Molecular Weight ($g\ mol^{-1}$)	Uptake Rate ($cm^3\ min^{-1}$)	Theoretical Uptake Rate ($ng\ ppm^{-1}\ min^{-1}$)	Experimental Uptake Rate ($ng\ ppm^{-1}\ min^{-1}$)
Benzene	0.0932	78.10	0.59	1.90	0.70 ± 0.09
Toluene	0.0849	92.10	0.54	2.04	1.03 ± 0.26
o-Xylene	0.0688	106.20	0.44	1.90	1.46 ± 0.67
m,p-Xylene	0.0727	106.20	0.46	2.01	1.46 ± 0.67

Adapted from Markes International [37].

3.1 Selection and Quantity of Sampling Locations

Although passive samplers are cost-efficient and require minimal capital expense, consideration must be given to the selection of sampling locations. In general, fewer well-placed samplers are more beneficial than numerous poorly placed samplers. It is recommended that air quality studies focus primarily on capturing direct impacts on human or environmental health. Table 6 provides general guidelines on where samplers should be placed and the relative fraction of sampler locations that should be incorporated into air quality surveys. These guidelines are qualitative and intended only to provide initial guidance in the absence of other project specifications or community needs. For an initial study, a minimum of 12 samplers is desired to provide sufficient data for statistical analysis; however, the quality of locations should be considered. See Section 3.3 for further guidance.

TABLE 6 Guidance on the Location and Relative Quantity of Passive Samplers Deployed in an Air Quality Study

Purpose	Location	Relative Quantity of Total Samplers Deployed (Approximate)
Determining baseline and background concentrations	Locate away from known emission sources, including industry, agriculture, transportation and population centres	10%
Determining direct impacts on human health or ambient environment	Locate at vulnerable population centres, including schools, hospitals, public parks and nature reserves	50%
Determining gradients in region, either to known pollution sources or owing to local topography and climate	Locate between known sources of emissions and affected populations, or locate relative to topographic change/local climate (e.g. sea breeze)	25%
Collocation with existing regulatory networks and/or continuous monitors	Locate directly at site of existing or planned monitors	10%
Other sites of special concern	Locate at areas of key concern, such as those of cultural heritage or particularly vulnerable populations	5%

3.2 Sampling Frequency and Duration

There are two contradictory effects that influence the frequency and duration of passive sampling. Short-duration sampling is desired to assess direct air quality impacts; however, short exposures result in limited mass uptake, whereby analytical quantification is challenged and overall data quality is impacted owing to increased uncertainty. Long-duration exposures can saturate samplers and magnify experimental error owing to changes in temperature, pressure, interference, data loss or vandalism. These limiting factors can be calculated as follows:

- **Minimum period of deployment**: Calculate the minimum quantifiable mass absorbed on the sampler based on the analytical quantification limit (greater than twice the detection limit of the relevant analytical method). Compare this with the expected lower quantile (either 5th or 10th percentile) of ambient concentrations (using the effective diffusion rate as the sampled air volume), either from pre-existing data in the region or from related literature studies, ideally from regions of comparable emissions and climate. This establishes the lower threshold of exposure time to quantify concentrations above a detectable level. Although 'non-detects' are acceptable, pollutant concentrations should be quantified in at least 50% of the exposed samplers during an air quality study.
- **Maximum period of deployment**: Calculate the maximum absorbable mass of the sampler, either by theory (if the mass of sorbent is known) or based on information provided by the manufacturer. Compare this with the expected upper quantile (either 90th or 95th percentile) of ambient concentrations (using the effective diffusion rate as the sampled air volume), either from pre-existing data in the region or from related literature studies, ideally from regions of comparable emissions and climate. This establishes the upper threshold of exposure time to avoid sampler saturation. It is recommended that the maximum period of deployment not exceed 75% saturation. If some samplers are expected to be above 75% saturation, additional deployments for subsampling periods should be added at these locations.

Sampling frequency and duration depend on the study types (Table 7). For an initial study, a minimum of 10 periods of deployment is desired to provide sufficient data for statistical analysis. If sampling periods are not concurrent owing to logistics or cost, it is recommended that at one location samplers be deployed continuously during the full period of the study (at the same frequency of deployment as other samplers) to verify that there are no underlying data artefacts or significant pollution events that are missed.

3.3 Field Siting Criteria

The physical placement of samplers should ensure that sites are situated in strategic locations, such that proper airflow around the sampler is guaranteed

TABLE 7 Guidance on the Sampling Frequency and Number of Deployments for Air Quality Studies

Purpose of Study	Suggested Sampling Frequency	Suggested Number of Deployments (and Study Duration)
General air quality survey • Determine general human exposure • Determine placement of long-term, continuous monitors • Determine impacted area of existing emitters	Daily to weekly (depending on calculated minimum period of deployment)	Daily: 30 deployments (1 month) Weekly: 52 deployments (1 year)
Extending data from continuous stack or fence-line monitors	Weekly to monthly (depending on calculated maximum period of deployment)	Weekly: 100 deployments (2 years) Monthly: 20 deployments (or indefinitely)
Determining long-term trends in and across regions	Every two weeks to monthly (depending on calculated periods of deployment)	Minimum 3 years to indefinitely

and excessive turbulence is avoided. Samplers should be located away from buildings and other structures; the air sampled should represent the air found in the surrounding atmosphere, not just in the immediate vicinity of the sampling point. Samplers should be placed in such a way that there are no unwanted impacts, and they should not be placed next to houses, chimneys, highways, incinerators, rubbish dumps, etc. Suggested distance from objects and known emission sources ranges from 2 m to 20 km (see Figure 6).

Samplers are typically placed under a protective shield or shelter to limit exposure to wind and precipitation at a height of 1.5−2 m (see Figure 7). High wind speeds may perturb the normally quiescent air within the sampler and shorten the diffusion path length. Conversely, if wind speeds in the surrounding environment drop below 5−10 cm s^{-1} there tends to be a stagnant layer of air at the entrance to the sampler [35]. If the pollutant uptake rate of the sampler exceeds the convective delivery of the pollutant to the sampler's inlet, this static air layer will be 'starved' and the pollutant concentration at the sampler inlet will drop below that of the bulk air being sampled. Laminar diffusion rather than turbulent diffusion becomes significant in this static layer. In reality, this phenomenon effectively results in an increase of the diffusion path length, decreasing the mass of pollutant sorbed.

Many different types of shields can be used for the protection of the sampler, but the impact of the protective shield on the static air layer should be taken into account. The assumption that the gas is fully absorbed at the sorbent filter end of the sampler implies that the concentration at the sorbent surface is zero (Figure 8). As such, the concentration gradient to ambient air is influenced by the construction of the sampler (i.e. stainless-steel mesh and the Teflon filter shown in Figure 4) and the protective shield.

3.4 Collocation with Reference Measurements

Although passive sampling techniques are well-accepted, it should be noted that methods still lack accepted harmonized data validation processes and techniques. Nonetheless, the deployment of duplicate or triplicate passive samplers and the parallel comparison with active sampling at regular intervals has become general practice. This improves data quality and allows for the incorporation of correction factors if necessary. In general, intercomparison is carried out at regular intervals, both within networks, and across collaborating networks [3,21,28]. Intercomparison with active samplers is essential, especially where a pollutant of interest is not directly measured (e.g. O_3 in some cases is linked directly to the 'oxidation capacity' in the atmosphere).

Where feasible, passive samplers should be collocated with active air pollution monitoring devices, e.g. continuous monitors or denuders. (Refer to Chapters 5 and 6 for more detail on these sampling methods). For example, the comparison of passive samplers for gaseous NH_3 (Radiello, RAD1201/RAD168) with active denuders in the NADP/AMoN showed good agreement,

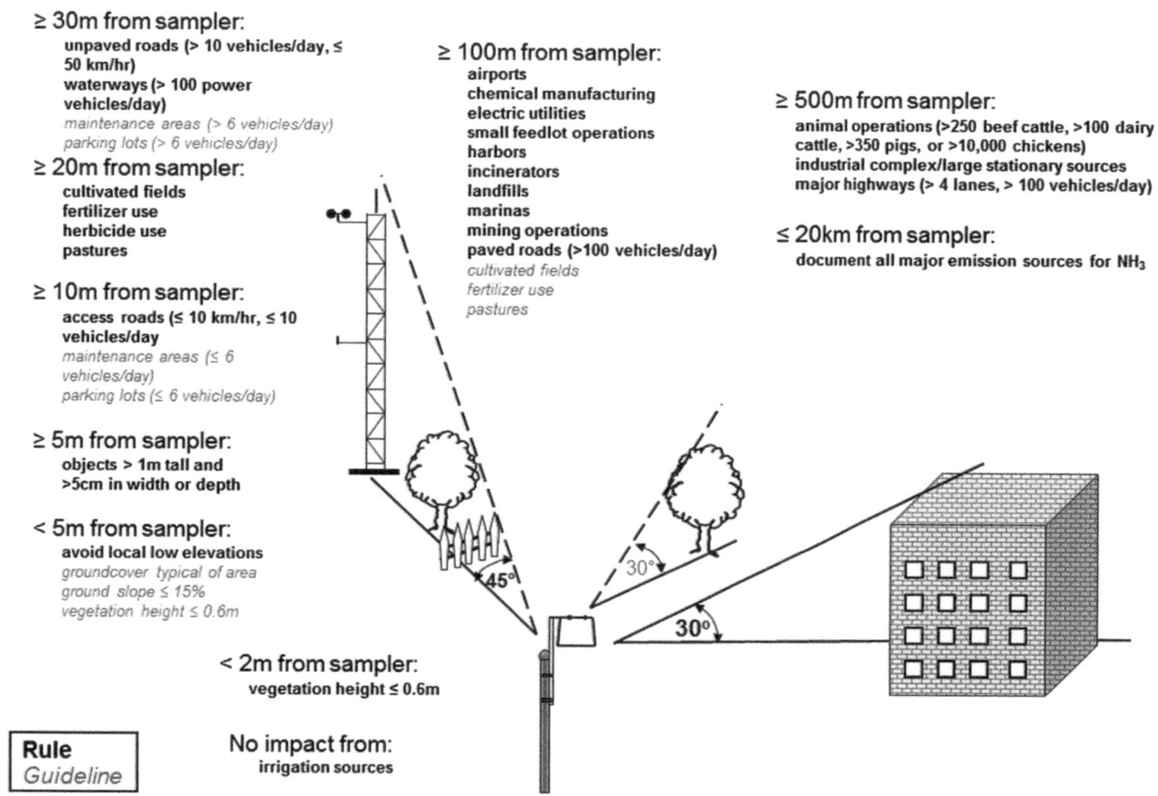

FIGURE 6 Suggested siting criteria for passive sampler placement for sampling of ambient ammonia [23].

FIGURE 7 Examples of passive sampler holders and shields as used by (a) the Atmospheric Chemistry Monitoring Network in Africa (IDAF) programme and (b) U.S. National Park Service.

with a relative percent difference typically within 20% (Figure 9). Similarly, the passive samplers and active denuders also agreed well with a continuous analyser (Pranalytic Nitrolux). Significant peaks observed with the continuous analyser were not captured by the denuders or passive samplers, although there was general agreement in the central tendency of concentrations (Figure 9).

For large-scale studies, validation tests can be performed by exposing samplers to controlled doses of the air pollutant of concern [9,38]. In addition, chamber and laboratory tests are useful to assess the influence of wind speed, sampling and storage times, temperature, humidity and the effects of other co-pollutants (see Figure 10 for examples of laboratory based intercomparisons). Site-specific international comparisons have also been undertaken. One of the most comprehensive in terms of global coverage was undertaken by the WMO [21]. In all these studies, the value and validity of passive samplers has been proven, when used in accordance to prescribed procedures.

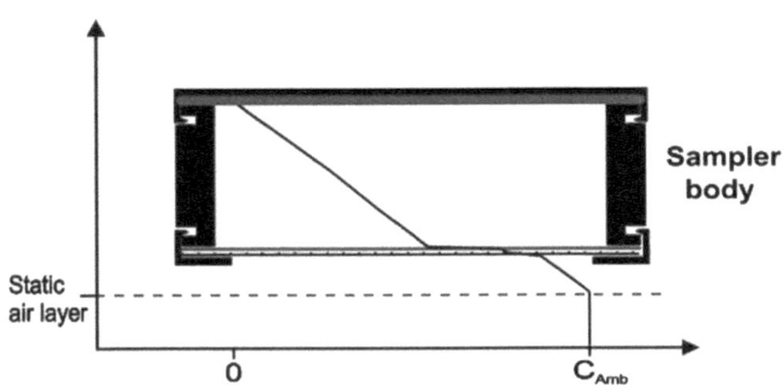

FIGURE 8 Concentration profile of pollutant around and within a passive sampler. The vertical distance represents the diffusion path length, and the horizontal axis represents the concentration of pollutant, which in ambient air is C_{Amb}.

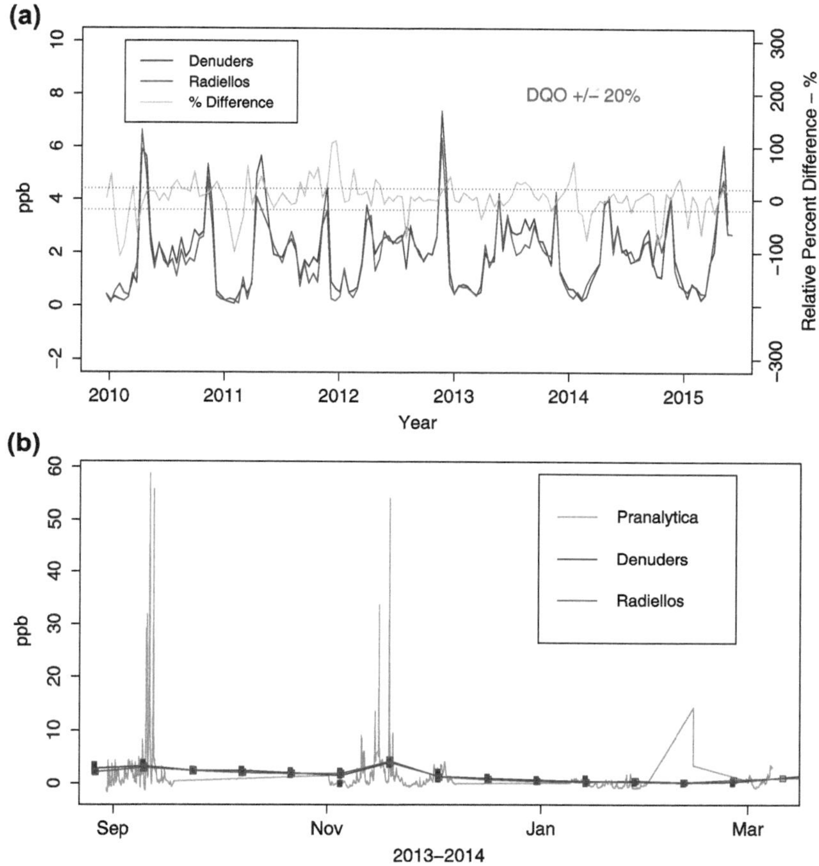

FIGURE 9 Comparison of active and passive measurements in the U.S. National Atmospheric Deposition Program/Ammonia Monitoring Network (NADP/AMoN): (a) Comparison of passive samplers with denuder measurements for NH_3 at Bondville, Illinois, USA, the data quality objective (DQO) benchmark of ±20% is also shown; and (b) Comparison of passive samplers with denuder measurements and continuous monitoring at Bondville, Illinois, USA.

4. METHODOLOGY

Most of the problems experienced with the accuracy and reproducibility of passive samplers are related to sampler preparation, transportation and storage. Trace amounts of contamination on sampler body parts, in storage vials, etc. can contribute to significant error in the final results. Paying meticulous attention to cleanliness during preparation and analysis is essential, and use of dedicated laboratory facilities and clean-air benches or glove boxes for sampler preparation is recommended to avoid cross-contamination. (See Section 5.1 for further details on Laboratory QA.) Furthermore, analytical log books should be maintained for the duration of

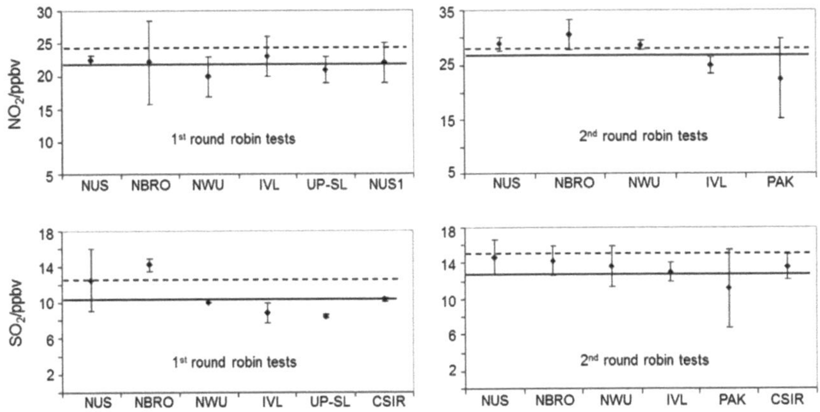

FIGURE 10 Interlaboratory comparisons between passive and active samplers conducted by the University of Singapore [39]. The error bar indicates the standard deviation based on three independent measurements, the dotted line is the average of the active sampler and the solid line is the grand mean of all the passive samplers (CSIR, South Africa; Swedish Environmental Research Institute (IVL), Sweden; National Building Research Organisation (NBRO), Sri Lanka; National University of Singapore (NUS); North-West University (NWU), South Africa; Pakistan Environmental Protection Agency (PAK), Pakistan; University of Peradeniya (UP-SL), Sri Lanka).

studies, noting the lot number of all supplies and reagents used, and the dates and times of sampler processing.

4.1 Network Operations

All supplies for deployment should ideally be prepared in one analytical facility using the same supplies for all stations. If feasible, this should include shipment containers, gloves supplied to field stations, and reagent-grade water and laboratory wipes used to clean the field shelter. During transit to field stations, due precaution should be taken to protect samplers, avoiding exposure to excess heat or contamination owing to negative pressures in air-freighting of samplers. If available, colour-change ‘temperature change’ tape or min/max thermometers should be shipped with the samplers. Samplers should be shipped in coolers with cold packs, via the fastest practical shipment means possible. Travel blanks should be shipped with every shipment initially to determine a baseline level of expected bias. Further details are provided in Section 5.2. Residence time of samplers at field stations should be kept to a minimum. If samplers are held for more than 1 week, samplers should be refrigerated (<4 °C) if feasible.

In general, handling of samplers at the field station should be kept to an absolute minimum. Samplers should be checked upon receipt to ensure there was no damage in shipment. Field site operators should use clean gloves to deploy and remove samplers. Operators should use precautions not to touch samplers with bare skin, breathe on them or expose them to cleaning agents or other harsh chemicals. The site operator should inspect the field shelter before deployment, ensuring that it is intact; if necessary the inside and outside of the shelter may be cleaned, but only after samplers are removed and secured. When not deployed, samplers should be stored in the packaging provided by the analytical laboratory.

A field sheet, with identifying information corresponding to the sampler(s) deployed, should be maintained for each deployment. As a minimum, deployment information should include:

- Station identifier
- Operator name and contact information
- Identification code of sampler (i.e. code permanently inscribed on sampler)
- On and off dates and times (hour, minutes and time zone) of sampler deployment and removal
- Operator identification of potential nearby sources of contamination during deployment (e.g. agricultural or animal activity, strong odours, fires, construction or other sources of emissions)
- Operator identification of anomalous meteorological events (heavy rain or fog, atmospheric inversion, or pollution event) and general state of vegetation (leaf emergence, greenness)
- Operator notes about the deployment, including issues during the sampling period (e.g. damage in shipment or vandalism)

A copy of all records should be retained at the field site, and samplers should be returned via traceable shipment methods. Particular care should be used if samplers are shipped across international borders. Coordination with customs officials is critical to avoid sampler contamination during shipment inspections. For this purpose, samplers should be packaged with see-through packaging, clearly labelling their contents (and warning about limited exposure, ideally with tamper-evident seals) and enclosing instructions on a letterhead explaining the purpose of the study in the local language.

4.2 Laboratory Methods for Sampling of Inorganic Gases

To ensure data quality, the following procedures are recommended, as followed in the DEBITS/IDAF programme [28]; similar procedures are also used in AMoN [40].

4.2.1 Sampler Preparation

Reagents used to prepare absorbing solutions and calibration standards should be the highest purity chemicals generally available. Ultra-high purity de-ionised water (specific resistance of 18.2 MΩ cm or more) should be used for sampler preparation, including rinsing of all apparatus and glassware. Plastic surgical gloves, rinsed with de-ionised water prior to use, should be worn whenever samplers are handled. Tweezers with plastic pad tips should be used to handle sampler components. All reagents must be freshly prepared prior to use. Wherever possible, one single lot of chemicals and supplies should be used for the duration of the study.

The cleaning procedures for sorbent filters (Whatman paper in the case of DEBITS) require that filters should be rinsed, soaked and sonicated (with the use of an ultrasonic bath) at least four times in de-ionised water and methanol before being dried in a vacuum desiccator prior to storage in air-tight plastic bags that have undergone similar washing procedures. All other sampler components are soaked overnight in a 3% (v/v) solution of detergent (EXTRAN MA 02 Neutral), rinsed thoroughly, sonicated in de-ionised water and dried in a clean environment before storage. The plastic vials used for the analysis of samples are soaked in a 0.2% (v/v) cleaning solution (EXTRAN AP 21 acidic) for 2+ hours and then soaked overnight in 3% (v/v) detergent (EXTRAN MA 02 Neutral), thoroughly rinsed and sonicated in de-ionised water before being dried in a vacuum desiccator.

Precautions to avoid contamination should also be taken when 'loading' the sampler. In the DEBITS samplers, the Whatman filter paper is loaded against the snap-on end cap using sterile plastic surgical gloves and tweezers with plastic pads and then attached to the main plastic ring. Then 50 mm^3 of the absorbing solution is pipetted evenly over the surface with the use of a micropipette that is able to accurately measure small volumes. Similar pipettes are also used during the preparation of calibration standards. The punched out snap-on cap containing the Teflon filter and steel mesh is attached to the ring. The samplers are marked with an identifying code and immediately placed in an airtight vial. Sampler vials are stored inside sealed plastic bags to further minimise any potential contamination by air leakages, etc. Samples should be kept refrigerated (<4 °C for up to 1 month) before shipment to the stations, or kept frozen (<0 °C) indefinitely.

4.2.2 Sampler Extraction and Analytical Preparation

Samplers and extract solutions should be handled in a clean-air bench or glove box. Samplers for inorganic gases are typically extracted using ultra-high purity de-ionised water (specific resistance of 18.2 MΩ cm or more) or buffer in an ultrasonic bath, depending on the pollutant measured. Laboratory spiking studies should be performed to evaluate the efficiency of extraction methods.

In the DEBITS programme, samples are prepared for ion chromatography (IC) analysis by removing the Whatman filter from the sampler and placing it in a clean vial. Predetermined volumes (based on the expected concentrations of pollutants) of de-ionised water are added to the samples, and the samples are extracted in an ultrasonic bath for 15 minutes.

A quantitative repipettor should be used to measure extraction solutions, with the volume used for extraction verified using a tared analytical balance. Retention time of extract solutions should be kept to a minimum (ideally

<24 h) before analysis, as they will readily scavenge pollutants from the laboratory air, and they can be easily contaminated in the laboratory. Extract solutions should be capped tightly after extraction and refrigerated (<4 °C) overnight or frozen (<0 °C) if held for more than 24 h before analysis.

4.2.3 Sample Analysis

Standard analytical techniques for inorganic analysis are used to analyse the sample extracts. Typical techniques include IC and UV–visible spectrophotometery, as described by Martins et al. [2]. In the DEBITS programme, SO_2, NO_2, NH_3 and O_3 samples are analysed on a suppressed Ion Dionex Chromatography System (ICS-300) fitted with an Ionpac AS18 (2 × 50 mm) analytical column, Ionpac AG18 (2 × 50 mm) guard column and ASRS Ultra II 2-mm self-regulating suppressor. In the NADP/AMoN, NH_3 sample extracts are analysed for ammonium ion by colourimetry [40].

Typical IC-chromatographs of sulphate (SO_4^{2-}) and NO_3^- samples collected during field campaigns are given in Figure 11. A five-point IC calibration curve is used for SO_2 analysis, by preparing standards at 1, 2.5, 5, 10 and 50 ppm; while O_3 is calibrated at 0.1, 1, 2, 5 and 10 ppm. The calibration range should be adjusted to match the expected range of concentrations in the study of interest.

Due to potential interferences, NO_2 sampler extracts can also be analysed using UV–visible spectrophotometry. The samples are leached in 5 cm^3 of deionised water and then treated with 5 cm^3 of the diazotising agent before measurement of absorbance at 540 nm. The diazotising agent consists of sulphanilamide, naphthylethylenediamine dihydrochloride (NEDA) and H_3PO_4. A five-point calibration curve is normally also used in the case for NO_2 analyses, by preparing standards at 0.1, 1.0, 2.5, 5.0 and 12.5 ppm.

4.2.4 Precautions and Known Interferences

Several uncertainties must be considered when using passive samplers to monitor ambient pollutant concentrations [27]. These issues include the linearity in the reaction response of the sampler [24], interferences from other chemical compounds in the atmosphere [29] and environmental factors prevailing during the sampling period.

The use of Fick's law assumes steady-state conditions, while ambient pollutant concentrations can vary widely. The assumption whether steady-state conditions can be assumed and whether passive samplers provide a truly integrated response to short-lived transients has been investigated [1]. Provided that the sampling period is well in excess of the molecular diffusion time (typically 1–10 s), this assumption is typically valid.

Certain sorbent chemicals for inorganic gases lack specificity to target pollutants, owing to interferences and reaction with other atmospheric constituents. With O_3 measurements, peroxyacetyl nitrate (PAN) and other atmospheric oxidants may result in an overestimation of O_3 concentrations.

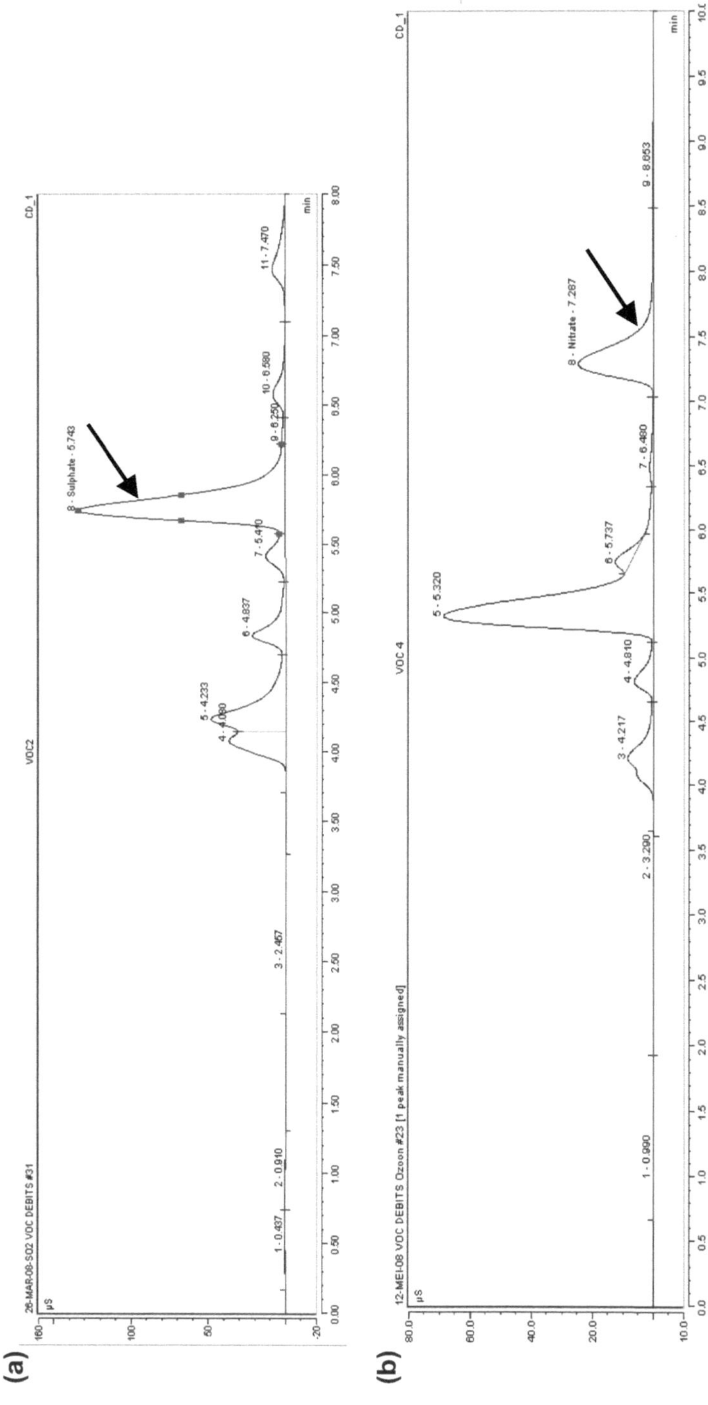

FIGURE 11 Example passive sampler chromatogram for the determination of (a) SO_4^{2-} (adsorbed product of SO_2) and (b) NO_3^- (adsorbed product of NO_2).

Chemical instability of sorbents may result in back- or reverse diffusion of pollutants, which will result in an underestimation of ambient concentrations. This can occur after an extended exposure if the vapour pressure of the analyte at the sorbent surface is greater than the external concentration, or if ambient concentrations drop to lower levels after initial high levels [1].

Environmental factors such as temperature and pressure influence the diffusion coefficient of a gas. In addition, humidity can affect the sorption capacity of hydrophilic sorbents and very high humidity may even cause condensation in tube-type samplers or dissolution of the sorbent chemicals, which alters the functioning of the sampler. Wind speed and turbulence may change the static air layer and thus influence the effective diffusion path. Nevertheless, extensive field studies in large networks (including IDAF) indicate that uncertainties in inorganic gas measurements are much better than that of passive samplers for organic gases. See Section 4.3.

The IDAF and DEBITS studies follow the approach and methodologies recommended by a WMO interlaboratory study [41]. These methodologies have been optimised after extensive testing by the original developers, with interferences caused by other co-pollutants either minimised to such an extent that they cease to be significant or completely eliminated.

4.3 Laboratory Methods for Sampling of Organic Gases

4.3.1 Sampler Preparation

The adsorbent tubes must be tested for leaks and preconditioned with an inert gas to activate them before use. For Tenax® tubes, the preconditioning is typically done by flushing the tube with helium for 30 min at 350 °C at a flow of 50 $cm^3 min^{-1}$. After preconditioning, the tubes need to be sealed with suitable caps (e.g. Swagelok) and stored at temperatures $<$18 °C. Tubes may be reused and have a maximum lifetime of 2 years when used frequently.

4.3.2 Sampler Deployment

The adsorption tubes for organic gases are exposed under similar protective shields as those described in Section 3.3. Sampler caps (Swagelok) are removed from the tubes before exposure in the field and replaced with diffusive caps that allow for effective air movement into the sampler. The tubes are placed facing upside down under the shield; after the sampling period, the diffusive caps are removed and the tubes sealed with the Swagelok caps. The exposed tubes are then wrapped individually in aluminium foil and maintained at room temperature or below ($<$18 °C) until they are analysed.

4.3.3 Sampler Desorption and Analysis

The analytical procedures for organic gases collected using passive tube-type samplers are described in the EPA TO-17 compendium and National Institute of Safety and Health (NIOSH) 2549 manuals [33,42]. The adsorbed gases are desorbed using a thermal desorber connected to a gas chromatograph (GC) for the separation and quantification of the sampled gases. The GC is coupled to a time-of-flight mass spectrometer (TOF-MS) for the identification of the desorbing gases (Figure 12).

The tubes are normally analysed by a two-stage thermal desorption method. The samples are first tested for leaks and then purged with zero air to reduce the risk of adsorbent oxidation during desorption. Thereafter, the tubes are heated for 5 min. The volatiles in the tubes are swept by nitrogen gas into a cold trap (−30 °C) where they are reconcentrated for 4 min. At the end of this period, the cold trap is heated rapidly for 5 min to transfer the concentrated

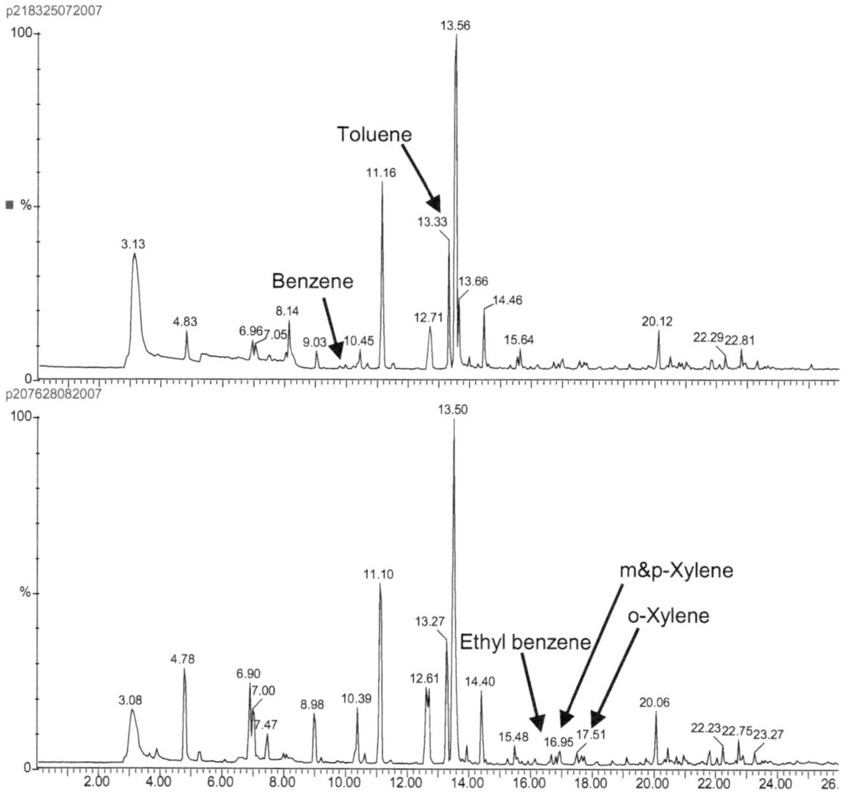

FIGURE 12 Example gas chromatographs of ambient volatile organic compound samples collected on Tenax® tubes; benzene is the small peak at an elution time 9:45, which is not always detected at this resolution, with the toluene peak following at 13:22−13:27. Ethyl benzene elutes from 16:40 to 16:41, m,p-Xylene peak is at 16:58−17:00 and o-Xylene at 17:40. In this example, the peaks were so closely spaced that a compound, such as benzene, could easily be mistaken for the compound eluting at 10:45 and 10:39, had high resolution GC−MS analysis not been used (Note: MS spectra are not included).

VOCs from the cold trap into the GC for separation and detection with the TOF-MS.

4.3.4 Precautions and Known Interferences

Samplers for organic gases require certain precautions owing to the low levels typically found in the ambient environment. The US NIOSH has set the accepted accuracy of ±25% and a bias of ±10% for air quality monitoring using organic passive samplers [42].

Analytical precautions should be taken to ensure that desorption is efficient, and that there are no leaks or oxidation during desorption. Analytical quantifications should be verified to ensure there are no overlapping peaks or interferences (i.e. as shown in Figure 12). In large-scale studies, spike recovery studies of compounds of interest should be performed for verification purposes to ensure efficiency of desorption and correct analytical quantification.

5. QUALITY ASSURANCE AND QUALITY CONTROL

QA/QC procedures are an essential component of good field and laboratory practices for passive sampling. Overall data quality objective (DQO) benchmarks should be established during project planning. In the absence of project specifications or prior data, the following data quality indicator targets for precision, bias, sensitivity and comparability may be used as initial goals:

- **Precision:** ±20%, measured as the median absolute relative percent difference of replicate (or triplicate) field measurements using multiple samplers
- **Bias:** +20%, measured as the median travel blank compared to the median of all laboratory blanks
- **Sensitivity**: >50%, measured as the percentage of deployed samplers above the analytical quantification limit
- **Comparability**: ±20%, measured as the regression slope of colocated passive and active/continuous measurements

5.1 Laboratory QA and QC

Laboratory blanks should be prepared with each batch of samplers following the same methodology of preparation and retained at the laboratory to be extracted and analysed with the deployed samples. The purpose of the blanks is to take into account and compensate for any matrix effects, systematic errors and contaminants present in the reagents, vials or filters. At a minimum, one of each of the following blanks is recommended for each deployment and/or analytical batch:

- Extraction/reagent solutions (for inorganic gas extractions)
- Extraction blank (of prepared, but not deployed sampler)

Calibration standards should be prepared for each analytical run, corresponding to the expected range of sample concentrations. The standards should be prepared from stock solutions that are verified with each analytical run and maintained for the duration of the project. The standards should be handled exactly as the samples, and a calibration blank prepared exactly as the standards but without the analyte of interest. The purpose of the calibration blank is to compensate for any contributions from the reagents and/or apparatus effects. The normal analytical procedure for determining the analyte concentrations is then followed. The signal of the calibration blank is subtracted from the signal of the standards. A straight (or in some cases curved) line is fitted to a plot of concentration versus instrument response signal for all the standards, using a least squares regression.

5.2 Field QA and QC

It is strongly recommended that samplers be exposed in duplicate or triplicate and accompanied by travel blanks, at a minimum at the initiation of studies and notably if samplers are shipped across international borders. If initial data prove that precision of replicate measurements and the bias of travel blanks are acceptable, their frequency may be reduced. However, the following replicate and travel blank deployment frequency is recommended as a minimum level throughout the duration of studies:

- Replicate measurements (duplicate or triplicate): 25% of deployments, as a percentage of total deployed samples (i.e. 25 replicate measurements per 100 deployed samples)
- Travel blanks: 10%, as a percentage of total deployed samples (i.e. 1 travel blank per shipment of 10 samplers)

Travel blanks are assembled, shipped and analysed exactly the same way and together with the samples but are not exposed to ambient air. They may remain in the sealed vials and plastic bags throughout the sampling period and stored at the field station.

5.3 Comparison with Reference Methods

As noted in Section 3.4, at least one passive sampler station should be collocated with an active and/or continuous reference method, such as an ambient air pollution monitoring station. Data from the reference station can be used to verify or correct passive sampler data and can be used to quantify uncertainty in the passive data (see Figure 13).

5.4 Data Validation and Reporting

Final ambient concentration data should be reported with laboratory and field blank data for each analytical batch. Data should be flagged to indicate

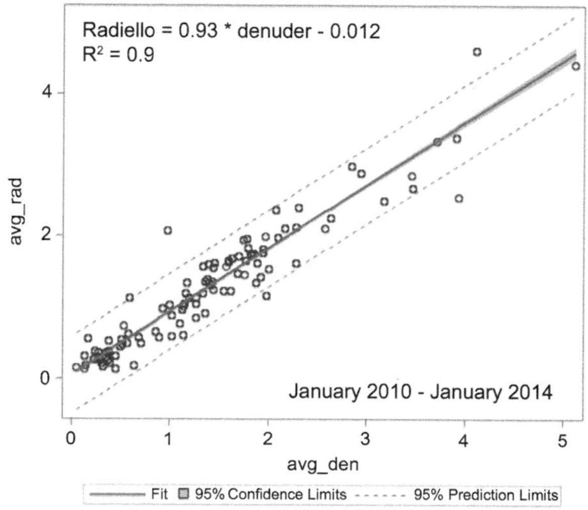

FIGURE 13 Regression of Radiello passive samplers (vertical axis) against an active denuder method (horizontal axis) for measurement of ambient ammonia at the NADP/AMoN field station, Bondville, Illinois. The plot includes all valid data from January 2010 to January 2014.

potential field sampling and analytical issues. The following factors should be taken into consideration for their potential bias on the resulting data:

- Local, anomalous sources of potential contamination during the sampling period
- Major deviation from field sampling protocol (e.g. vandalism)
- Potential laboratory analytical issues (e.g. sample handling)
- Major deviation from laboratory analytical protocols (e.g. spilt sample)
- Concentration below laboratory quantification limit

6. CASE STUDIES

The unique role and place of passive sampling in ambient air quality studies is illustrated by the following case studies from the DEBITS programme and the NADP/AMoN.

6.1 Mapping Regional Pollutant Concentrations

6.1.1 Determining Exceedance of Critical Loads in South Africa

Emissions of nitrogen (N) and sulphur (S) in many developing countries around the world are increasing owing to an increase in energy demand [2]. This is true in South Africa, where the cumulative impacts of both wet and dry S and N deposition are of concern owing to their adverse impacts on soil and surface water. A study was initiated in 2005 to assess the effects of these trace gases on a regional scale for an extended period [43]. A 2-year long programme of passive monitoring at 37 sites, distributed on a 1-degree grid from the Highveld to the Indian Ocean coast, Mozambique, Zimbabwe and Botswana borders, was initiated to assist decision-making processes and policies regarding the acidification risk in this region (Figure 14). The specific aim of the study was to evaluate whether or not acidic atmospheric deposition posed a threat to soils, plants and water bodies of South Africa [43].

Monthly mean SO_2 and NO_2 concentrations were measured using duplicate passive samplers prepared and analysed as described in Section 4.2. The results of two concurrent years of monitoring SO_2 and NO_2 (September 2005–August 2007) showed a consistent spatial pattern in elevated concentrations (Figure 15). These data, together with wet deposition measurements, have been used for the assessment of potential critical load exceedances in this region [44,45].

These results showed that interannual differences for both SO_2 and NO_2 were not substantial, although SO_2 showed larger interannual variations. Potential critical load exceedance was determined using these results in combination with soil sensitivity maps.

6.1.2 The DEBITS Network

DEBITS is the longest running task of the IGAC programme within the International Geo-Biosphere Program. This programme was initiated as a cooperative programme involving the GAW Project of the WMO in the 1990s, due to the global lack of information on wet deposition in the tropics. It was

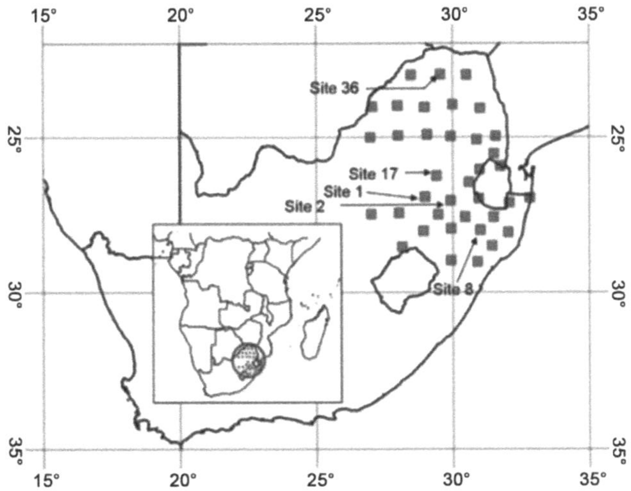

FIGURE 14 Passive sampler network operated during a 2-year period within South and sub-Saharan Africa.

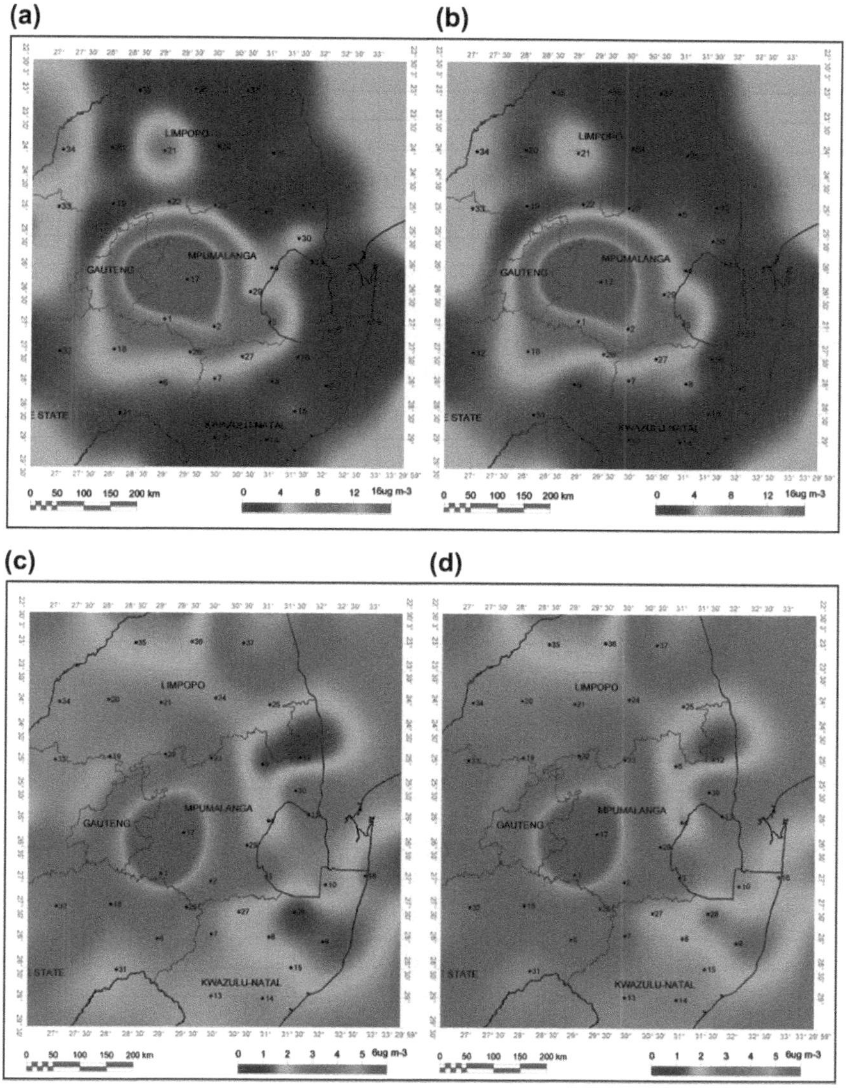

FIGURE 15 Average of monthly mean annual concentrations of: (a) SO_2, September 2005–August 2006; (b) SO_2, September 2006–August 2007; (c) NO_2, September 2005–August 2006; and (d) NO_2, September 2006–August 2007.

determined that wet deposition alone could not explain the global cycles of atmospheric species, and the programme was expanded to measure gaseous species of S, N and O_3. The remoteness of sites and unavailability of electricity at most sites necessitated a non-conventional method of monitoring gaseous species. Ferm-type passive samplers were initially deployed in Asia and later fully rolled out in the African component of the programme called IDAF (**I**GAC, **D**EBITS, **Af**rica). Slight changes in absorbents were made to improve their performance in dry and hot environments, and intensive intercalibrations with active and other commercially available samplers were performed. As mentioned in Section 4.2.4, correction factors (for NO_2, NH_3 and O_3) were necessary due to back reactions and interferences by other compounds (especially in the case of O_3).

The use of passive samplers enabled researchers to obtain accurate information on the concentration of SO_2, NO_2, NH_3 and O_3 at sites over a long time period at a fraction of the cost of active samplers. Passive samplers for NO_2, HNO_3, NH_3, and SO_2 were deployed in this network at 10 regionally representative transects through several African ecosystems, from dry savanna to forest to wet savanna (see Figure 16 and Table 8). The resulting information [3] was combined with modelled results to provide global estimates of N deposition (Figure 17, [46]).

6.2 Determining Long-Term Ammonia Trends: The NADP/AMoN

The NADP/AMoN was established in the United States in October 2007 to routinely measure gaseous atmospheric NH_3 using Radiello®-type passive samplers [17,40]. The purpose of AMoN was to provide data to assess the long-term trends in ambient NH_3 concentrations and deposition of reduced N species; validate atmospheric models; better estimate total N inputs to

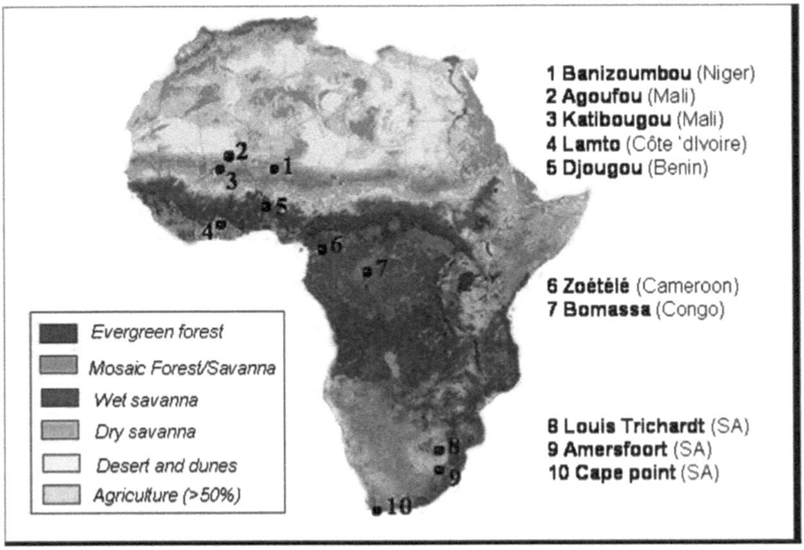

FIGURE 16 Vegetation and location map of the 10 measurements stations of the IDAF network.

TABLE 8 Mean Annual Concentrations (ppb) at the IDAF Sites [3]

Station (period)	NO_2	NH_3	O_3[a]	SO_2[b]	HNO_3[c]
West and Central African sites					
Banizoumbou (98–07)	2.4 ± 0.4 *(113)*	6.3 ± 2.0 *(112)*	11.9 ± 2.3 *(76)*	0.6 ± 0.2 *(66)*	0.5 ± 0.2 *(112)*
Katibougou (98–07)	1.9 ± 0.3 *(117)*	6.6 ± 1.0 *(113)*	12.6 ± 2.2 *(73)*	0.6 ± 0.2 *(68)*	0.4 ± 0.1 *(116)*
Agoufou (05–07)	1.8 ± 0.4 *(28)*	7.4 ± 0.8 *(25)*	12.2 ± 2.6 *(23)*	1.0 ± 0.2 *(26)*	0.5 ± 0.1 *(28)*
Lamto (98–07)	1.0 ± 0.3 *(115)*	4.0 ± 1.2 *(112)*	10.9 ± 1.8 *(78)*	0.5 ± 0.2 *(67)*	0.3 ± 0.1 *(116)*
Djougou (05–07)	1.2 ± 0.1 *(30)*	3.9 ± 1.4 *(27)*	14.0 ± 2.8 *(29)*	0.8 ± 0.3 *(29)*	0.4 ± 0.2 *(30)*
Zoetele (98–07)	0.9 ± 0.2 *(112)*	4.2 ± 0.9 *(103)*	4.8 ± 1.0 *(73)*	0.3 ± 0.1 *(65)*	0.2 ± 0.1 *(113)*
Bomassa (98–06)	1.4 ± 0.4 *(83)*	4.7 ± 1.7 *(76)*	4.0 ± 0.4 *(45)*	0.4 ± 0.2 *(44)*	0.3 ± 0.1 *(81)*
Southern African sites					
Amersfoort (97–05)	2.5 ± 1.0 *(105)*	1.2 ± 0.7 *(85)*	27 ± 8.3 *(89)*	2.8 ± 1.1 *(10)*	0.9 ± 0.5 *(26)*
Louis Trichardt (95–05)	0.7 ± 0.4 *(71)*	1.2 ± 0.7 *(65)*	35 ± 8.6 *(72)*	0.8 ± 0.7 (70)	0.2 ± 0.1 *(27)*
Cape Point (95–05)	1.2 ± 0.6 *(114)*	1.5 ± 0.7 *(120)*	27 ± 7.3 *(114)*	0.7 ± 0.4 *(106)*	0.5 ± 0.2 *(29)*

[a]*In West and Central Africa O_3 monitoring started in 2001 and.*
[b]*SO_2 in 2002.*
[c]*In South Africa HNO_3 started in 2003.*

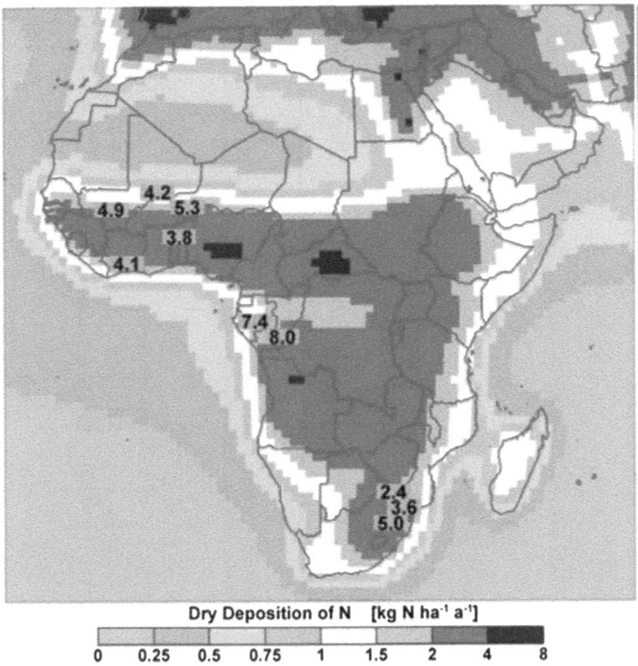

FIGURE 17 2001 modelled dry deposition of $N_{oxidized}$ and $N_{reduced}$ with measurement-based inferential estimates from 2000 to 2007 superimposed [46].

ecosystems; assess changes in atmospheric chemistry due to S and N dioxide reductions; and assess compliance with fine particulate ($PM_{2.5}$) standards [40]. The AMoN complements the NADP's National Trends Network (NADP/NTN), which has monitored the wet deposition of ammonium and other major ions across the US since 1978 [47]. Further information about AMoN methods and data are available at URL: nadp.isws.illinois.edu/AMoN.

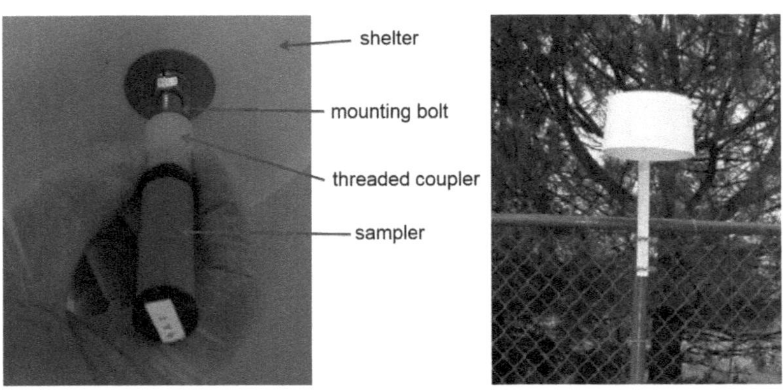

FIGURE 18 Radiello passive samplers and field housing used in the NADP/AMoN [40].

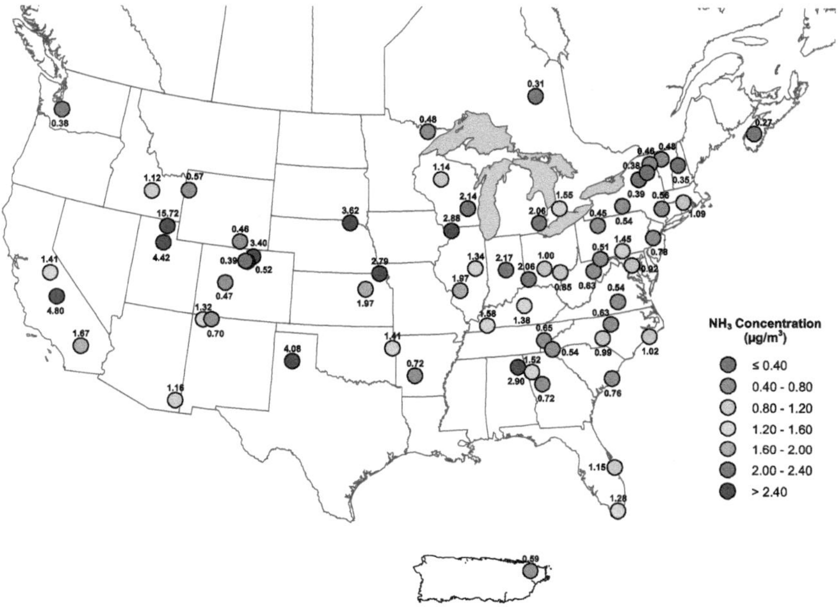

FIGURE 19 Time-weighted annual mean ambient ammonia concentration for 2014 (µg m^{-3}) as reported by the National Atmospheric Deposition Program/Ammonia Monitoring Network (NADP/AMoN).

Samplers in the NADP/AMoN are deployed continuously over 2 week periods at each site at a standardized height (~2 m) using a field housing provided by the NADP (Figure 18). Triplicate samples and travel blanks are deployed randomly as an indicator of data quality. All site operators comply with NADP established standard operating procedures, using only sampling materials provided by the NADP/AMoN's Central Analytical Laboratory. Concentrations measured from AMoN samplers are also compared to reference denuders and other passive samplers at the NADP field station, Bondville, Illinois, USA (see Figures 9 and 13). Time-weighted ambient NH_3 data during 2014 showed a tendency for higher concentrations in the central and western regions of the US (Figure 19).

7. CONCLUSIONS

The studies outlined in this chapter demonstrate the unique role of passive samplers in ambient monitoring of air quality. Their usage in both screening and spatial mapping studies is of great value. Further, they can provide high quality data; however, accuracy and reproducibility is strongly linked to sound operational and laboratory practices. Passive samplers should be tested and validated against reference active samplers to ensure reliability, and good QA procedures need to be followed. Site maintenance and calibration against known references are as important as for any other monitoring device. Meticulous attention must be given to cleanliness during preparation and analysis of samplers. Further, storage and transportation conditions are as important as preparation and analysis to ensure good data quality.

REFERENCES

[1] R.H. Brown, Monitoring the ambient environment with diffusive samplers: theory and practical considerations, J. Environ. Monit. 2 (2000) 1–9.

[2] J.J. Martins, R.S. Dhammapala, G. Lachmann, C. Galy-Lacaux, J.J. Pienaar, Long-term measurements of sulphur dioxide, nitrogen dioxide, ammonia, nitric acid and ozone in southern Africa using passive samplers, S. Afr. J. Sci. 103 (2007) 336–342.

[3] M. Adon, C. Galy-Lacaux, V. Yoboué, C. Delon, J.P. Lacaux, P. Castera, E. Gardrat, J. Pienaar, H. Al Ourabi, D. Laouali, B. Diop, L. Sigha-Nkamdjou, A. Akpo, J.P. Tathy, F. Lavenu, E. Mougin, Long term measurements of sulfur dioxide, nitrogen dioxide, ammonia, nitric acid and ozone in Africa using passive samplers, Atmos. Chem. Phys. 10 (2010) 7467–7487.

[4] J. Namieśnik, B. Zabiegala, A. Kot-Wasik, M. Partyka, A. Wasik, Passive sampling and/or extraction techniques in environmental analysis: a review, Anal. Bioanal. Chem. 381 (2005) 279–301.

[5] C.S. Gordon, J.T. Lowe, US Patent 1.644014 (1927).

[6] E.D. Palmes, A.F. Gunnison, Personal monitoring device for gaseous contaminants, Am. Ind. Hyg. Assoc. J. 34 (1973) 78–81.

[7] P.W. West, G.C. Gaeke, Fixation of sulphur dioxide as disulfitomercurate(II) and subsequent colorimetric estimation, Anal. Chem. 28 (1956) 1816–1819.

[8] M. Ferm, A Sensitive Diffusional Sampler, IVL publication B-1020, Swedish Environmental Research Institute, 1991. Box 47086, 402 58 Goteborg, Sweden.

[9] E.D. Palmes, A.F. Gunnison, J. DiMattio, C. Tomczyk, Personal sampler for nitrogen dioxide, Am. Ind. Hyg. Assoc. J. 37 (10) (1976) 570–577.

[10] J.F. Mazur, R.L. Bamberger, G.E. Podolak, Development and evaluation of an ammonia dosimeter, Am. Ind. Hyg. Assoc. J. 39 (1978) 749–753.

[11] C. Monn, M. Hangartner, Passive sampling for ozone, J. Air Waste Manage. Assoc. 40 (1990) 357–358.

[12] Rose, Perkins, Passive dosimetry – state of the art review, Am. Ind. Hyg. Assoc. J. 43 (1982) 605–621.

[13] R.H. Brown, Position paper on their environmental monitoring potential and the need for further research. In the use of diffusive samplers for monitoring of ambient air, IUPAC Commission on atmospheric chemistry (VI.4) Report, Pure Appl. Chem. 65 (8) (1993) 1859–1874.

[14] SKC Inc., Passive Samplers, http://www.skcinc.com/, (accessed 17.08.15).

[15] Swedish Environmental Research Institute, Air Quality Monitoring – Diffusive and Passive Sampling, http://www.diffusivesampling.ivl.se/, (accessed 17.08.15).

[16] Gradko International, NO_2 and Nitrogen Oxides Diffusion Tubes, http://www.gradko.com/environmental/products/no2-and-nitrogen-oxides-diffusion-tubes.shtml, (accessed 17.08.15).

[17] Fondazione Salvatore Maugeri, radiello, http://www.radiello.com/, (accessed 17.08.15).

[18] PerkinElmer Inc., Sample Tubes for Automated Thermal Desorber (ATD) 400, http://www.perkinelmer.com/catalog/product/id/n9307005, (accessed 17.08.15).

[19] IGAC Newsletter No. 27 (DEBITS Special Issue), January 2003, http://www.igacproject.org/sites/all/themes/bluemasters/images/NewsletterArchives/Issue_27_Jan_2003.pdf, (accessed 17.08.15).

[20] M. Ferm, H. Rodhe, Measurements of air concentrations of SO_2, NO_2 and NH_3 at rural and remote sites in asia, J. Atmos. Chem. 27 (1997) 17–29.

[21] G.R. Carmichael, M. Ferm, N. Thongboonchoo, J.H. Woo, L.Y. Chan, K. Murano, P.H. Viet, C. Mossberg, R. Balasubramanian, J. Boonjawat, P. Upatum, M. Mohan, S.P. Adhikary, A. B. Shrestha, J.J. Pienaar, E.B. Brunke, T. Chen, J. Tang, G.A. Ding, C.P. Leong, S. Dhiharto, H. Harjanto, A.M. Jose, W. Kimani, A. Kirouane, J.P. Lacaux, S. Richard, O. Barturen, J.C. Cerda, A. Athayde, T. Tavares, J.S. Cotrina, E. Bilici, Measurements of sulfur dioxide, ozone and ammonia concentrations in Asia, Africa, and South America using passive samplers, Atmos. Environ. 37 (2003) 1293–1308.

[22] A.S.M. Lourens, J.P. Beukes, P.G. van Zyl, G.D. Fourie, J.W. Burger, J.J. Pienaar, C.E. Read, J.H.L. Jordaan, Spatial and temporal assessment of gaseous pollutants in the Mpumalanga highveld of South Africa, S. Afr. J. Sci. 107 (1/2) (2011) 1–8.

[23] NADP, National Atmospheric Deposition Program, Site Selection and Installation Manual, NADP Program Office, Illinois State Water Survey, Champaign, IL, 2014. http://nadp.isws.illinois.edu/lib/manuals/NADP_Site_Selection_and_Installation_Manual_2014_11.pdf (accessed 17.08.15).

[24] J. Zhou, S. Smith, Measurement of ozone concentrations in ambient air using a badge-type passive monitor, J. Air Waste Manage. Assoc. 47 (1997) 697–703.

[25] P. Bruno, M. Caputi, M. Caselli, G. de Gennaro, M. de Rienzo, Reliability of a BTEX radial diffusive sampler for thermal desorption: field measurements, Atmos. Environ. 39 (2005) 1347–1355.

[26] R.G. Lewis, J.D. Mulik, R.W. Coutant, G.W. Wooten, C.R. McMillin, Thermally desorbable passive sampling device for volatile organic chemicals in ambient air, Anal. Chem. 57 (1985) 214–219.

[27] S.V. Krupa, A.H. Legge, Passive sampling of ambient, gaseous air pollutants: an assessment from an ecological perspective, Environ. Pollut. 107 (2000) 31–45.

[28] R.S. Dhammapala, Use of diffusive samplers for the sampling of atmospheric pollutants (M. Sc. thesis), Potchefstroom University for Christian Higher Education, South Africa, 1996.

[29] P.K. Koutrakis, J.M. Wolfson, A. Bunyaviroch, S.E. Froehlich, K. Hirano, J.D. Mulik, Measurement of ambient ozone using a nitrite-coated filter, Anal. Chem. 65 (3) (1993) 209–214.

[30] M. Ferm, Private Communication, 1994.

[31] Atmospheric Chemistry Monitoring Network in Africa, Gas Collection Procedure, idaf.sedoo.fr/spip.php?rubrique38, (accessed 17.08.15).

[32] Markes International, Thermal desorption sorbent tubes for air sampling, http://www.markes.com/Products/Sampling-accessories/Sorbent-tubes/default.aspx, (accessed 17.08.15).

[33] EPA, Compendium Method TO-17, Determination of Volatile Organic Compounds in Ambient Air Using Active Sampling onto Sorbent Tubes, U.S. Environmental Protection Agency, Cincinnati, OH 45268, January 1999.

[34] J. Crank, The Mathematics of Diffusion, second ed., Oxford University Press, New York, 1975.

[35] M. Harper, C.J. Purnell, Diffusive sampling – a review, Am. Ind. Hyg. Assoc. J. 48 (1987) 214–218.

[36] S.E. Schwartz, P. Warneck, Units for use in atmospheric chemistry, Pure Appl. Chem. 67 (1995) 1377–1406.

[37] Markes International Ltd. http://www.ingenieria-analitica.com/downloads/dl/file/id/593/product/61/td8_principles_of_diffusive_monitoring.pdf.

[38] D. Grosjean, M.W.M. Hisham, A passive sampler for atmospheric ozone, J. Air Waste Manage. Assoc. 42 (2) (1992) 169–173.

[39] J. He, B. Rajasekhar, Final Report on Passive Sampler Inter-comparison under Malé Declaration, 2009 (unpublished results).

[40] NADP, National Atmospheric Deposition Program, Ammonia Monitoring Network, http://nadp.isws.illinois.edu/amon/, (accessed 09.07.15).

[41] G.R. Carmichael, Report on Passive Samplers for Atmospheric Chemistry Measurements and Their Role in GAW, WMO Report no. 122, 1997, pp. 1–21.

[42] NIOSH Manual of Analytical Methods (NMAM) 2549: Volatile Organic Compounds (Screening), 4th rev. 1996. Cincinnati, OH.

[43] M. Josipovic, Acid deposition emanating from the South African Highveld – a critical levels and critical loads assessment (Ph.D. thesis), University of Johannesburg, 2009.

[44] M. Josipovic, H.J. Annegarn, M.A. Kneen, J.J. Pienaar, S.J. Piketh, Atmospheric dry and wet deposition of sulphur and nitrogen species and assessment of critical loads of acidic deposition exceedance in South Africa, S. Afr. J. Sci. 107 (2011) 1–10.

[45] M. Josipovic, H.J. Annegarn, M.A. Kneen, J.J. Pienaar, S.J. Piketh, Concentrations, distributions and critical levels exceedance assessment of SO_2, NO_2 and O_3 in South Africa, Environ. Monit. Assess. 171 (2010) 181–196.

[46] R. Vet, R.S. Artz, S. Carou, M. Shaw, Chul-Un Ro, W. Aas, A. Baker, Van C. Bowersox, F. Dentener, C. Galy-Lacaux, A. Hou, J.J. Pienaar, R. Gillett, M.C. Forti, S. Gromov, H. Hara, T. Khodzher, N.M. Mahowald, S. Nickovic, P.S.P. Rao, N.W. Reid, A global assessment of precipitation chemistry and deposition of sulphur, nitrogen, sea salt, base cations, organic acids, acidity and pH, and phosphorus, Atmos. Environ. 93 (2014) 3–100.

[47] NADP, National Atmospheric Deposition Program, Governance Handbook, NADP Program Office, Illinois State Water Survey, Champaign, IL, 2008.

Chapter 3

Biomonitors

Patricia B.C. Forbes,*[,1] Leandri van der Wat[1] and Eve M. Kroukamp[1,2]
[1]*Department of Chemistry, Faculty of Natural and Agricultural Sciences, University of Pretoria, Pretoria, South Africa;* [2]*Spectrum Analytical Facility, University of Johannesburg, Johannesburg, South Africa*
**Corresponding author: E-mail: patricia.forbes@up.ac.za*

Chapter Outline

1. Introduction **54**
2. Lichens **56**
2.1 Introduction 56
2.2 Lichens as Biomonitors of Inorganic Air Pollutants 58
2.3 Lichens as Biomonitors of Organic Air Pollutants 59
3. Mosses **65**
3.1 Introduction 65
3.2 Mosses as Biomonitors of Air Pollutants 65
4. Conifer Needles **67**
4.1 Introduction 67
4.2 Conifer Needles as Biomonitors of Organic Air Pollutants 73
4.3 Conifer Needles as Biomonitors of Inorganic Air Pollutants 74
5. Tree Bark **76**
5.1 Introduction 76
5.2 Tree Bark as a Biomonitor of Air Pollutants 76
6. Other Plants **78**
7. Sampling Considerations **78**
7.1 Sampling of Biomonitors for Organic Analytes 83
7.2 Sampling of Biomonitors for Inorganic Analytes 84
8. Analyte Extraction Methods **85**
8.1 Soxhlet Extraction 85
8.2 Microwave-Assisted Extraction 87
8.2.1 Extraction of Organic Analytes 87
8.2.2 Extraction of Inorganic Analytes 88
8.3 Ultrasound-Assisted Solvent Extraction 88
8.3.1 Introduction 88
8.3.2 Extraction of Inorganic Analytes 91
8.3.3 Extraction of Organic Analytes 93
8.4 QuEChERS 93
8.4.1 Introduction 93
8.5 Sample Cleanup 95
9. Analytical Techniques **98**
9.1 Organic Analytes 98
9.2 Inorganic Analytes 100
10. Conclusions, Challenges and Future Outlook **100**
References **102**

Comprehensive Analytical Chemistry, Vol. 70. http://dx.doi.org/10.1016/bs.coac.2015.09.003

1. INTRODUCTION

Biomonitors are living organisms used to obtain quantitative and qualitative information about certain aspects of the environment, as they are able to assimilate contaminants over an extended period of time and are therefore reflective of environmental conditions [1]. For this reason, various biomonitors are useful in monitoring air pollution in a given region. They are commonly used in studies where the high cost of technical equipment, which would measure the air pollution profile over an extended period of time for a range of contaminants, cannot be afforded. Biomonitors provide an integrated pollution profile as a result of their ability to accumulate a range of air pollutants [2]. Most air pollution biomonitoring studies have used plants such as fungi, lichens, mosses, pine needles, tree bark and leaves [3,4]. This is because plants have been shown to bioaccumulate air pollutants through a number of different mechanisms such as entrapment of particulates, gas exchange through stomata and surface ion exchange. Moreover, plants are capable of accumulating these pollutants in concentrations which far exceed their basic metabolic requirements.

One of the earliest biomonitoring studies was on the decline of lichens in Paris and its surrounding areas as a result of air pollution [5]. Numerous subsequent studies have shown the effect of air pollution on various plants [6–8]. Applications such as these are known as *bioindicator* studies, where plants and other organisms are assessed for changes in physiology, abundance and behaviour [1]. Typically the visual health of an organism is evaluated across a geographical region by noting morphological changes, as shown in Figure 1, for example, where a study on the photosynthetic activity of lichens exposed to diesel exhaust fumes was investigated [9].

On the other hand, *biomonitoring* studies provide quantitative information obtained from measurements of accumulated compounds, reflecting the environmental profile from which the organism was sampled. As instrumentation capable of the quantification of pollutants became more accessible, an

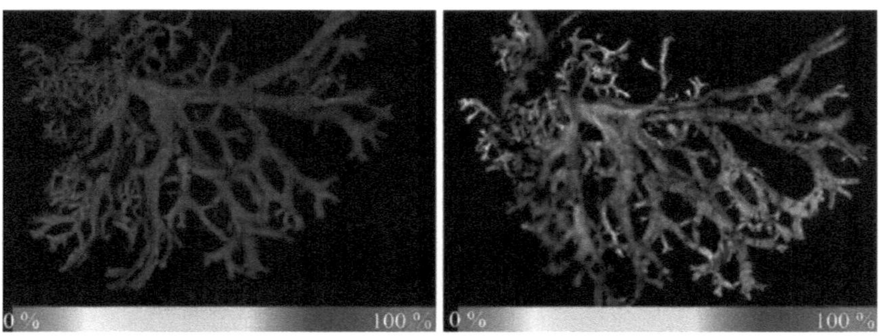

FIGURE 1 False colour imaging of *Pseudevernia furfuracea* lichen before (left) and after (right) exposure to diesel exhaust, confirming the loss of photosynthetic activity (dark blue (dark grey in print versions) indicates high photosynthetic activity) [9]. *Image courtesy of* the Journal of Environmental Protection.

increase in the number of biomonitoring studies linking environmental concentrations to concentrations in biological populations has occurred [1]. A study by Blasco et al. combined the use of lichens as biomonitors and bioindicators to assess the air quality in the Pyrenees, which included both the mapping of the lichen biodiversity as well as an investigation into the polycyclic aromatic hydrocarbon (PAH) content of lichens [10]. Typically, biomonitors are either removed from their original location and analysed directly (passive biomonitors) or transplanted from a lesser impacted to a contaminated site for a specific period of time (transplant/active biomonitors) [11].

Mosses were used to measure fluorine emitted from industrial sources in the 1950s [12] and since the 1960s, mosses have been used to chart heavy metal deposition in Scandinavia [13]. Lichens have been used as bioindicators, for sulphur dioxide, nitrogen dioxide, ozone and hydrofluoric acid [14—18].

Target analytes in biomonitoring studies have included organic compounds such as hydrocarbons as they accumulate within organisms because they are lipophilic. PAHs and their halogenated derivatives, which are potentially hazardous to living organisms [19], have been studied using lichens, moss, pine needles and bark. Other organic analytes of interest include phenols [20], polychlorinated biphenyls (PCBs) and polychlorinated dibenzo-*p*-dioxins and furans (PCDD/Fs). Inorganic air pollutants have been extensively studied using biomonitors, including isotopes [21], metals [1] (such as Fe, Zn, Pb, Cu, Mn, Co, Cd, Ni, Cr and Hg), lanthanides and rare earth elements [22], ammonia and nitrogen. In order to study these air pollutants, different types of organisms have been used, including the bark of various tree species, lichens, pine needles, moss, tobacco, wheat, barley, maize, grass, leaves of trees, leaf cuticles, wild chard and various shrubs.

When using biomonitors for air pollution monitoring, the selection of an appropriate biomonitor is vital. The chosen species should be widely dispersed throughout the sampling area [22], which may allow for national and international pollution monitoring initiatives [23]. Target organisms should be chosen based on their abundance, bioaccumulation potential and ecological relevance to the specific air pollution study. Different biomonitors may also be representative of the system for different time periods, and so the desired monitoring period and pollutant retention should be established prior to sampling. Mosses, for example, exhibit concentrations which are representative of longer accumulation periods than can be found in leaves [4]. Species selection is also important because studies have shown that different plant species within a group may experience selectivity towards specific contaminants [4,24]. In addition, if the relationship between concentrations of contaminants within the monitor and the environmental exposure levels are poorly defined, the information gained is purely qualitative [1]. As a result, a clear link between the amount of contaminant from different sources to which a biomonitor organism is exposed, and the concentration which is bioconcentrated needs to be determined [25]. Tree bark and leaves, for example, may be exposed to a variety of pollutant sources such as soil, air and groundwater whereas others, such as lichens, rely almost entirely upon atmospheric sources. It is also for this reason that higher order organisms are not as suitable as plants for air biomonitoring purposes, due to the wide range of sources to which they are exposed.

The accumulation of airborne metals and organic pollutants by biomonitor organisms is expected to be related to a number of physical and chemical factors of the pollutants, such as water solubility, hydrophobicity, vapour pressure, particle size and metal oxidation state. Environmental aspects such as temperature, as well as biomonitor species, also affect uptake rates [4]. Early work in the field of biomonitors focused on the establishment of dose–response relationships, in terms of correlations between elemental levels in lichen biomonitors and sampled air particulate matter (such as Ref. [26]). Hanssen et al. [27], Ross [28], Berg et al. [29], as well as Berg and Steinnes [24], all demonstrated the correlation between metal concentrations in moss and wet deposition. Later on, studies by Augusto et al. [30] compared the concentrations of PAHs accumulated by lichens, pine needles, soil and direct air sampling. The correlation of the air samples to both pine needles and lichen PAH data suggested that both biomonitors were effective at providing an understanding of atmospheric PAH levels.

This chapter focuses on the application of various organisms as biomonitors in air pollution monitoring, with emphasis on the more commonly utilised biomonitor species; namely lichens, moss, conifer needles and tree bark. The extraction of target analytes from within the organism's matrix as well as common analytical techniques employed to determine the identities and quantities of analytes extracted from the biomonitors are outlined. Sampling considerations, as well as general analytical challenges, are also discussed. Biomonitoring has been applied in numerous air pollution studies, and some examples of these are included in this chapter.

2. LICHENS

2.1 Introduction

Lichens are the product of a symbiotic relationship between algae (phycobiont) and a fungus (mycobiant) [31]. They are perennial, represent uniform morphology over their life span, and are slow growing [32]. Annual radial growth of 0.1–1.0 mm has been reported for crustose lichens, 1.0–2.5 mm for foliose lichens and 2.0–6.0 mm for fruticose lichens, respectively [33]. This growth rate varies between species and also between geographical locations for one species, and has been correlated to rainfall [33]. Prevalent in a variety of biomes, lichens are found in extremes of temperature, from the arctic tundra to the rainforests. Lichens grow on a range of substrata, both natural, such as rock, trees, soil, wood and leaves, as well as anthropogenic, including rubber, plastic, glass, stonework, concrete, plaster, ceramic, tiles and bricks [33]. The microclimate has been found to significantly impact on lichen colonisation and biodiversity, particularly factors such as precipitation, light intensity, humidity, wind speed and air quality [33].

The lack of a cuticle in lichens allows them to gain most of their nutrients from the air [32], and since this movement is unhindered, the hyper-accumulation of metals and organic pollutants is possible. Certain species of lichens which persist on rocks are able to supplement their nutritional requirements by releasing chemicals onto the rock substrate, thereby accelerating breakdown of the rock surface [34]. Lichens which persist on trees, however, simply use trees as a substrate and are not believed to acquire any nutrients from the tree or bark [35].

Some lichens are particularly sensitive to pollution where high atmospheric pollutant concentrations may result in the reduction or complete absence of a particular lichen species in an environment. An example of this was found in a study by Dongarrà and Varrica [36] where lichens (*Parmelia* spp.), which were found in abundance in other areas, were absent within close proximity to a volcano. Such traits are desirable in bioindicator studies. Other types of lichens are more robust; withstanding high levels of atmospheric pollutants (e.g. *Parmelia sulcata*) and some may possess detoxification mechanisms capable of degrading or excreting pollutants [34]. In situations where pollutant loading is high, however, lichens may experience reduced growth and altered morphology [37].

It was Hawksworth and Rose [38] as well as Rogers [39] who first applied lichens as biomonitors with respect to the uptake of sulphur dioxide from the atmosphere. Sloof et al. [40] compared the concentration of pollutants in lichens to the atmospheric concentrations of suspended and deposited particulate matter. Elemental analyses of Zn, Co and Sc were performed on lichens by Sloof and Wolterbeek [41], which correlated well with determined atmospheric concentrations. Migaszewski et al. [20] showed that lichens were also suitable for the study of atmospheric PAHs, followed by Guidotti et al. [42] who performed a lichen transplant study, targeting PAHs. Augusto et al. studied the PCDD/F profile for the first time in lichens sampled in Portugal [43]. For a full review on lichens as biomonitors for organic air pollutants, the reader is referred to a paper by van der Wat and Forbes [44]. Lichens have also been used to study ammonia and nitrogen deposition [45,46]. Over the years, lichen biomonitoring research has been conducted across a wide spectrum of inorganic analytes including Hg, most transition metals, radionuclides, fluoride, S, N and acid rain, as a result of bioaccumulation in the lichen thallus [2,31,47–50].

Lichens found in nature exist in different growth forms, namely foliose (leaflike, Figure 2), crustose (form a crust), squamulose (flattened-pebble appearance) and fruticose (branching tubes).

Proposed mechanisms for accumulation of air pollutants by lichens include the trapping of particles on the surface, diffusion of gases across the cell

FIGURE 2 A foliose lichen suitable for air pollution biomonitoring.

membrane or ion exchange. Factors such as surface water (rain) have been found to facilitate the movement of metals into the lichen thalli via ion-exchange processes [51]. Extracellular uptake of metal ions is determined by ligands in the walls of the fungus component of the lichen, whereas intracellular uptake is subject to the nature of the ion, the cell membrane permeability and the affinity of extracellular ligands for cations [52]. As a result of the lipophilic nature of the surface of lichens, the uptake of hydrophobic compounds, such as semi-volatile organic compounds (SVOCs), is easily accommodated by the surface, where the compounds' incorporation into cells is metabolically controlled [53,54]. Organic air pollutants are typically lipophilic, thus uptake tends towards dry deposition [55,56].

2.2 Lichens as Biomonitors of Inorganic Air Pollutants

Metal concentrations in lichens have been shown to correlate well with atmospheric concentrations [3]. It was thought that mosses are more likely to accumulate metals than lichens due to their growth structure and porosity [3]; however, contradictory evidence later found that lichens possessed higher concentrations of heavy metals than mosses [57]. Foliose lichens have primarily been used as biomonitors [1,3,22,31,32,36,57], due to their ease of removal from the substrate. They also present a large surface area and are widely distributed in a variety of different habitats. Crustose [58], fruticose [59] and squamulous [58] lichens have also been used in air pollution monitoring, although to a lesser extent. Crustose lichens are often difficult to remove from the substrate without contamination of the sample taking place. Certain types of *Parmelia* species growing on rock surfaces have been used to assess air pollution [36], although uptake of minerals from the rock substrate may impact on the results [34]. Contamination from soil and groundwater are also potential limitations in such studies. It should be noted that *Diploschistes muscorum* is capable of sequestering metals as oxalate crystals or lichen acid complexes [34].

Species differences have been found to affect the bioconcentration potentials for certain elements, as was shown in a study by Mendil et al. [59]. In this study Fe, Zn and Cr were found to accumulate highest in *Physcia adscendes*, Mn and Cu in *Flavoparmelia caperata*, Ni and Cd in *Xanthoria parietina* and Pb in *Ramalina polymorpha*. Species differences have also been observed for the monitoring of nitrogen where *P. sulcata* was found to accumulate the lowest amount of nitrogen, and higher concentrations were observed in both *X. parietina* and *Physcia* spp. [21]. Active biomonitoring using lichen transplants has the advantage of allowing their use in areas where they may be naturally absent and prevent the influence of substrate concentrations on *in situ* samples [60].

In terms of the analysis of inorganic pollutants in lichens, a wide range of techniques has been employed, such as inductively coupled plasma–mass spectrometry (ICP-MS), inductively coupled plasma–optical emission spectrometry (ICP-OES), X-ray absorption spectroscopy, flame atomic absorption spectroscopy and instrumental neutron activation analysis. Some of the most common inorganic analytes which have been assessed in lichen biomonitoring studies, and the range of concentrations thus determined, are shown in Table 1.

2.3 Lichens as Biomonitors of Organic Air Pollutants

Lichens were first applied as biomonitors to the investigation of inorganic pollutants, but in time their application to the monitoring of organic air pollutants such as PAHs [42], phenols [20], PCDD/Fs [63] and PCBs [64–66] was realised, for which various lichen species have been employed [67–70]. Table 2 shows examples of the organic analytes studied using lichens, as well as the corresponding sample preparation techniques, methods of analysis and concentrations thus determined.

As with mosses, lichens are believed to accumulate organic air pollutants by both dry and wet deposition, in a passive fashion [40]. Lipid metabolites released onto the lichen surface are suspected to behave in a similar way to the cuticle in plants [71]. Bauer et al. [72] and Baur et al. [73] showed that diffusion rates of heavier persistent organic pollutants (POPs) are lower than those of the lower molecular weight POPs, therefore heavier POPs are most often associated with particles that remain on the surface of lichens. Augusto et al. [74] also showed that the concentrations of PCDD/Fs remain stable after periods of wet conditions (rain or fog, for example), indicating that some POPs are incorporated into the lichen thallus and are not easily rinsed off. The trend has not been established for all classes of organic air pollutants, but a similar trend is expected [74,75].

TABLE 1 Examples of Inorganic Air Pollutants Determined Using Lichens as Biomonitors

Lichen Species	Location	Analytes	Sample Mass (g)	Extraction Method	Detection Method	Concentration Detected	References
Parmelia sulcata	The Netherlands	Al, As, Br, Ca, Cd, Co, Cr, Cs, Cu, Fe, Ga, Hf, Hg, I, K, La, Lu, Mg, Mn, Na, Ni, Pb, Pb, Sb, Sc, Se, Sr, Th, Tl, U, V, W, Yb, Zn	0.030–0.1 g	N/A	INAA	Range (μg g^{-1}): Lu 0.009–Fe 30,000	[1]
Hypogymnia physodes	Finland	Cr, Ni	0.2 L fresh volume	Wet digestion	ICP-OES	Range (μg g^{-1}): Cr 4–5, Ni 3.3–3.8	[3]
H. physodes	Finland	Cr, Pb, Ni, Cd, Cu, Fe, Zn	Whole thalli	Wet digestion	ICP-OES	Average (μg g^{-1}): Cd 0.69, Cr 2.1, Cu 7.1, Fe 540, Ni 2.5, Pb 17, Zn 84	[57]
H. physodes	Slovenia	Ag, As, Ba, Br, Ce, Cd, Co, Cr, Cs, Fe, Ga, Hf, Hg, K, La, Mo, Na, Rb, Sb, Sc, Se, Sr, Sm, Tb, Th, U, W, Zn	0.1–0.2 g	N/A	INAA	Range (μg g^{-1}): Tb 0.001–K 8644	[22]
Parmelia sp.	Italy	Au, As, Ba, Br, Ca, Co, Cr, Cs, Fe, Hf, K, Mo, Na, Rb, Sb, Sb, Sc, Th, U, Zn, Cu, Pb, Ni, Mn, Sr, V, P, Mg, Ti, Al, Y	Not specified	N/A	INAA, ICP	Range (ppm): Au 0.05–Ca 14,667	[36]
Parmelia caperata	Italy	Cr, Cu, Fe, Mn, Zn, Al, Cd, Ni, Pb, Hg	Not specified	Pressurised digestion system	FAAS and GFAAS	Range (μg g^{-1}): Cr 0.1–Al 277	[32]

Parmotrema austrosinense	South Africa	Cr, Mn, Co, Ni, Zn, Sn, Pb	0.25 g	Microwave	FAAS	Average (μg g^{-1}): Pb 41.5–181.1, Mn 97.1	[31]
Ramalina polymorpha, Xanthoria parietina, Physcia stellaris, Flavoparmelia caperata, Physcia adscendens	Turkey	Fe, Mn, Zn, Pb, Ni, Cr, Cu, Cd	1 g	Microwave	FAAS, GFAAS	Range (μg g^{-1}): Fe 103.8–455.5, Mn 31.6–170.5, Zn 10.4–77.6, Pb 0.9–6.5, Ni 1.0–10,1, Cr ND-3.8, Cu ND-25.6, Cd ND-1.5	[59]
P. sulcata	Ghana	Sb, Mn, Cu, V, Al, Co, Hg, As, Cd, Th	0.2 g	N/A	INAA	Range (ppm): Co 0.12–Mn 1097	[61]
Cladonia furcata, Hypocenomyce scalaris, Lepraria incana	Poland	Mn, Pb	0.005–0.020 g	Ultrasound	FAAS	Range (μmol g^{-1}): Zn 1.28–17.16, Pb ND-2.66	[58]
X. parietina	Turkey	Al, As, Cd, Co, Cu, Fe, Hg, K, Mn, Ni, Pb, S, Ti, Tl, V, Zn	1 g	Wet digestion	ICP-MS	Range (μg g^{-1}): Tl 0.02–Fe 13,550	[62]
P. sulcata, P. adscendens, Physcia tenella, X. parietina	Germany	N isotopes	0.003–0.004 g	N/A	Element analyser coupled with isotope ratio MS	Range (%): 28–41	[21]

ND, not detected; FAAS, flame atomic absorption spectroscopy; ICP-OES, inductively coupled plasma-optical emission spectrometry; ICP-MS, inductively coupled plasma-mass spectrometry; INAA, instrumental neutron activation analysis; GFAAS, Graphite furnace atomic absorption spectroscopy.

TABLE 2 Various Lichen Biomonitoring Studies for the Analysis of Organic Air Pollutants

Lichen Species	Location	Analytes	Sample Mass (g)	Extraction Method	Analysis Method	Concentration Detected	References
Pseudevernia furfuracea (L.) *Zopf*	Dolomites, Italy	PAHs	0.6–0.8	Automated Soxhlet extraction	GC-MS	758.0 ng g^{-1}	[76]
Parmotrema hypoleucinum	Portugal	PAHs	2	Soxhlet	HPLC-FLD-DAD/V-UV	95.5–873.8 ng g^{-1}	[30]
P. hypoleucinum (Steiner) Hale	Portugal	PAHs	2	Soxhlet	HPLC-FLD-DAD/V-UV	93.56–599.24 ng g^{-1}	[69]
P. hypoleucinum (Steiner) Hale	Portugal	PAHs	2	Soxhlet	HPLC-FLD-DAD/V-UV	58–536 ng g^{-1}	[77]
Hypogymnia physodes	Poland	PAHs	Not reported	Soxtec liquid–solid extraction	GC-MS	3–1887 ppb	[20]
Parmelia sulcata	Spain	PAHs	0.2	DSASE	GC-MS	352–1654 ng g^{-1}	[78]
Evernia prunastri	Pyrenees, Spain	PAHs	0.2	DSASE	GC-MS	692–6420 ng g^{-1}	[10]
P. sulcata	Somport tunnel, Spain	PAHs	0.2	DSASE	GC-MS	0.91–1.92 μg g^{-1}	[68]
P. sulcata, E. prunastri, Ramalina farinacea, P. furfuracea, Usnea sp. and *Lobaria pulmonaria*	Pyrenees, Spain	PAHs	0.2	DSASE	GC-MS	238–6240 ng g^{-1}	[79]
Xanthoria parietina	Zaragoza city, Spain	PAHs	0.2	DSASE	GC-MS	+/– 330 ng g^{-1}	[80]

Rinodina sophodes (Ach.) Massal	Kanpur city, India	PAHs	2	Soxhlet	HPLC-V-UV	0.189–0.494 $\mu g\ g^{-1}$	[81]
Phaeophyscia hispidula	Himalayas, India	PAHs	2	Soxhlet	HPLC-V-UV	3.38–25.0 $\mu g\ g^{-1}$	[82]
Pyxine subcinerea	India	PAHs	2	Soxhlet	HPLC-V-UV	1.25–187.3 $\mu g\ g^{-1}$	[67]
Dermatocarpon vellereum Zschacke	Himalayas, India	PAHs	2	Soxhlet	HPLC-V-UV	0.136–4.96 $\mu g\ g^{-1}$	[83]
P. furfuracea	Italy	PAHs	2	USAE	GC-MS	56–159 $\mu g\ kg^{-1}$	[42]
P. furfuracea	Italy	PAHs	2	USAE	GC-MS	168.0–395.0 $ng\ g^{-1}$	[84]
Xanthoria Parientina	Portugal	PCDD/Fs	Not reported	Solvent extraction	GC-HRMS	427.74 $ng\ kg^{-1}$	[43]
X. Parientina	Portugal	PCDD/Fs	Not reported	Soxhlet	HRGC/HRMS	197.5–1218.7 $ng\ kg^{-1}$ dw	[63]
X. Parientina and *Ramalina canariensis Steiner*	Portugal	PCDD/Fs	Not reported	Soxhlet	HRGC/HRMS	170.8–1058.6 $ng\ kg^{-1}$	[70]
X. Parientina	France	PCDD/Fs, PCBs	Not reported	Not reported	GC-HRMS	Only toxic equivalents reported	[85]
Pyxine coralligera	Venezuela	PAHs	2	USAE	HPLC-FLD-DAD	0.24–9.08 $\mu g\ g^{-1}$	[86]
Cladonia alpestris	Sweden	PCBs, DDT, DDE and chlorinated hydrocarbons	Not reported	Soxhlet	GC-ECD	–	[65]

GC-MS, gas chromatography-mass spectrometry; HPLC-FLD-DAD/V-UV, high performance liquid chromatography fluorescence detector-diode array detector/visible ultraviolet detector; HPLC-V-UV, high performance liquid chromatography visible ultraviolet detector; GC-HRMS, gas chromatography - high-resolution mass spectrometry; HRGC/HRMS, high resolution gas chromatography/high resolution mass spectrometry; GC-ECD, gas chromatography-electron capture detector.

One of the limitations of using lichens as biomonitors is that for a long time, there existed no studies reporting the translation of the concentrations of organic analytes found in lichens into atmospheric equivalent concentration values. However, a study by Augusto et al. [74] investigated PAH concentrations in lichens and compared the results to those from an active outdoor air sampler. This study showed a good correlation for the sum of the 16 U.S. Environmental Protection Agency (EPA) PAHs, benzo[*a*]pyrene and the higher molecular weight PAHs. Seasonal fluctuations in the PAH profile were also observed in this study, suggesting that the season in which lichens are sampled should be kept consistent [74].

According to Jensen [87], who studied pine needles in Scandinavia and compared the results to an ongoing moss and lichen study, lichens were not able to accumulate high levels of the more volatile SVOCs, such as hexachlorocylohexane (HCH). HCH levels were seen to have large, erratic variations amongst lichen samples, compared to pine needles that showed good reproducibility. It was therefore concluded that lichens are not suitable for the analysis of HCH, as a result of the large variations after rainfall, suggesting that lichens are not capable of accumulating these specific analytes sufficiently [75,87].

In contrast, a study by Augusto et al. [30] showed that lichens accumulate PAHs more effectively than soil and pine needles. It was similarly demonstrated in a study by Schrlau et al. [70] that lichens accumulate PAHs more effectively than pine needles and mosses. Migaszewski et al. [20] investigated PAHs in lichens and sampled the host bark in addition to lichens, concluding that the lichen contained higher levels of PAHs, compared to its host bark. The concentrations were also found to be consistent throughout various lichen samples (of the same species), despite growing on different tree species. These studies have provided convincing evidence of the potential use of lichens as biomonitors for the study of atmospheric pollution, particularly in terrain that is difficult to access with bulky, expensive sampling equipment.

Interspecies differences in lichens have been investigated for both PCDD/Fs [88] as well as PAHs [79], where differences in the profile of both classes of pollutants were observed, which highlights the importance of appropriate choice of lichen species, as well as consistent sampling to minimise variation. Three different lichen classes have been studied to identify which lichens absorb organic air pollutants most effectively. The lichen class absorbing low molecular weight POPs most effectively is reported to be fruticose lichens as a result of their high surface area to volume ratio, whereas foliose lichens appear to favour the absorption of the higher molecular weight organic pollutants [69,70,79,88]. However, in general, despite the lichen class studied, similar POP profiles have been reported; where two- and four-ring PAHs are primarily found in PAH studies [10,20,30,42,68,76,82] with phenanthrene, naphthalene, fluoranthene and benzo[*a*]anthracene usually being the

PAHs found at the highest concentrations [30,79]. PCDD/F profiles often are dominated by TeCDD/PeCDDs and PeCDFs [43,63,88].

3. MOSSES

3.1 Introduction

Mosses are flowerless, nonvascular plants that form part of the phylum Bryophyta. They are often found in damp, shady locations and grow in dense mats. Mosses can be divided into two predominant categories: *Pleurocarpous* mosses which are prostrate, highly branched and matted [89] and *Acrocarpous* mosses which are tufted, upright, slightly branched and are more tolerant to dryness and air pollution than the *Pleurocarpous* spp. [89]. The four main growth types of mosses are sheets, cushions, haircaps and rockcaps. Mosses have a wide distribution and can be found from sea level to some of the highest altitudes occupied by plants [90]. They grow on a variety of different substrates including trees, rocks, buildings and in wetlands.

Similar to lichens, Bryophytes do not possess a root system or waxy cuticle and have poorly developed vascular bundles [4]. Mosses are therefore reliant upon the air as a major source of nutrients, although some Bryophytes may also acquire metals from the substrate upon which they are growing through rising capillary water [24]. Mosses possess a large surface area to mass ratio [4] and their lack of a cuticle layer allows minerals, water and pollutants to easily penetrate the plant cells, and the cationic exchange capacity of mosses is high [91]. Moreover, the absence of well-developed vascular bundles allows mosses to absorb higher concentrations of metals than other vascular plants [92]. As outlined by Spagnuolo et al. [25], the major factors affecting accumulation of metal contaminants by mosses are the amount and type of metal deposition as well as the form of metal (soluble or insoluble). In addition, meteorological conditions are of importance [93]. For inorganic air pollutants, the cation-exchange properties of mosses are important in terms of analyte uptake, whilst for organic analytes, the capacity of mosses to absorb large organic molecules is key [94]. The Antarctic mosses have been used as biomonitors to assess POPs and PAHs [95]. An observed decrease in concentrations of these pollutants was observed over a period of 8 months after a fire at the Brazilian Antarctic station, indicating detoxification mechanisms of mosses.

3.2 Mosses as Biomonitors of Air Pollutants

Mosses from the *Pleurocarpous* species are used most frequently in air biomonitoring studies because of their easy isolation from the substrate, and therefore contamination by soil particles is minimal [4,96]. Due to their slow growth and large surface area, accumulated metal concentrations are often high, increasing the ease of chemical analysis and reducing the effect of small amounts of contamination which may occur [24]. In terms of sampling, ground growing mosses are generally collected in areas between canopies and therefore do not reflect deposits which may occur below the canopy of trees.

In terms of metal accumulation, mosses have been shown to reflect temporal trends of atmospheric deposition over longer periods than leaves, whilst for organic air pollutants, mosses such as *Leptodon smithii* have been found to accumulate higher concentrations of atmospheric metals than holm oak leaves but lower concentrations of low molecular weight PAHs [4]. In a study by Kansanen and Venetvaara [3], mosses were shown to have a clear relationship between deposition and uptake, exhibited good reproducibility and were found to accumulate higher concentrations of metals in comparison to lichens, although this was disputed by Lippo et al. [57]. The discrepancies in these findings could be due to a number of factors such as cleaning procedures and the part of the plant used for analysis as Lippo et al. [57] used the youngest segments of the moss for analysis whereas Kansanen and Venetvaara [3] used a mixture of young and old segments. It is not always possible to find mosses growing in industrial areas, which are often the areas of interest for air pollution studies [97]. Moreover, certain species of mosses may be absent in dry areas and therefore may not be suitable for global sampling campaigns [98]. The porous structure of mosses is also particularly difficult to clean of detritus therefore they cannot be used to accurately determine bioconcentrated pollutants such as metals [25].

Species differences in mosses have been observed in biomonitoring studies, where *Tortula muralis* was found to accumulate higher concentrations of Fe, Cu and Cd than *Tortula intermedia*, and the latter was found to accumulate higher concentrations of Mn and Zn [59]. *Barbula unguiculata* from this study was also found to exhibit a degree of selectivity for certain target elements, specifically Pb, Ni and Cr. Species differences have also been observed in studies by Berg and Steinnes [24] where *Hylocomium splendens* accumulated higher concentrations of Cr, Fe, Co, Ni, Cu, Ga, Nb, Mo, Sb, Eu, Gd, Tb, Dy, Er, Tm, Lu, W, Tl, Pb and Th than *Pleurozium schreberi*, which exhibited selective accumulation for V, Mn, Rb and Cd. Uyar et al. [91] assessed a number of different species and found favourable accumulation in *Hypnum lacunosum* for Fe and Cr, *B. glaerosum* for Pb, *Hypnum cupressiforme* for Ni and *S. purum* for As, Cu and Co. Species differences have even been observed in N isotope studies where concentrations were found to be lowest in *P. schreberi* and the highest in *H. cuppresiforme*. Such selectivity exhibited by species is an important consideration in moss species selection. These differences also imply that interspecies calibrations cannot be effectively used in biomonitoring initiatives.

A range of inorganic and organic air pollutants have been determined in moss substrates from as early as 1950 [94]. In some cases both inorganic and organic analytes have been targeted in the same study, such as by Gerdol et al. [99], where PAHs and metals were determined in urban and rural sites in

Northern Italy. It is possible to use mosses that are naturally available in the study area, or to transplant them from 'clean areas' or background sites to sampling sites of interest. This overcomes the restrictions of the lack of natural occurrence and also allows for well-defined exposure times [94]. Moss biomonitors have been used to determine the fate and atmospheric transport of air pollutants, such as the study by Liu et al. [100], which provided insights into the role of high mountains in South China in the atmospheric transport of PAHs. Mosses such as *Polytrichum formosum* have been used as biomonitors for Pb source apportionment studies using isotope ratios [101]. Mosses have also been employed in national surveys, such as in Hungary, where heavy metals in moss growing at 116 sites were determined [102]; as well as in long-term studies, such as that of Poikolainen et al. [103], which investigated the temporal variation in heavy metal content of mosses from a sample site over the period 1985–2000 [103]. Examples of some of the most commonly assessed inorganic and organic pollutants in mosses are included in Table 3, which also includes the analytical methods employed and the range of concentrations found.

4. CONIFER NEEDLES

4.1 Introduction

As a result of the waxy layer on conifer needles, they are able to accumulate gaseous and particulate-bound pollutants [111]. It is believed that pine needles accumulate organic air pollutants by an uptake process via their stomata or via diffusion through the wax layer [112]. The accumulation ability of pine needles is important since they take up metals at concentrations much higher than the required essential levels. As a result of their large surface area, they are able to trap airborne elements, which are further improved by their high cation-exchange capacity [113]. Conifers are evergreen; therefore they can accumulate air pollutants for several years. In addition, needles ranging from 1 to 3 years can be found on the same branch and the age of pine needles can be easily determined [114].

The first study using pine needles as biomonitors was targeted towards trace elements (Hg, Cd, Cu, Zn, Fe and Co) in Italy [115]. In later years, various SVOCs including chlorobenzenes (1,2,3,4-tetra-, penta-, hexachlorobenzene (HCB)), HCH isomers (α-, β-, γ-, δ-, ε-HCH), DDX (*p,p'*-DDT, -DDE, -DDD), PCB congeners and PAHs [116] were analysed in pine needles. PAHs have also been investigated using pine needles in other studies [117].

Due to the prevalence of pine species across the world, they are a good choice of biomonitor since they allow a global understanding of pollutant levels in pine needles to be determined, as well as to establish pollutant transport patterns between regions. They have therefore been employed in intercontinental studies, such as in a comparison of atmospheric PAH levels

TABLE 3 Examples of Biomonitoring Studies Focusing on Inorganic and Organic Air Pollutants in Mosses

Moss Species	Location	Analytes	Sample Mass (g)	Extraction Method	Detection Method	Concentration Detected	References
Hylocomium splendens and Pleurozium schreberi	Finland	Cr, Ni	2 L fresh moss	Wet digestion	FAAS	Range ($\mu g\ g^{-1}$): Cr 1–6, Ni 3–6	[3]
H. splendens	Norway	Li, Be, B Na, Mg, Al, Ca, V, Cr, Mn, Fe, Co, Ni, Cu, Zn, Ga, As, Rb, Sr, Y, Mo, Cd, Sb, Te, Cs, Ba, La, Hg, Tl, Pb, Bi, Th, U	0.5	Wet digestion in pressurised vessels	ICP-MS	Range ($\mu g\ g^{-1}$): Be 0.02–Ca 3100	[104]
H. splendens	Norway	Na, Mg, Al, Cl, K, Ca, Sc, V, Cr, Mn, Fe, Co, Ni, Cu, Zn, As, Se, Br, Rb, Sr, Zr, Mo, Ag, Sb, I, Cs, Ba, La, Sm, Eu, Tb, Yb, Hf, Ta, W, Au, Th, U	0.5	N/A	INAA	Range (ppm): Sc 0.03–Fe 72,100	[105]
H. splendens, P. schreberi	Finland	Cr, Pb, Ni, Cd, Cu, Fe, V, Zn	Green parts	Wet digestion	ICP-OES	Range ($\mu g\ g^{-1}$): Cd 0.38–Fe 380	[57]
Hypnum cupressiforme	Turkey	Pb, Cd, Zn, Cr, Mn, Fe	2	Hotplate wet digestion	FAAS	Range ($\mu g\ g^{-1}$): Cd 6.87–Fe 16,500	[106]

H. splendens, P. schreberi	Norway	Li, Be, Mg, Ca, V, Cr, Mn, Fe, Co, Ni, Cu, Zn, Ga, Ge, As, Se, Rb, Sr, Y, Zr, Nb, Mo, Cd, Sn, Sb, Cs, Ba, La, Ce, Pr, Nd, Sm, Eu, Gd, Tb, Dy, Ho, Er, Tm, Yb, Lu, Hf, Ta, W, Tl, Pb, Th, U	0.5	Pressurised hotplate wet digestion	ICP-MS	Range ($\mu g\ g^{-1}$): Tm 0.000095–Ca 3700	[24]
Polytrichum formosum	Germany	Pb isotopes	0.15–0.2	Microwave	ICP-MS	Range of ratios: 1.131–1.154	[101]
Pleurocarpous mosses	Turkey	Fe, Pb, Ni, Cr, As, Cu, Co	1	Hotplate wet digestion	GFAA[a]	Range ($\mu g\ g^{-1}$): Ni 0.11–Fe 4166.5	[91]
Tortula muralis, Barbula acuta, Barbula unguiculata, Tortula ruralis, Bryum capilare, Tortula intermedia	Turkey	Fe, Mn, Zn, Pb, Ni, Cr, Cu, Cd	1	Microwave digestion	FAAS and GFAA	Range ($\mu g\ g^{-1}$): Cu 1.6–Fe 468.1	[59]
H. cupressiforme	Spain	V, Cr, Ni, Cu, Zn, As, Cd, Hg, Pb and total nitrogen	Not specified	Microwave digestion	ICP-MS	Range ($\mu g\ g^{-1}$): Hg 0.03–Zn 1087	[93]
H. cupressiforme	Spain	PAHs and NPAHs	0.2	Ultrasound	APGC-Q-TOFMS[b]	Sum of PAHs ($ng\ g^{-1}$) 188–1733, sum of NPAHs ($ng\ g^{-1}$) 41–315	[107]

Continued

TABLE 3 Examples of Biomonitoring Studies Focusing on Inorganic and Organic Air Pollutants in Mosses—cont'd

Moss Species	Location	Analytes	Sample Mass (g)	Extraction Method	Detection Method	Concentration Detected	References
H. cupressiforme	Italy	Cd, Cu, Zn, As, Pb, Ni, V, Pt, Pd, Cr, Mn, Al, Fe	0.25	Microwave	ICP-MS	Concentrations not reported. Bioconcentration in order of: Al > Fe > Mn > Zn > Pb > Cu > V > Cr > Ni > As > Cd > Pb > Hg	[97]
H. cupressiforme	Romania	PCBs, PBDE, OCPs, DDT and metabolites, HCH isomers	3	ASE system[c]	ECNI-GC-MS[d]	Range (ng g^{-1}): sum HCH 8.9–133, sum DDT 5.8–95.3, sum Chlordanes: ND-0.4	[108]
H. cupressiforme, P. schreberi, Pseudoscleropodium purum	Germany	N isotopes	0.003–0.004	N/A	Elemental analyser coupled with isotope ratio mass spectrometer	Range: 18–49%	[21]
Leptodon smithii	Italy	Ca, Cd, Cr, Cu, Fe, K, Mg, Mn, Na, Ni, Pb, Zn and 16 PAHs	Not specified	Microwave digestion	FAAS. GFAA, GC-MSD[e]	Total PAHs (ng g^{-1}): 72.33–380.98	[4]

Sanionia uncinata, Warnstorfia sarmentosa	Antarctica	PCBs, HCB, PAHs	3 g	Soxhlet extraction	GC-ECD[f]	Sum PCBs (ng g^{-1}): ND-1.64, DDTs: ND-1.93, HCHs: ND-2.91, sum PAHs: 126–955	[95]
Brachythecium rutabulum	Spain	PAHs	10–30 mm long moss shoots	Ultrasonic bath	GC-MS	Not specified	[109]
T. muralis	Italy	PAHs, N, Cd, Co, Cr, Cu, Fe, Ni, Pb, V, Zn	5–0.5 g	Ultrasound, wet digestion	GC-MS, elemental analyser, ICP-MS	Metals: range (μg g^{-1}): Cd 0.35–Fe 15,605	[99]
Sphagnum	The Netherlands	PAHs	20 g	Soxhlet	LC-fluorescence detection	Range (ng g^{-1}): 15 (benzo(k) fluoranthene, rural site)–13,000 (fluoranthene, impacted site)	[94]
H. cupressiforme	Hungary	PAHs	5 g	Ultrasound	HPLC-fluorescence detection	Total PAHs (μg kg^{-1}): $0.1567–10.45 \times 10^4$	[110]
H. cupressiforme	Hungary	Cd, Cr, Cu, Fe, Ni, Pb, V and Zn	0.2 g	ASE	ICP-AES	Mean (mg g^{-1}): Cd 0.9–Fe 2070	[102]
H. splendens and P. schreberi	Finland	Cd, Cr, Cu, Fe, Ni, Pb, V and Zn (and As and Hg for some samples)	~2 g	Wet digestion	ICP-AES; CV-AFS[g] (Hg); GFAA (As)	Range (μg g^{-1}): Cd 0.01–Fe 3150	[103]

Continued

TABLE 3 Examples of Biomonitoring Studies Focusing on Inorganic and Organic Air Pollutants in Mosses—cont'd

Moss Species	Location	Analytes	Sample Mass (g)	Extraction Method	Detection Method	Concentration Detected	References
Hypnum plumaeformae	China	PAHs	5 g	ASE	GC-MSD	Sum PAHs (ng g^{-1}): 310–1340	[100]

ICP-AES, inductively coupled plasma-atomic emission spectroscopy; FAAS, flame atomic absorption spectroscopy; ICP-OES, inductively coupled plasma-optical emission spectrometry; ICP-MS, inductively coupled plasma-mass spectrometry; INAA, instrumental neutron activation analysis; GC-MS, gas chromatography-mass spectrometry.
[a] *Graphite furnace atomic absorption.*
[b] *Atmospheric pressure gas chromatography–quadrupole time-of-flight mass spectrometry.*
[c] *Accelerated solvent extraction.*
[d] *Electron capture negative ion gas chromatography–mass spectrometry.*
[e] *Gas chromatography–mass selective detector.*
[f] *Gas chromatography–electron capture detector.*
[g] *Cold vapour–atomic fluorescence spectroscopy.*

between Korea, Mexico and the United States [118]. Not only are the needles used as biomonitors, but also often the bark of pine trees is studied in addition to the needles as a comparison [119]. Mosses have also been studied in conjunction with pine needles for comparative purposes to determine persistent organic air pollutants at local and regional scales [120].

In terms of trace elements, where the elements exceed the essential (background) concentration it is indicative of additional uptake either via the soil or via atmospheric wet and dry deposition [121]. Using enrichment factors, elemental source apportionment is possible if soil or bedrock concentrations are determined concurrently with the pine needle concentrations [122], as in a study by Lehndorff et al. [121].

4.2 Conifer Needles as Biomonitors of Organic Air Pollutants

Pine needles have been assessed as a tool for long-term trend determinations of atmospheric levels of organic pollutants, specifically PAHs, PCBs and organochlorine pesticides (OCPs) [114], where it was found that the needles provided very similar information to high-volume air sampling.

Pine needles have been used to study POPs, as it has been shown that they accumulate these pollutants [70]. However, it was observed that lichens accumulated higher concentrations of POPs, compared to the pine needles. This was also observed by Migaszewski et al. [20] in Poland, who found that the lichen *Hypogymnia physodes* bioaccumulated higher levels of organic analytes than pine needles. This is most likely due to the fact that lichens have longer (and mostly undefined) exposure periods than that of pine needles [77]. Pine needles are also known to accumulate gas-phase POPs preferentially over particle-bound POPs [70], possibly due to the smaller surface area of the needles.

As a result of a DDT spraying program conducted in the Southern regions of East Germany during 1984, a study was conducted by Eriksson et al. [123] to monitor the extent of pesticide pollution, which showed that pine needles are appropriate biomonitors for DDT and its derivatives. Around the same time, Stratchen et al. [124] studied organochlorine compounds in pine needles across West Germany, Denmark, Sweden and Norway. Successfully targeted compounds included HCB, pentachlorophenol, HCH isomers (α-HCH and γ-HCH (lindane)), DDT and DDE. The study found that there was no difference amongst pine needles sampled from trees of different ages. This has not been the case for studies of other air pollutants, such as that by Bargagli et al. [115] which used pine needles to monitor Hg, as well as a study by Lehndorff et al. [117] who found that fresh and old needles varied in total PAH concentration. The study by Lehndorff et al. [117] also identified large seasonal variations, where the older pine needles showed large increases in total PAH concentration during the winter months. A study conducted in Portugal [125] investigated the PAH profile across several geographic regions in

Portugal; in cities, industrial areas, rural areas and remote locations. The PAH concentrations found in the needles ranged from 96 ng g^{-1} dry weight (dw) (remote sites), 118 ng g^{-1} dw (rural sites), 337 ng g^{-1} dw (urban sites) to 866 ng g^{-1} dw (industrial areas).

Pine needles were used to study the PAH profile in Portugal, Greece and Spain, where Ratola et al. [126] found PAH levels to be similar in all countries (279, 301, 294 ng g^{-1} dw, respectively) with a general dominance of three- and four-ring PAHs. This study also used the diagnostic ratios of phenanthrene/anthracene and fluoranthene/pyrene that indicated mixed petrogenic and pyrogenic sources of PAHs, demonstrating the applicability of using pine needles as biomonitors in this regard. A recent study by Silva et al. [127] investigated brominated flame-retardants (BFRs), PCBs, OCPs, PAHs and one class of contaminant of emerging concern, the synthetic musk fragrances (SMCs) in a multicomponent study using pine needles. Various determined concentrations across sites ranged 0.45–0.87 ng g^{-1} dw (BFRs), 0.35–1.01 ng g^{-1} dw (PCBs), 0.26–12.2 ng g^{-1} dw (HCBs), 245.7–967.8 ng g^{-1} dw (PAHs) and around 277.5 ng g^{-1} dw (SMCs). A selection of various studies using pine needles as monitors for organic air pollutants are summarised in Table 4.

4.3 Conifer Needles as Biomonitors of Inorganic Air Pollutants

The use of vascular plant leaves such as conifer needles has successfully been applied to the biomonitoring of heavy metals [132], such as a study by Bargagli et al. [133] which found elevated levels of mercury in older pine needles and a similarity between the concentration in pine needles and lichens was observed. It was suggested that pine needles are accumulators of mercury, and that the release of metals from pine needles may play an important role in the mercury cycle [133].

In a study by Ceburnis and Steinnes [134], correlations between the concentration of heavy metals in both needles and mosses suggested that mosses were more appropriate as biomonitors for metals due to higher elemental concentrations. Their study therefore recommends the use of mosses when a multi-component analysis is to be made, but does not disregard the use of pine needles when moss is not suitable or available for use as a biomonitor. It has also been suggested that pine needles are only feasible as biomonitors for trace elements when sampled close to the pollution source under investigation [134].

Lehndorff et al. [135] sampled pine needles in an urban area in Cologne, Germany, for the biomonitoring of atmospheric Fe, Pb, Cd, Zn, Cr, Cu, Ni and sulphur, and correlated the results to PAH data. The concentrations found for Fe, Pb and Cd were 132, 1.1 and 0.06 mg g^{-1}, respectively, showing accumulation above essential nutrient levels. Sulphur concentrations were found to be between 868 and 2076 mg g^{-1}, and the elevated levels of Fe, Pb and Zn were suggested to be indicative of high traffic volumes, as supported by the PAH results [135]. A study by Dmuchowski and Bytnerowicz [136] mapped

TABLE 4 Various Studies Using Pine Needles to Monitor Organic Air Pollutants

Pine Species	Location	Analytes	Sample Mass (g)	Extraction Method	Analysis Method	Total Concentration Detected	References
Pinus pinea L.	Portugal	PAHs	10	Soxhlet/USAE	GC-MS	21.86–339.28 ng g^{-1}	[126]
P. pinea L. needles and bark	Portugal	PAHs	2	USAE/MAE	GC-MS	213–1773 ng g^{-1}	[119]
P. pinea L.	Portugal	Pesticides	5	USAE	GC-MS	18.76–56.05 ng g^{-1}	[128]
Wet needles (mixed species)	Korea, USA, Mexico	PAHs	10	Column	GC-MSD	31–563 ng g^{-1}	[118]
Pinus sylvestris L.	Germany	PAHs	25	ASE/Soxhlet	GC-MS	Up to 55 ng g^{-1}	[129]
Pinus pinaster	Portugal	PCBs, BFRs, HCBs, PAHs, SMCs	5	USAE	GC-MS ion trap	0.26 (HCBs)–967.8 ng g^{-1} (PAHs)	[127]
Pinus nigra	Germany	PAHs	10	ASE	GC-MS	51–410 ng g^{-1}	[117]
Pinus thunbergii Parl., Pinus densiflora Sieb et Zucc	Japan	PCDD/Fs, PCBS, PCNs	Not reported	Soxhlet	HRGC-HRMS	0.17 (TCDDs)–2100 pg g^{-1} (PCNs)	[130]
P. pinaster, P. pinea	Portugal	PAHs	1/5	MA-HS-SPME[a]	GC-MS	111–1320 ng g^{-1}	[131]

GC-MS, gas chromatography-mass spectrometry.
[a]*Microwave-assisted headspace solid-phase microextraction.*

the distribution of sulphur, Zn, Cd, Pb, Cu, and As across Poland using Scots pine needles. Deposition zones were therefore identified and were used to identify sources of air pollution, as well as the directions in which pollutants were being dispersed over the sampling years [136].

A consideration that needs to be borne in mind when using pine needles to monitor inorganic air pollutants is whether the fraction related to compounds deposited on the needle surface (and not incorporated into the organism) is desired in the final extract. This is necessary, since a study by Wyttenbach et al. [132] found that the fraction washed off can contribute as much as 60% to the total concentration of an element. This has implications on how the results of a study are interpreted and its associated uncertainties. Examples of studies involving conifer needles for inorganic air pollutant biomonitoring are provided in Table 5.

5. TREE BARK

5.1 Introduction

The accumulation of pollutants by tree bark is mostly due to the ability of the tree to absorb pollutants from soil, where water serves as a solvent for a number of ionic materials [139]. Tree bark has, however, been used in air biomonitoring studies because bioconcentrated fractions have been found to correlate with atmospheric values for 46 elements [98]. Trees are also easily identified, widely dispersed and easily sampled [98]. Pine bark in particular has shown to be effective for the assessment of spatial and temporal trends in air pollution studies because it is very porous and inert, and pine trees are widely distributed [119,140], as discussed in Section 4.

5.2 Tree Bark as a Biomonitor of Air Pollutants

The high lipid content and large surface area of tree bark contribute to hyperaccumulation of organic pollutants [141] such as POPs [119], where the estimated residence time of PCBs and PCDD/Fs in tree bark is >10 years [142]. In a study by Türkan et al. [106] a reduction in pollution was observed at increasing distances from the source, however, these differences were not as marked as what was observed for the moss, *H. cupressiforme*. In contrast, a study by Tarcau et al. [108] found that DDT and its metabolites as well as HCH isomers were in higher concentrations in tree bark samples than in moss. Species-pollutant selectivity and differences in sample collection and sample preparation may contribute to these differences. In terms of PAH biomonitoring, Ratola et al. [119] found that pine needles were more effective bioaccumulators of PAHs than tree bark, with concentrations up to 17 times higher in the needles than in the bark. Inorganic analytes commonly assessed in bark biomonitoring studies are SO_2 and heavy metals, where the SO_2 concentrations largely depend upon the pH of the bark [143]. Tree bark is used

TABLE 5 Various Studies Using Pine Needles as Biomonitors for Trace Metal Pollution Monitoring

Pine Species	Location	Analytes	Sample Mass (g)	Extraction Method	Analysis Method	Total Concentration Detected	References
Pinus halepensis L.	Jordan	Pb, Cd, Cu and Zn	2	Ashed in furnace, dissolved in $HNO_3 + HCl$ (1:1)	ICP-AES	0.12–262 μg g^{-1}	[137]
Pinus Eldarica Medw.	Iran	Pb, Zn, Cu, Ni, Cr	1	Microwave digestion in concentrated HNO_3	FAAS	0.39–62.3 ppm	[138]
Picea abies	Lithuania	As, Cd, Cr, Mn, Pb, V, Zn	1.5	Digested in hot concentrated HNO_3 for 3 days	AAS	0.018–684 μg g^{-1}	[134]
Pinus nigra	Germany	Ba, Ca, Cd, Fe, Mo, Na, Pb, Sb, Ti, V, Zr		Hot aqua regia digestion	HR-ICP-MS	Ca (1.5–12.1 g kg^{-1}), Fe (32–350 mg kg^{-1}), Na (2–676 mg kg^{-1}), Ba, Ti and Pb (0.1–4.5 mg kg^{-1}). Mo and Sb (20–1000 mg kg^{-1}), Cd and Zr (500–300 mg kg^{-1})	[121]

ICP-AES, inductively coupled plasma-atomic emission spectroscopy; FAAS, flame atomic absorption spectroscopy; AAS, atomic absorption spectroscopy; HR-ICP-MS, high resolution-inductively coupled plasma-mass spectrometry.

less commonly in the biomonitoring of air than other biomonitors such as mosses and lichens [1,57,106]. This is because the potential sources of pollutants are not limited to the air. In comparison to lichens and mosses, tree bark poorly reflects regional differences of inorganic atmospheric pollutant concentrations [57]. Bark has also been shown to accumulate lower concentrations of heavy metals than mosses or lichens [3]. An additional challenge in using bark as a biomonitor of air pollution is the lack of standardisation of techniques [106], therefore future studies employing the use of bark will need to address these challenges. Table 6 shows some examples of studies which have used tree bark in the assessment of inorganic and air organic pollutants.

6. OTHER PLANTS

A number of other plant species have been employed as biomonitors of air pollution, although to a somewhat lesser extent than the species discussed in the previous sections of this chapter. The leaves of various species of trees and shrubs have been used to determine levels of organic pollutants, where PAHs have been studied in *Ficus benghalensis* leaves [145]; in leaf cuticles and leaf tissue [146]; in leaves of *Quercus ilex* L. [147] and in maize [148], for example. In addition, Barriada-Pereira et al. [149] determined pesticide levels in various plants.

These matrices have similarly been used for inorganic air pollutants, such as heavy metals [150]. Celik et al. [151] utilised the leaves of *Robinio pseudoacacia* L. to monitor airborne Fe, Zn, Pb, Cu, Mn and Cd; whilst Calzoni et al. [152] employed the perennial shrub, *Rosa rugosa*. Studies have also investigated the impacts of inorganic air pollutants on the growth and viability of agricultural crops, an example being the effects of sulphur dioxide pollution on wheat [153].

A concern with these biomonitors is that the levels of pollutants determined therein may not solely reflect the contribution arising from exposure to atmospheric concentrations, but also from uptake from the soil by the root system (a portion of which may have arisen from deposition of the pollutant from the air), and from groundwater and irrigation water. This limitation needs to be taken into consideration when interpreting the results from such studies [98].

7. SAMPLING CONSIDERATIONS

When using biomonitors, a number of factors should be carefully considered in order to minimise variations caused by species differences, specificity of the organism towards the target analytes and meteorological conditions.

1. A sufficient amount of the organism must be present in the ecosystem under study to contribute towards a representative sample amount (unless a transplant study is planned).

TABLE 6 Examples of Biomonitoring Studies of Inorganic and Organic Analytes Using Tree Bark

Common Name	Plant Species	Location	Analytes	Sample Mass	Extraction Method	Detection Method	Concentration Detected	References
Pine	Not mentioned	Finland	Cr, Ni	0.2 L Fresh sample volume	Wet digestion	FAAS	Range ($\mu g\ g^{-1}$): Cr 0.5–3.5, Ni 1.1–2.2	[3]
Turkish pine	*Pinus brutia*	Turkey	Pb, Cd, Zn, Cr, Mn, Fe	2 g	Hotplate wet digestion	FAAS	Range ($\mu g\ g^{-1}$): Cd 2.25–Fe 1850	[106]
Scots pine	*Pinus sylvestris* L.	Finland	Cr, Pb, Ni, Cd, Cu, Fe, V, Zn	Outermost 3 mm	Wet digestion	ICP-OES	Range ($\mu g\ g^{-1}$): Cd 0.31–Fe 110	[57]
Pine tree	*P. sylvestris*	Germany, Norway, Russia, Poland	Benzo[*a*]pyrene, fluoranthene, pyrene, α-hexachlorocyclohexane (α-HCH), DDT, Al, As, B, Ca, Cd, Ce, Cr, Cu, Fe, Hg, Mo, NH_4^+, Ni, NO_3^-, PO_4^{3-}, Pb, Sr, SO_4^{2-}, Ti, V, W, Zr, Zn	0.5 g for Heavy metals, 5 g for PAHs	Microwave and solvent extraction device	IC, HPLC-MS, GC-ECD, ICP-OES, ICP-MS,	Range ($\mu g\ g^{-1}$): Hg 0.001–Ca 27,000, benzo(*a*)pyrene 0.0004–0.0210, fluoranthene 0.0074–0.1332, pyrene 0.005–0.083, α-HCH 0.008–0.120, DDT 0.019–0.304	[140]

Continued

TABLE 6 Examples of Biomonitoring Studies of Inorganic and Organic Analytes Using Tree Bark—cont'd

Common Name	Plant Species	Location	Analytes	Sample Mass	Extraction Method	Detection Method	Concentration Detected	References
Olive tree	*Olea europaea*	Portugal	46 trace elements	<0.001 g	N/A	INAA and PIXE	–	[98]
Indian liliac, mango, sacred fig, cashew, cassod tree, gum arabic tree, sea almond tree	*Azadirachta indica, Mangifera indica, Ficus religiosa L., Anacardium occidentale L. Cassia siamea lam., Acacia senegal. Terminalia catappa* used for validation of method	Nigeria	Cd, Cr, Cu, Fe, Ni, Pb, V, Zn	1 g	Ashing and wet digestion	FAAS	Range (μg g^{-1}): Ni 5.0–Fe 695	[144]
Cluster pine, stone pine	*Pinus pinaster and Pinus pinea*	Portugal	PAHs	1,2 and 5 g	Microwave and ultrasound	GC-MS	Sum PAHs (ng g^{-1}): 22–1773	[119]
Not specified	*Tili, Pinus, Fagus sylvatica*	France and Germany	PCDD/Fs and PCBs and trace metals As, Sb, Cd, Co, Sn, Ni, Cr, Pb, Fe	5–30 g	ASE	GC-ECD, ICP-OES and ICP-MS	GC-EDC (ng g^{-1}): PCBs 4–140 HRGC-MS (ng g^{-1}): PCBs 0.006–41.01 metals (ppm): As 1.7–Fe > 10,000 ppm	[142]

Oak, cherry, hornbeam, alderwood, ashwood	Not mentioned	Romania	DDT and metabolites, and HCH isomers	1.5 g	ASE system	GC-MS, GC-ECNI/MS	Range (ng g^{-1}): sum HCH 12.4–131.4, sum DDT 510.8–442.6, sum chlordanes: ND-0.6	[108]
Black alder, silver birch, Scots pine, sweet cherry, plum, English oak, sessile oak, willow, small-leaved lime, large-leaved lime, Norway maple, Sycamore, European hornbeam, common hazel, European ash, apple, Canadian poplar, checker tree	*Alnus glutinosa*, *Betula pendula*, *P. sylvestris*, *Prunus avium*, *Prunus domestica*, *Quercus robur*, *Quercus petraea*, *Salix* spp., *Tilia cordata*, *Tilia platyphyllos*, *Acer platanoides*, *Acer pseudoplatanus*, *Carpinus betulus*, *Corylus avellana*, *Fraxinus excelsior*, *Malus domestica*, *Populus canadensis*, *Sorbus torminalis*	Germany	N isotopes	0.003–0.004 g	N/A	Element analyser coupled with isotope ratio mass spectrometer	Range (%): 1–21	[21]

GC-ECD, gas chromatography-electron capture detector; FAAS, flame atomic absorption spectroscopy; ICP-OES, inductively coupled plasma-optical emission spectrometry; ICP-MS, inductively coupled plasma-mass spectrometry; INAA, instrumental neutron activation analysis; PIXE, particle-induced x-ray emission; GC-ECNI/MS, gas chromatography-electron capture negative ion mass spectrometry; GC-MS, gas chromatography-mass spectrometry.

2. The biomonitor must be widely distributed throughout the desired sampling region.
3. The biomonitor should be easy to identify or alternatively access to an expert who is able to correctly identify the organism is required.
4. Good tolerance of the organism towards the target pollutants.
5. The population of the organism should be well established so that future sampling campaigns can be undertaken to establish trends over a period of years and so that the biodiversity of the sampling region is not negatively impacted.
6. There should be evidence of a correlation between the concentration of the target pollutants in the atmosphere and the concentration of the pollutants in the tissue of the chosen organism.
7. A similar bioaccumulation factor (the biomagnification of the analyte concentration in the tissue of the biomonitor compared to the ambient levels as established in point 6) should be maintained for samples collected at different sites by keeping the species, substrate from which it is sampled, meteorological conditions and sample handling parameters as consistent as possible.

It is also necessary to establish, prior to analysis, whether the presence of the target pollutants in the tissues is as a result of atmospheric pollution only (thus showing specificity towards the uptake mechanism) or if the pollutants are accumulated by other means, introducing a bias [154]. For this reason, lichens and mosses are popular choices, since it is believed that the only contribution of pollutants to their tissue is as a result of atmospheric sources (via dry and wet deposition) [155].

Augusto et al. [74] performed a study on PAHs in lichens and found seasonal variations (highest PAH values during the dry winter) and demonstrated that meteorological parameters such as temperature, air pressure, humidity as well as wind speed affected the PAH concentrations in both lichens and sampled air. A study by Jensen [87] showed erratic variations in the SVOC concentrations in both lichens and mosses, and linked these results to weather conditions; where HCH concentrations were higher after rainy spells compared to dry, hot weather. The concentrations of HCHs remained nearly constant in the pine needles, which were analysed alongside the moss and lichen samples, suggesting that the hydration state of lichens and mosses may affect the accumulation of organic air pollutants therein [75]. Degradation of photosensitive analytes, which occurs more significantly during the summer months, has also been noted in biomonitoring studies [156], and humidity and air pressure may contribute towards the absorption and retention of organic air pollutants [77].

A study by Wolterbeek and Bode [157] proposed that when biomonitors are to be utilised in trace metal studies, the pH, NH_4^+, NO_3^- and SO_4^{2-} levels within the biomonitors are also relevant, as the retention of metals, such as Ca and Hg are affected by acidic precipitation, which in turn is influenced by micro- and macro-climatic factors. It has also been proposed by Tretiach and Carpanelli that when considering lichens and their assimilation of pollutants, it might be more meaningful to represent the assimilation rates based on the chlorophyll content rather than on the principle of dry weight [158]. They recommended that the concentration of chlorophyll in samples should always be reported.

7.1 Sampling of Biomonitors for Organic Analytes

A standardised approach has not yet been formulated with respect to sampling biomonitors used to target organic analytes. However, recommended best practise which minimises uncertainties in the results generated are included in this section.

The geographical location (co-ordinates) where sampling takes place should be recorded, along with the sampling altitude, since atmospheric pressure has been shown to influence results [74]. The location of nearby air pollutant emission sources should be noted. Sampling should be conducted during a dry spell, in order to minimise any variations introduced by rain wash-off. The weather (wind direction, temperature, humidity and air pressure) during sampling should be noted. Acquiring longer-term meteorological data for the sampling sites (such as annual rainfall and temperature) may also assist with data interpretation. The species of the substrate from which the biomonitors are to be removed must be identified, as well as the total number of individual substrates from which biomonitors were sampled.

No smoking should be permitted near the sampling site in order to prevent sample contamination. Gloves should be worn, and sterilised stainless steel tweezers or a sterilised blade is recommended for the removal of the biomonitors for organic air pollutant monitoring. For lichens, sampling is typically performed at a height of at least 50 cm above the ground, on the branches and trunk of trees [68]. In the case of pine needles, whole needles are removed, typically from the bottom and outer branches of the trees. Moss samples are collected away from trees (on dead tree stumps and over the ground only) and stored in amber jars. If bark is the biomonitor of choice, it is to be sampled from a height >150 cm above the ground, on the trunk, removing only the most external layer [159]. Samples may be stored in brown paper bags, but amber vials are more commonly used to reduce contamination and analyte loss, and to prevent photodegradation of the analytes. Wrapping the samples in foil prior to placing them into plastic bags is an alternative [160].

The samples should be kept cool to reduce any losses, and manually cleaned of all dirt, bark, leaves or stones once in the laboratory, although most organic air pollutant biomonitoring studies do not mention a washing step. Lichens are typically oven dried at 35 °C for 3–4 days to remove all moisture [68], before being ground with an agate mortar and pestle, and stored in the freezer until use. Pine needles are either cut into 1-cm-long pieces or left whole and stored (wrapped in foil and placed in a plastic bag) in the freezer without drying. The water content of the needles may be established by drying ~5 g of needles at 80 °C until a constant weight is obtained [131]. Moss samples may be dried at 40 °C for 2–3 days prior to being ground in a blender, and stored in glass jars in the freezer [107].

7.2 Sampling of Biomonitors for Inorganic Analytes

The basic best practice principles of sampling biomonitors for inorganic air pollutants are similar to those for organic target analytes, as described in Section 7.1, therefore only the differences in methods are presented here. Samples should be taken using acid-washed plastic [161], nylon, ceramic or Teflon tweezers [36] in order to prevent metal contamination thereof, and powder-free latex, nitrile or vinyl gloves should be worn.

Epiphytic lichens should be collected 1–2 m above the ground [3], and sampling should be done all around the tree to limit the effect of prevailing wind conditions (unless this factor is to be studied). Lichens of a similar diameter should be sampled to reduce variations in age of the lichens, although it is recommended that the outermost 3–4 mm of the foliose lichen thallus be detached for analysis if the aim of the study is to assess concentrations which have been accumulated over the period of a year [32], however, these restrictions may lead to reduced sample mass. Epiphytic moss samples should be collected at 1–2 m above the ground and should not be older than 2–3 years [4].

Samples should be stored in acid-washed polyethylene or polypropylene bottles or paper bags and stabilised as soon as possible. Samples should be cleaned of extraneous material prior to storage to ensure that these materials do not contaminate the biological monitor. This cleaning step should take place as soon as possible after sampling so as to limit changes which may occur due to microbial activity and surface dust deposition. There has been little standardisation with respect to sample pre-extraction preparation. Some authors recommend washing the samples with water to remove surface dust deposits [3], however washing may result in the dissolution of ionic compounds. Ultrasonic baths have been used to clean foliose lichens [162]. Lichens are not washed, however, in the determinations of metal concentrations in adhering particles [22] or isotopes [21], but if the bioconcentrated portion is of interest, then the samples should be washed [109].

Mosses are particularly difficult to clean effectively due to their porous nature [25]. Techniques such as nitrogen air jets and ultrasonic probes have been used to remedy this, however, it was found that nitrogen air jets are too aggressive (at 4–8 bars) and could not effectively remove extraneous materials at lower pressures [25]. The results from such pressurised jet systems were also found to be irreproducible and unpredictable. Attempts to remove particulate matter from mosses by using an ultrasonic probe were found to be an unviable option since it required the presence of a solution, which leached analyte from the sample [25].

It has been recommended in a number of studies that samples be ground and homogenised [163], but this may increase the risk of contamination [161]. Drying or freezing steps aid homogenisation and stabilise the sample in terms of microbial activity, thereby preventing changes to the analyte concentration which microbes may induce through their metabolic processes. Commonly used techniques are air-drying [32,36], oven-drying [164], freezing to between −20 and −80 °C [101], freeze-drying [165] and cryogenic freezing with liquid nitrogen [22,166]. Pre-ashing of samples has been used [3]; however, the use of this technique may liberate volatile elements such as Bi, Cd, S, Sb and Se [163].

8. ANALYTE EXTRACTION METHODS

The most commonly employed techniques utilised to extract air pollutants from biomonitors are discussed in this section.

8.1 Soxhlet Extraction

Soxhlet has been used for over a century to leach analytes of interest from their solid matrices into a solvent of choice. The Soxhlet apparatus allows the continual dripping of condensed extracting solvent onto the sample, held in a porous thimble, encouraging extraction of the analytes into the solvent [167,168]. A siphon aspirates the solvent dripping into the thimble when an overflow volume level is exceeded, and the liquid is poured back into the distillation flask. This process allows the transfer equilibrium to be continually disturbed, improving the extraction efficiency when compared to static leaching techniques [168].

Soxhlet extraction has historically been used to extract organic analytes from lichens including PAHs and other SVOCs [30,43,67,82,169] (Table 7). The method has also been applied with success to the extraction of SVOCs from bark and pine needles [30,126,159,170,171].

Advantages of using Soxhlet extraction include the utilisation of either large or small sample volumes, the omission of a filtration step, the lack of matrix dependence and ease of use. The technique has been automated in commercial systems, which are faster and use less solvent [167]. A serious disadvantage of using the Soxhlet methodology for sample preparation is the large volumes of wasted solvent that are expensive to purchase and dispose of, many of which are harmful to the environment. Another disadvantage is the long extraction times (2–24 h) which decreases sample throughput and utilises more energy [76,88]. In addition, boiling solvent over the period of a few hours may degrade thermolabile species [168]. The introduction of newer techniques such as microwave-assisted extraction (MAE) and accelerated solvent extraction (ASE) are increasing in popularity, however, the use of Soxhlet for routine analysis remains relevant as a result of the relative

TABLE 7 The Different Soxhlet Extraction Conditions Applied in Various Biomonitor Studies Targeted Towards Organic Air Pollutants

Biomonitor	Analytes	Sample Mass (g)	Solvent Scheme	Solvent Volume (mL)	Extraction Time (h)	References
Lichen	PCDD/Fs	10	Methylene chloride:hexane (1:1)	Not reported	18–24	[43,88]
Lichen	PAHs	2	Dichloromethane	100	16	[67,82,83,172]
Lichen	PAHs	2	Acetonitrile	200	24	[30,88,173]
Lichen	PAHs	0.2	Dichloromethane	250	6	[80]
Lichen	PAHs	0.6–0.8	Hexane:acetone (1:1)	150	2	[76]
Lichen	PCDD/Fs	Not reported	Toluene	200	24	[63,88]
Lichen	Chlorinated hydrocarbons	Not reported	Hexane	Not reported	8	[65]
Pine needles	PCDD/Fs, PCBs, PCNs	Not reported	Toluene and dichloromethane:methanol (1:1)	300 each	7 each	[130]
Pine needles	OCPs	Not reported	Hexane:acetone (1:1)	80	20	[171]

affordability of the method as well as its robustness, therefore, it is often used as a reference technique when new extraction methods are developed.

8.2 Microwave-Assisted Extraction

The principle of MAE is that the analytes of interest adsorbed or absorbed by the substrate are partitioned into the extraction solvent by the application of microwave energy (in the range of 0.3–300 GHz). Microwaves heat the contents of the vessel simultaneously and nearly instantaneously, without heating the vessel, reducing extraction times significantly.

The extraction efficiency is heavily dependent on the choice of extraction solvent, since the solvent should readily absorb microwave energy [174]. As a result, non-polar solvents produce poor extraction efficiencies, unless they are used in conjunction with solvents that have large dipole moments [175]. The solvent should selectively and efficiently extract the analytes without overheating, which would lead to degradation. The water content of samples should also be controlled to minimise losses due to heating [176], thus it is recommended that the sample is air dried, oven dried at low temperatures, or freeze dried, as applicable.

MAE can be considered as either pressurised MAE, where the vessels are closed, with the pressure and temperature being controlled (by elevating the pressure), or as focused MAE, where the extraction vessel is kept open, and the extraction takes place at atmospheric pressure and the boiling point of the solvent [177,178].

8.2.1 Extraction of Organic Analytes

The earliest use of microwave energy to assist in the extraction of organic analytes dates back to the 1980s, where household equipment was used [179,180]. Specialised laboratory equipment has since been developed. Guo and Lee demonstrated in their study on PAHs in soils that higher temperatures during extraction increases the diffusion rates of the solvent which then permeate the solid matrix more effectively. The partition coefficient is also affected such that the analytes are released from the solid more readily at higher temperatures [174].

Typically, a cleanup and concentration step is necessary after MAE before the extract can be analysed. The filtration and cleanup procedure would also ideally serve to remove interferences and thus reduce matrix effects [181]. A study by Ratola et al. [119] compared the use of MAE and ultrasound-assisted extraction (USAE) to determine the PAH content of pine needles in Portugal. The study utilised a modified domestic microwave oven and dichloromethane:hexane (1:1, v/v) for the MAE extraction. The recovery of PAHs using MAE varied widely between 10% and 120%, with the recoveries of PAHs using USAE reported as being very similar although it was observed that MAE had better recoveries for the heavier molecular weight PAHs compared to USAE. A study on PAHs in spruce needles and pollen [182] found MAE to be superior over USAE when n-hexane:acetone (3:2, v/v) was used, as shown in Table 8.

In general, when compared to traditional Soxhlet extraction, MAE offers a reduction in extraction time (between 5 and 20 min compared to >6 h for Soxhlet), reduced solvent consumption (10−50 mL in MAE compared to 250−500 mL in Soxhlet) as well as enhanced reproducibility and reduced solvent wastage [175,183].

A new microwave-assisted headspace solid-phase microextraction (MA-HS-SPME) technique was developed by Ratola et al. [131] for the analysis of PAHs in pine needles. The technique tested both polydimethylsiloxane and divinylbenzene fibres, and the optimised extraction technique produced clean extracts with no need for further cleanup steps. In particular, improved recoveries of higher molecular weight PAHs were observed, compared to previously developed USAE and MAE techniques [126,131].

8.2.2 Extraction of Inorganic Analytes

Microwave-assisted digestion is one of the most commonly used sample preparation techniques for the analysis of total metal concentrations from a range of matrices, as it is fast and effective. Samples are digested under high temperatures and pressures with minimum loss of sample and with substantially less acid than what is required with conventional methods. The analytically pure acids used depend upon the analytes and matrix of interest. Acid mixtures which are commonly employed in air biomonitor digestions are HNO_3 and H_2O_2 [113]; HNO_3, H_2O_2 and HF [164,184]; HCl and HNO_3 [185]; and HCl, HNO_3 and $HClO_4$ [186] with further examples provided in Table 9. Sample digestion runs should include spiked samples and method blanks so as to ensure that analyte recoveries are optimal and to check for sample contamination. The filtration of samples after digestion is necessary, where the composition of the filter and its pore size are important considerations.

Despite the fact that less acid is required in microwave digestion techniques than in conventional methods, high concentrations and limited dilution allowances in a number of techniques require that the acids are removed or reduced prior to analysis. Samples containing HF may be evaporated in an open vessel on a hotplate and then resuspended in a weak acid solution; however, the loss of volatile analytes may be incurred. Another option is the use of boric acid which complexes to F^- ions [187].

8.3 Ultrasound-Assisted Solvent Extraction

8.3.1 Introduction

USAE has gained popularity in the last few years as an attractive alternative to microwave digestion techniques [25,189]. Benefits of USAE over other techniques, such as Soxhlet extraction, accelerated solvent extraction and

TABLE 8 Examples of Microwave-Assisted Extraction Techniques for the Analysis of Organic Air Pollutants in Biomonitors

Biomonitor	Analytes	Sample Mass (g)	Solvent Scheme	Solvent Volume (mL)	Extraction Time	References
Pine needles	PAHs	2	Hexane:dichloromethane (1:1)	90	30 min at 513 W	[119]
Pine needles	PAHs, SMCs	2	Water and ethanol	50 water, 0.9 ethanol	60 min at 560 W	[131]
Pollen and spruce needles	PAHs	5	Hexane:acetone (3:2, v/v)	50	20 min	[182]

TABLE 9 Examples of the Use of Microwave Digestions for the Analysis of Inorganic Air Pollutants in Biomonitors

Biomonitor	Acids Used	Sample Mass (g)	References
Moss	4 mL HNO_3 (65%), 1 mL H_2O_2 (30%)	0.15–0.2	[101]
Lichen	Mixture of HNO_3, H_2O_2 and HF	0.05–0.1	[161]
Lichen	HNO_3	0.15	[166]
Garlic and Indian mustard	1.5 mL HNO_3 and 1.5 mL of H_2O_2	0.2	[188]
Lichen	3 mL HNO_3, 3 mL H_2O_2, 0.8 mL HF and 2 mL deionised water	0.25	[31]
Lichen and mosses	6 mL HNO_3 (65%), 2 mL H_2O_2 (30%) and 2 mL H_2O	1	[59]
Pine needles and bark	90 mL hexane:dichloromethane	1, 2, 5	[119]
Moss	Not specified	Not specified	[93]
Moss	2.5 mL H_2O_2 (30%), 5 mL HNO_3 (65%)	0.25	[97]
Tree leaves and mosses	HNO_3 and HF (2:1, v/v)	Not specified	[4]

supercritical fluid extraction, are that it is comparatively simple, efficient, cheap, is available in most laboratories and does not require high temperatures, pressures or concentrated harmful chemicals [190]. USAE techniques are commonly used in biomonitoring applications due to the fact that smaller sample sizes can be accommodated, less solvent is used, and the extractions are relatively fast compared to traditional liquid extraction techniques [42,80].

Ultrasonic techniques rely on the principle that once a solid sample in solution is subjected to ultrasonic energy, bubbles are formed and implode [189]. This causes high localised temperatures and pressures [189] which result in the extraction of compounds or elements from the sample matrix [25]. The major parameters affecting USAE are particle size, sample mass to extractant ratio [80,162,191], extractant concentration, sonication time and the use and choice of surfactants [191]. Surfactants are sometimes used to ensure that the sample is distributed homogenously and that small particles remain in suspension. Tween 85 and Triton X-100 were found to be ideal surfactants for use in biomonitoring studies, as opposed to surfactants such as glycerol that resulted in high backgrounds and poor reproducibility [191].

USAE can be used for a wide range of biomonitoring applications; from extraction of surface material, cleaning of samples [25] and the extraction of metallic species [188]. USAE requires fewer analytical steps than a number of other extraction techniques and is therefore less susceptible to contamination [162]. Furthermore, ultrasonic methods can also be used to accelerate sequential extraction steps that may be used in the partitioning of metals [190], and to facilitate the suspension and homogenous distribution of materials in slurry extracts [191]. Ultrasonic extraction is also applicable to the extraction of organic compounds from matrices to which they are weakly bound [190], as long as the heat is maintained at a temperature below which analytes are susceptible to degradation or volatilisation (for example, below 35 °C for PAH extraction from lichens [68]).

Sonication by an ultrasonic probe is up to 100 times more efficient than ultrasonic baths [162]. This is because the reproducibility and repeatability of ultrasonic baths is negatively influenced by non-uniformity of ultrasound energy distribution throughout the bath, as well as a loss of power with time [192]. The use of ultrasonic probes for the cleaning of moss samples was found to be ineffective, however, since the analyte of interest was leached from the sample into the solution [25]. Ultrasonic baths, on the other hand, are much gentler methods and are thus often preferred for speciation analysis [193] and may be a suitable alternative to probes for the removal of dust from the surface of samples such as moss.

8.3.2 Extraction of Inorganic Analytes

Compared to microwave digestions, USAE is used less frequently and requires a substantial amount of method development compared to the currently available microwave digestion methods. Table 10 provides examples of the use

TABLE 10 Examples of Biomonitoring Studies Which Have Used Ultrasonic Methods for Inorganic Analytes

Use	Species	Analytes	Optimal Extraction Medium	Sample Mass (g)	Optimal Sample Size	Optimal Temperature (°C)/Sonication Amplitude (%)/Extraction Time	References
Total extraction of metals	*Parmelia sulcata*, IAEA lichen 336, NIST SRM 2976	Na, K, Ca, Mg, Fe, Al, Cr, Mn, Co, Ni, Cu, Zn, Ge, As, Se, Rb, Sr, Zr, Ag, Cd, In, Sb, Cs, Ba, Pb, Bi	5 mL of 1% HNO_3	0.1	200–300 mesh	20–40%	[162]
Extraction of selenium from plants for selenium speciation	*Allium sativum* and *Brassica juncea*	Se species	0.1 M HCl, 25 mM ammonium acetate buffer (pH 5.6) and protease	0.2	Not specified	3 min	[188]
Total extraction of metals	Tobacco leaves	Al, As, Cd, Ni, Pb	10 mL of 15–20% HNO_3	0.2	<35 μm	60–80 °C	[189]
Suspension of extraction slurries	*Cupressus lawsoniana*, *Cupressus sempervirens*, *Quercus ilex*	Cd, Cu, Pb, Ni	5% HNO_3 with either Tween (0.5%) or Triton X (1%)	1–15 mg mL^{-1}	<32 μm		[191]

of ultrasonic extraction techniques in a number of studies for a variety of inorganic air biomonitoring applications.

8.3.3 Extraction of Organic Analytes

The USAE technique has grown in popularity because it is a fast method that uses less solvent and less energy, with satisfactory results. Furthermore, a sonication-assisted extraction procedure has been detailed by the EPA (Method 3550B) and USAE was also found to be superior over Soxhlet and pressurised liquid extraction in a study by Ratola et al. [126] that focused on PAHs in pine needles. Examples of experimental conditions when using USAE on various biomonitor matrices are given in Table 11. Concern has been raised about the possibility that high ultrasound frequencies may increase the degradation of SVOC analytes. A study by Manariotis et al. showed that phenanthrene, naphthalene and pyrene were degraded at frequencies above 582 kHz, particularly when the extractions were performed for long periods of time (>30 min) [194]. This should be carefully considered when developing an experimental method utilising USAE.

Whereas Guidotti et al. [42,84] used cyclohexane as the solvent for USAE of PAHs from lichens (sample mass 2 g), combining only two extractions of 30 mL each, Domeño et al. [80] extracted a 0.2 g lichen sample using portions of 15 mL dichloromethane in each of four subsequent extractions. A study, in which a novel dynamic ultrasonic-assisted extraction technique for lichens was developed by Domeño et al. [80], found that hexane extracted PAHs most efficiently and the new dynamic sonication-assisted solvent extraction (DSASE) technique has since been popular in numerous PAH studies using lichens [68,78,79], which utilises a continuous flow of solvent, resulting in reduced total solvent volumes, minimised extraction time and minimal sample handling compared to traditional USAE techniques, with a reduced possibility of analyte degradation.

8.4 QuEChERS

8.4.1 Introduction

This sample extraction technique was developed to be quick, easy, cheap, effective, rugged and safe. QuEChERS, as it was therefore named, was developed by Anastassiades et al. [195] and quickly found popularity as an effective sample preparation technique for pesticide analysis in foodstuffs [196]. The method serves as an alternative to traditional liquid–liquid and solid-phase extraction (SPE) techniques, reducing the number of steps as well as the total extraction time and solvent consumption [197].

The QuEChERS methodology involves two steps. The first step is the extraction of homogenised samples using an organic solvent of choice with a solution of extraction salts, often made up of magnesium sulphate (absorbs water and increases ionic strength), sodium chloride (improves partitioning to organic

TABLE 11 Experimental Conditions Used When Applying USAE to Biomonitors for the Study of Organic Air Pollutants

Biomonitor	Analytes	Sample Mass (g)	Solvent Scheme	Solvent Volume (mL)	Extraction Time (min)	References
Parmelia sulcata	PAHs	0.2	Hexane	2	10	[78]
Evernia prunastri	PAHs	0.2	Hexane	2	10	[10]
P. sulcata	PAHs	0.2	Hexane	2	10	[68]
P. sulcata, E. prunastri, Ramalina farinacea, Pseudevernia furfuracea, Usnea sp. and *Lobaria pulmonaria*	PAHs	0.2	Hexane	2	10	[79]
Xanthoria parietina	PAHs	0.2	Hexane	2	10	[80]
P. furfuracea	PAHs	2	Cyclohexane	30	30	[42]
P. furfuracea	PAHs	2	Cyclohexane	30	30	[84]
Pyxine coralligera	PAHs	2	Cyclohexane: dichloromethane (4:1 v/v)	30	30	[86]
Pinus pinea L.	PAHs	10	Hexane: dichloromethane (1:1)	30	10 (×3)	[126]
P. pinea L.	PAHs	2	Hexane: dichloromethane (1:1)	30	10 (×3)	[119]
P. pinea L.	Pesticides	5	Acetonitrile	30	10 (×3)	[128]
Pinus pinaster	PCBs, BFRs, HCBs	5	Hexane: dichloromethane (1:1)	100	10 (×3)	[127]
P. pinaster	PAHs, SMCs	5	Hexane: dichloromethane (1:1)	100	30	

phase) and buffer salts (trisodium citrate dihydrate and disodium hydrogen citrate sesquihydrate), followed by a centrifugation step. The addition of the extraction salts (Na_2SO_4 or $MgSO_4$) is an exothermic process and therefore increases the temperature of the reaction vessel, improving the extraction efficiency [195]. The second step then involves the cleanup of the supernatant using dispersive SPE (dSPE) which differs from SPE since it does not use a fixed solid bed. The addition of internal standards, buffers or acids can be done in the first step in order to improve the extraction efficiencies and track recoveries. It has been noted that water should be added to the extract, so that the extraction efficiency is improved as a result of the partitioning of polar compounds between the organic and aqueous phases [198,199]. The dSPE step involves the use of a drying salt (magnesium sulphate) as well as SPE sorbent packing materials such as primary–secondary amine (PSA), often supplemented with graphitised carbon black (GCB) or C_{18} material to clean the extracts and remove moisture. The dSPE step also removes sugars, fats, organic acids, proteins, pigments and sterols; thereby simplifying what would be many consecutive cleanup steps into one cleanup procedure. Instrumental analysis using either gas or liquid chromatography (GC or LC) follows, with or without the addition of a pre-concentration step after the cleanup.

Analysis of pesticides in many matrices has been undertaken, and QuEChERS has become a popular choice for extracting PAHs from foodstuffs [200–202]. Although no studies have been published on the use of QuEChERS to extract organic air pollutants from biomonitors, it certainly has potential to do so. Sadowska-Rociek and co-workers have evaluated the QuEChERS methodology for the determination of PAHs in fresh herbs [203]. The study found that acetonitrile gave recoveries between 71.6% and 116.9%. The sample extracts were cleaned up using PSA, GCB and $MgSO_4$, followed by GC-MS analysis. Included in the methodology was a 15 min shaking step, which deviates from the original QuEChERS methodology of extractions between 1 and 5 min [198], but remains, however, a considerably faster extraction technique compared to Soxhlet, USAE and some MAE methods.

In a QuEChERS study by Albinet et al. [198] on particle-bound PAHs in ambient air, analytes were extracted from particles collected onto filters with dichloromethane, acetonitrile or acetonitrile:water (5:2, v/v). The extraction with acetonitrile was the solvent of choice due to instrumental compatibilities; the different solvent schemes showed negligible differences in extraction efficiencies. Although the work by Albinet et al. [198] was not based on biomonitoring, the successful extraction of PAHs suggests that with the optimisation of extraction solvent, agitation time and cleanup methodology, analytes of interest could be extracted from biomonitors using the QuEChERS approach.

8.5 Sample Cleanup

Regardless of the extraction technique used to prepare the biomonitors for analysis, a cleanup step is necessary to remove lipids, proteins, pigments, chlorophyll, carotenoids and other compounds that may cause matrix effects for organic analyte determinations or cause damage to expensive analytical instrumentation. Clean extracts are imperative to producing reliable qualitative and quantitative results, particularly of compounds present at low levels, as in the case of air pollutants.

The most popular cleanup technique for organic analytes in biological samples remains classical liquid adsorption chromatography [176] in which the prepared extracts are passed through adsorbent columns or solid-phase cartridges (Figure 3). The EPA has approved cleanup methods using silica gel, alumina and florisil (EPA methods 3630C, 3610B and 3620B, respectively) for organic extracts of solid environmental samples, including biomonitors. Ideally, the column material should allow for the selective separation of the target analytes from the matrix background. Additional parameters requiring optimisation are sorbent mesh size and amount (this can vary from small SPE cartridges to large columns filled with activated silica gel). If the biomonitor has a high lipid content, saponification may be necessary using an alkaline solution or a sulphuric acid wash prior to cleanup using a lipid-specific column that would separate the analytes from the lipids in the sample.

Sometimes combinations of adsorbent phases are required in order to efficiently remove unwanted compounds from the extract. In a tobacco study targeted towards OCPs, the cleanup step involved a florisil column cleanup step, followed by silica column chromatography [204]. In a similar manner, florisil cartridges with 5% deactivated alumina were used to purify extracts from grass and various plant species for the study of OCPs [149,171]. It has also been reported that PCBs, PCDD/Fs and polychlorinated naphthalenes (PCNs) were selectively eluted from a series of multilayer silica columns in a study of pine needles [205] and plant extracts containing DDT were passed through a florisil column, after being treated with sulphuric acid [206]. The cleanup of plant extracts for PAH-targeted analysis has also been performed

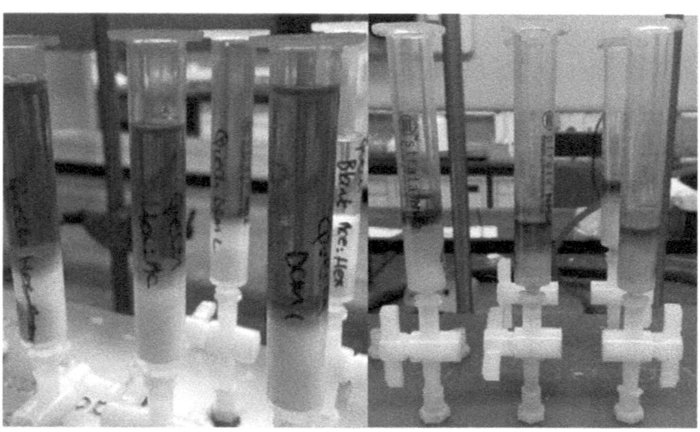

FIGURE 3 Solid-phase extraction of lichen extracts.

using florisil columns [207]. In order to clean up tree leaf extracts for the analysis of OCPs, Barriada-Pereira et al. [208] found that a carbon sorbent cleaned the extracts most selectively and efficiently. In a study on pine needles, Ratola et al. [126] determined that an alumina cartridge cleanup provided the best recoveries (between 72% and 100%) and limits of detection as low as 0.22 ng g^{-1} dw. A gel permeation cleanup procedure has also been developed for PAHs extracted from pine needles [129].

Silica column chromatography has been employed extensively to lichen extracts, with the use of different solvent schemes by different groups [42,70,82]. A direct comparison between the results of these studies is a challenge as a result of the different affinities solvents have for the different PAHs [80], leading to biased elutions. The mass of sorbent is also not clearly defined in most studies, which influences the extent to which the analytes are recovered.

More recently, SPE has been used to clean up lichen extracts targeted towards PAHs with mainly normal phase $-NH_2$ columns being used, after a study by Blasco et al. [78] found that these columns, combined with an elution solvent of hexane:dichloromethane (65:35) yielded the best recoveries. SPE remains a favourable choice in comparison to traditional column chromatography since it uses less solvent, reducing the likelihood of loss of SVOCs since the blow down and pre-concentration steps are shortened.

In terms of the QuEChERS technique, a cleanup step has been worked into the methodology, however, this cleanup step needs to be optimised for each application. In the study on emission samples by Albinet et al., simple filtration by the use of a polyethylene terephthalate syringe filter was considered the 'cleanup' step [198]. Sadowska-Rociek et al. [209] tested various sorbent types for the cleanup procedure for tea extracts, including C_{18}, PSA, strong anion exchange (SAX), $-NH_2$ and GCB. PSA is used to remove sugars, fatty acids and sterols, GCB is used to remove pigments (in particular, chlorophyll), C_{18} is used to remove non-polar compounds such as lipids, NH_2 is a weak anion exchanger and thus extracts acids and charged compounds, and finally SAX is a strong anion exchanger, typically used to extract organic acids. Sadowska-Rociek et al. found that SAX was the most appropriate sorbent for the tea samples, whilst the use of GCB resulted in losses of PAHs, in particular the high molecular weight PAHs [209]. This could be as a result of intercalation between the graphite structure and the planar structure of PAHs, thus caution should be exercised when it comes to the use of GCB, depending on the target analytes.

Size exclusion chromatography is primarily used to remove lipids and larger interfering compounds (typically larger than 500 Å). The gels most commonly used are Bio-Beads S-X3 [176]. The main advantage of using gel permeation chromatography over other techniques is that it is entirely non-destructive, unlike some of the liquid adsorption chromatography techniques that use sulphuric acid or alkaline solutions to remove lipids. As with the liquid adsorption chromatography cleanup technique, the EPA has established a gel permeation chromatography cleanup method for organic extracts obtained from environmental samples (EPA Method 3640A GPC Cleanup). Hubert et al. [129] used size exclusion chromatography to clean up pine needles, mosses and tree leaves. An S-X3 Bio-Beads column was used for the separation of PAHs in spruce needle extracts, using chloroform as the mobile phase [210]; for the cleanup of pasture vegetation extracts targeted towards the analysis of PAHs [211]; as well as the analysis of PCDD/Fs in lichens [43,88]. Often size exclusion techniques are followed by liquid adsorption chromatography, since not all the interferences are removed as a result of the complex matrix of biomonitors.

9. ANALYTICAL TECHNIQUES

After target analytes have been extracted from biomonitors and the extract cleaned up, if necessary, it may then be pre-concentrated by evaporation procedures prior to analysis. An important consideration in biomonitoring studies is the use of internal standards (including deuterated standards of the analytes of interest), which may be added to the sample at various stages of the analysis in order to determine extraction recoveries and analytical efficiencies. Certified reference materials for some target analytes and some biomonitor species are commercially available, and these are valuable in method development and validation. The analytical techniques most commonly utilised in biomonitor studies have been included in many of the tables provided in this chapter. They are briefly discussed here, however, more detail on these techniques is provided in the chapters of Section III of this book.

9.1 Organic Analytes

GC and LC are common techniques used to analyse biomonitor extracts for organic air pollutants derived from a range of sample preparation methods. As an example, the EPA PAH-related methods all recommend the use of either GC or LC in the analysis of sample extracts from a range of matrices including ambient air and foodstuffs [212].

When analysing lichen extracts for PCDD/Fs and PAHs, the main analytical techniques used are high performance liquid chromatography (HPLC) (using reverse phase C_{18} columns) and GC, using a variety of column configurations with column lengths ranging from 15 to 60 m. GC-MS and LC with fluorescence detection both have their advantages. If a particular PAH is of concern, such as benzo[*a*]pyrene or anthracene, the sensitivity and selectivity of using LC with fluorescence detection will be far greater at trace levels, due to the highly fluorescent characteristics of these PAHs [212]. On the other hand, benzo[*ghi*]perylene has poor fluorescence sensitivity and is therefore better suited to analysis using GC-MS [212]. Sacrificing peak separation, particularly of the benzo fluoranthenes, as well as benzo[*a*]anthracene and chrysene, for the sake of faster run times should be avoided to ensure an accurate representation of the PAH profile, owing to the different toxicities of PAH compounds. In general when deciding which method should be applied to a sample, a number of criteria guide the decision including:

- The type of matrix and analytes of interest.
  If the matrix is complex with a high level of interferences, GC-MS is a more appropriate technique due to the enhanced separation achievable using GC techniques [213]. Depending on how efficient the cleanup procedure is, it is likely that the resolving power of GC-MS would be advantageous over LC techniques, as long as the target analytes are amenable to GC analysis.
- The required sensitivity.
  Both GC-MS and HPLC coupled to a fluorescence detector (FLD) are known to be very sensitive, for example, detection limits as low as 0.1 ng g^{-1} dw for HPLC-FLD [214]. Other LC detectors such as the diode-array detector may not have the desired sensitivity for the analytes concerned. Comprehensive two dimensional GC–time-of-flight mass spectrometry (GCxGC-TOFMS) for environmental sample analysis allows full scan data at very fast scan rates, as well as enhanced peak capacity and reduced noise, leading to improved sensitivity [215–217].

Examples of analytical techniques for various biomonitors used for organic air pollutant monitoring are given in Table 12.

TABLE 12 Common Analytical Techniques Used to Analyse Extracts from Biomonitors Used for the Monitoring of Organic Air Pollutants

Biomonitor	Analytes	Instrument	References
Wild chard	PAHs	GC-FID	[218]
Pine needles	PAHs	GC-MS	[119,126]
Lichen	PAHs	HPLC-UV-V	[81]
Lichen	PAHs	HPLC-DAD-V-UV and FLD	[30]
Moss	PAHs and NPAHs	APGC-MS/Q-TOF	[107]
Lichen	PCDD/Fs	GC-HRMS	[43,85]
Lichen	PCDD/Fs	HRGC/HRMS	[63,70]
Lichen	PCBs	GC-HRMS	[85]
Pine needles	Pesticides	GC-ECD	[171]

APGC-MS/Q-TOF, atmospheric pressure gas chromatography coupled to a quadrupole hyphenated to a time-of-flight mass spectrometer; GC-HRMS: gas chromatography - high-resolution mass spectrometry; HRGC/HRMS, high resolution gas chromatography/high resolution mass spectrometry; GC-ECD, gas chromatography-electron capture detector.

9.2 Inorganic Analytes

ICP-OES and ICP-MS are the most common techniques used in the analysis of micro- and macro-elements from plant biomonitoring studies, as they are capable of high sample throughput and multi-element determinations, and they have large dynamic ranges [219]. Choice of nebuliser and spray chamber needs to be determined for each specific matrix, and argon humidifiers may be used in samples where salt concentrations are high. Standards and samples need to be closely matrix-matched as any deviations may result in signal drift [220].

ICP-MS is able to analyse concentrations of analytes in the ppb to ppt range, which is important when monitoring elements which may have environmental impacts at very low levels, such as Pb and Cd. This technique is also able to analyse isotopes and thus can be used in source appointment studies. ICP-MS is, however, not as robust as ICP-OES, and problems are experienced with samples having high total dissolved solids (TDS) [220]. The tolerance of ICP-MS for TDS is typically within the range of 0.1−0.2%, and samples are usually filtered prior to analysis. Additionally, high concentrations of acids in samples have a negative effect on the lifetime of the sampling cones [221], therefore when using HF for sample digestion, platinum cones should be used. Analysis of samples containing high concentrations of carbon and matrix ions may result in deposits forming on the cones leading to signal drift. Several techniques are employed to reduce this, such as sample dilution, the use of ion-exchange columns, and the introduction of oxygen to burn off excess carbon. Chloride ions which are present in HCl matrices are not only extremely corrosive but also pose a problem in the analysis of low concentrations of As, as $^{40}Ar^{35}Cl$ may form in the plasma which has the same nominal mass as ^{75}As. Unfortunately As is monoisotopic, and so the use of an alternative isotope is not an option. Correction equations, collision or reaction gases may therefore be applied to reduce or remove the interference.

10. CONCLUSIONS, CHALLENGES AND FUTURE OUTLOOK

One of the biggest challenges in biomonitoring of air pollution is that the measured concentrations are not necessarily fully reflective of ambient conditions. Real-time monitoring has advantage in this context; however it may incur significant costs.

The advantages of using biomonitors are a result of their ability to accumulate compounds and elements above their essential requirements. The application of biomonitors to monitor air pollution is favourable as a result of the low cost involved and ease of sampling since no bulky equipment is required. They also provide an integrated pollution profile over an extended period of time and allow for air pollutant mapping of large geographical regions. These features suggest that biomonitoring will continue to be applied in air pollution studies for the foreseeable future. The limitations of the use of biomonitors include that the comparison of results from different types of biomonitors cannot be made as a result of morphological and ecophysiological differences, resulting in different pollutant profiles being detected. There is, however, value in using the results in a complimentary way in order to obtain a comprehensive understanding of regional air pollutant profiles.

In order to establish the monitoring of air pollutants using biomonitors, an understanding of the uptake and release mechanisms of each individual sub-species is required. This is a result of various studies highlighting differences in accumulation rates, such as the study on PAHs in lichens by Blasco et al. [78]. In this context, plant species such as mosses and lichens may require expert identification in biomonitoring studies [98]. Furthermore, dose–response relationships need to be well defined so that the accumulation, release and preservation of pollutants within the organism and the effects thereof on the organisms can be better understood.

Biomonitoring has been successfully used to monitor the spatial behaviour of pollutants and to assess the variation in pollutant levels with changes in land use, along with the potential human and environmental health risks related to this. In order to meaningfully contribute towards risk assessment and management, the impact of various emission sources needs to be understood. Studies such as Augusto et al. [74], which correlate the PAH values found in lichens with atmospheric data, are important in this regard although biological processes in plants employed as biomonitors may complicate such studies [114]. One of the analytical challenges with biomonitoring studies is matrix effects due to chlorophyll and other biological compounds present in the sample extract, which require effective cleanup prior to analysis in the case of organic target analytes.

An area in which biomonitors show promise is in source apportionment of air pollutants. Caution must be exercised when applying any pollutant-specific diagnostic ratios to the results of biomonitoring studies, however, since photodegradation, varying analyte stabilities and differing sensitivities of biomonitors to analytes may impact on the results. Phase transport and degradation also need to be considered, since conservation of the pollutant from the source to the sampling point needs to be demonstrated in order to use existing diagnostic ratios [222]. Therefore, the development of new source diagnostic ratios with conditions specific to biomonitors would be beneficial. Multi-component studies that include trace metals and a range of organic air pollutants would also further develop the understanding of source apportionment in this field. Finally, standardisation in the methods and approaches employed with respect to biomonitoring studies is highly desirable.

REFERENCES

[1] M. De Bruin, IAEA Bull. 4 (1990) 22–27.

[2] K.J. Puckett, Bibl. Lichenol. 30 (1988) 231–267.

[3] P.H. Kansanen, J. Venetvaara, Water Air Soil Pollut. 60 (1991) 337–359.

[4] F. de Nicola, V. Spagnuolo, D. Baldantoni, L. Sessa, A. Alfani, R. Bargagli, F. Monaci, S. Terracciano, S. Giordano, Chemosphere 92 (2013) 1224–1230.

[5] W. Nylander, Bull. Soc. Bot. Fr. 13 (1886) 364–372.

[6] C. Crowther, A.E. Ruston, J. Agric. Sci. 4 (1911) 25–55.

[7] A. Wieler, Verhandlung Naturwissenschafthistorischen Vereins des Preusischen Rheinlandes und Westfalen 70 (1913) 387–400.

[8] D.L. Hawksworth, Int. J. Environ. Stud. 1 (1971) 281–296.

[9] U. Langmann, P. Madl, R. Türk, W. Hofmann, G. Brunauer, J. Environ. Prot. 5 (2014) 1331–1341.

[10] M. Blasco, C. Domeño, C. Nerín, Anal. Bioanal. Chem. 391 (2008) 759–771.

[11] J. Namiesnik, W. Wardencki, Pol. J. Environ. Stud. 11 (3) (2002) 211–218.

[12] W.H. MacIntyre, L.J. Harden, W. Hester, Ind. Eng. Chem. 44 (1952) 1365–1370.

[13] A. Rühling, G. Tyler, Water Air Soil Pollut. 1 (2001) 311–323.

[14] L.C. Pearson, E. Skye, Science 148 (1965) 1600–1602.

[15] D.J. Hill, New Phytol. 73 (1974) 1193–1205.

[16] D.H.S. Richardson, K.J. Puckett, in: B.W. Ferry, M.S. Baddeley, D.L. Hacksworth (Eds.), Air Pollution and Lichens, Athlone Press, London, 1973, pp. 283–299.

[17] T.H. Nash, B. Torrey, Bot. Club 98 (1971) 103–106.

[18] A. De Wit, Bibl. Lichenol. 5115 (1976).

[19] U. Varanasi (Ed.), Metabolism of Polycyclic Aromatic Hydrocarbons in the Aquatic Environment, CRC Press, Boca Raton, FL, 1989.

[20] Z.M. Migaszewski, A. Galuszka, P. Paslawski, Environ. Int. 28 (2002) 303–313.

[21] S.H. Boltersdorf, R. Pesch, W. Werner, Environ. Pollut. 189 (2014) 43–53.

[22] Z. Jeran, R. Jaćimovic, F. Baltič, B. Smodiš, H.Th. Wolterbeek, Fresenius J. Anal. Chem. 354 (1996) 681–687.

[23] T. Ellerman, J.K. Nøjgaard, C. Nordstrøm, J. Brandt, J. Christensen, M. Ketzel, S.S. Jensen, Dan. Air Qual. Monit. Programme 37 (2012) 1–66.

[24] T. Berg, E. Steinnes, Environ. Pollut. 98 (1997) 61–71.

[25] V. Spagnuolo, S. Giordano, A. Pérez-Llamazares, A. Ares, A. Carballeira, J.A. Fernández, J.R. Aboal, Sci. Total Environ. 463–464 (2013) 727–733.

[26] Z. Jeran, J. Jacimovic, B. Smodis, F. Batic, Epiphytic lichens as quantitative biomonitors for atmospheric element deposition, in: IAEA TECDOC-1152, Biomonitoring of Atmospheric Pollution (With Emphasis on Trace Elements)—BioMAP, Lisbon, 21–24 September 1997, IAEA, Vienna, 2000, pp. 22–28.

[27] J.E. Hanssen, J.P. Rambaek, E. Semb, E. Steinnes, Ecological impact of acid deposition in Norway, in: Proc. Intern. Conf., Sandefjord, Norway, Oslo, 1980, pp. 116–117.

[28] H. Ross, Water Air Soil Pollut. 50 (1990) 63–76.

[29] T. Berg, O. Royset, E. Steinnes, Atmos. Environ. 29 (1995) 353–360.

[30] S. Augusto, Environ. Pollut. 158 (2010) 483–489.

[31] P.B.C. Forbes, M. Thanjekwayo, J.O. Okonkwo, M. Sekhula, C. Zvinowanda, Fresenius Environ. Bull. 18 (2009) 609–614.

[32] S. Loppi, S.A. Pirintsos, Environ. Pollut. 121 (2003) 327–332.

[33] V. Shukla, D.K. Upreti, R. Bajpai, Lichens to Biomonitor the Environment, Springer, India, 2014.
[34] G. Sarret, A. Manceau, D. Cuny, C. van Haluwyn, S. Deruelle, J.L. Hazemann, Y. Soldo, L. Eybert-Berard, J.J. Menthonnex, Environ. Sci. Technol. 32 (1998) 3325–3330.
[35] R. Bargagli, F. Monaci, F. Borghini, F. Bravi, C. Agnorelli, Environ. Pollut. 116 (2002) 279–287.
[36] G. Dongarrà, D. Varrica, Sci. Total Environ. 212 (1998) 1–9.
[37] M. Hauck, C. Mulack, A. Paul, Environ. Exp. Bot. 48 (2002) 107–117.
[38] D.L. Hawksworth, F. Rose, Nature 227 (1970) 145–148.
[39] R.W. Rogers, Search 8 (1977) 75–77.
[40] J.E. Sloof, Atmos. Environ. 29 (1995) 11–20.
[41] J.E. Sloof, H. Wolterbeek, Lichenologist 23 (1991) 139–165.
[42] M. Guidotti, J. Chromatogr. A 985 (2003) 185–190.
[43] S. Augusto, J. Atmos. Chem. 49 (2004) 53–65.
[44] L. Van der Wat, P.B.C. Forbes, TrAC 64 (2015) 165–172.
[45] S. Jovan, B. Mccune, Water Air Soil Pollut. 170 (2006) 69–93.
[46] L. Frati, S. Santoni, V. Nicolardi, C. Gaggi, G. Brunialti, A. Guttova, S. Loppi, Environ. Pollut. 146 (2007) 311–316.
[47] J. Garty, in: B. Markert (Ed.), Plants as Biomonitors: Indicators for Heavy Metals in the Terrestrial Environment, VCH, New York, 1993, pp. 193–257.
[48] S. Gombert, J. Asta, M.R.D. Seaward, Environ. Pollut. 123 (2003) 281–290.
[49] Z. Jeran, T. Mrak, R. Jacimovic, F. Bati, D. Kastelec, R. Mavsar, P. Simonic, Environ. Pollut. 146 (2007) 324–331.
[50] A. True, N. Panichev, J. Okonkwo, P.B.C. Forbes, Clean Air J. 21 (2012) 19–25.
[51] E. Nieboer, P. Lavoie, R.L.P. Sasseville, K.J. Puckett, D.H.S. Richardson, Can. J. Bot. 54 (1976) 720–723.
[52] J. Hutchinson, D. Maynard, L. Geiser, USFS-Air Quality and Lichens, a Literature Review, 1996. http://gis.nacse.org/lichenair/index.php?page=literature, 1996. (accessed 01.04.15).
[53] C. Branquinho, F. Catarino, D. Brown, M.J. Pereira, A. Soares, Sci. Total Environ. 232 (1999) 67–77.
[54] C. Branquinho, in: M.N.V. Prasad (Ed.), Metals in the Environment: Analysis by Biodiversity, Marcel Dekker, New York, 2001, pp. 117–158.
[55] J.C. Duinker, F. Bouchertall, Environ. Sci.Technol. 23 (1989) 57–62.
[56] M.S. McLachlan, M. Horstmann, Environ. Sci. Technol. 32 (1998) 413–420.
[57] H. Lippo, J. Poikolainen, E. Kubin, Water Air Soil Pollut. 85 (1995) 2241–2246.
[58] B. Pawlik-Skowrońska, M. Bačkor, Environ. Exp. Bot. 72 (2011) 64–70.
[59] D. Mendil, F. Çelik, M. Tüzen, M. Soylak, J. Hazard. Mater. 166 (2009) 1344–1350.
[60] S. Ayrault, R. Clochiatti, F. Carrot, L. Daudin, J.P. Bennett, Sci. Total Environ. 372 (2007) 717–727.
[61] L.K. Boamponsem, J.I. Adam, S.B. Dampare, B.J.B. Nyarko, D.K. Essumang, Nucl. Instrum. Methods B 268 (2010) 1492–1501.
[62] A.D. Demiray, I. Yolcubal, N.H. Akyol, G. Çobanoğlu, Ecol. Indic. 18 (2012) 632–643.
[63] S. Augusto, M.J. Pereira, A. Soares, C. Branquinho, Int. J. Hyg. Environ. Health 210 (2007) 433–438.
[64] J. Garty, A.S. Perry, J. Mozel, Nord. J. Bot. 2 (1983) 583–586.
[65] J.P. Villeneuve, E. Holm, Chemosphere 13 (1984) 1133–1138.

[66] E. Bacci, D. Calamari, C. Gaggi, R. Fanelli, S. Focardi, M. Morosini, Chemosphere 15 (1986) 747–754.
[67] V. Shukla, D.K. Patel, D.K. Upreti, M. Yunus, Environ. Chem. Lett. 10 (2012) 159–164.
[68] M. Blasco, C. Domeño, C. Nerín, Environ. Sci. Technol. 40 (2006) 6384–6391.
[69] S. Augusto, Environ. Sci. Technol. 43 (2009) 7762–7769.
[70] J.E. Schrlau, L. Geiser, K.J. Hageman, D.H. Landers, S.M. Simonich, Environ. Sci. Technol. 45 (2011) 10354–10361.
[71] I. Oksanen, Appl. Microbiol. Biotechnol. 73 (2006) 723–734.
[72] H. Bauer, J. Schönherr, Pestic. Sci. 35 (1992) 1–11.
[73] P. Baur, A. Buchholz, J. Schönherr, Plant Cell Environ. 20 (1997) 982–994.
[74] S. Augusto, M.J. Pereira, C. Máguas, C. Branquinho, Chemosphere 92 (2013) 626–632.
[75] H. Kylin, H. Bouwman, Environ. Sci. Technol. 46 (2012) 10982–10989.
[76] J. Nascimbene, Sci. Total Environ. 475 (2014) 90–96.
[77] S. Augusto, C. Máguas, C. Branquinho, Environ. Pollut. 180 (2013) 330–338.
[78] M. Blasco, C. Domeño, K. Bentayeb, C. Nerín, Environ. Anal. Chem. 87 (2007) 833–846.
[79] M. Blasco, C. Domeño, P. Lopez, C. Nerin, J. Environ. Monit. 13 (2011) 2588–2596.
[80] C. Domeño, M. Blasco, C. Sánchez, C. Nerín, Anal. Chim. Acta 569 (2006) 103–112.
[81] D.K. Upreti, D.K. Patel, Environ. Monit. Assess. 184 (2012) 229–238.
[82] V. Shukla, D.K. Upreti, Environ. Monit. Assess. 149 (2009) 1–7.
[83] V. Shukla, D.K. Upreti, Environ. Chem. Lett. 11 (2013) 19–23.
[84] M. Guidotti, Bull. Environ. Contam. Toxicol. 83 (2009) 852–858.
[85] S. Denys, D. Gombert, K. Tack, Chemosphere 88 (2012) 806–812.
[86] R. Fernandez, F. Galarraga, Z. Benzo, G. Marquez, A.J. Fernández, M.G. Requiz, J. Hernandez, Int. J. Environ. Anal. Chem. 91 (2011) 230–240.
[87] S. Jensen, Use of Moss And/or Lichen to Monitor Airborne Organic Pollutants (In Swedish); Report to the Swedish Environmental Protection Agency, Stockholm, Sweden, 1990.
[88] S. Augusto, C. Máguas, C. Branquinho, Ecotoxicology 18 (2009) 1036–1042.
[89] J. Fabure, C. Meyer, F. Denayer, A. Gaudry, D. Gilbert, N. Bernard, Water Air Soil Pollut. 212 (2010) 205–217.
[90] L. Kent, http://www.bio.umass.edu/biology/conn.river/mosses.html, (accessed 22.04.15).
[91] G. Uyar, M. Ören, M. Ince, Fresenius Environ. Bull. 16 (2007) 1–9.
[92] S. Chakrabortty, G.T. Paratkar, Aerosol Air Qual. Res. 6 (2006) 247–258.
[93] L. González-Miqueo, D. Elustondo, E. Lasheras, J.M. Santamaría, Chemosphere 78 (2010) 965–971.
[94] J.W.M. Wegener, M.J.M. van Schaik, H. Aiking, Environ. Pollut. 76 (1992) 15–18.
[95] F.I. Colabuono, S. Taniguchi, C.V.Z. Cipro, J. da Silva, M.C. Bícego, R.C. Montone, Mar. Pollut. Bull. 93 (2015) 266–269.
[96] V. Spagnuolo, S. Terracciano, S. Giordano, Chemosphere 74 (2009) 1158–1164.
[97] M. Buonocore, C. Annicchiarico, N. Cardellicchio, A. Di Leo, S. Giandomenico, L. Spada, Environ. Eng. Manage. J. 12 (2013) 245–248.
[98] A.M.G. Pacheco, L.I.C. Barros, M.C. Freitas, M.A. Reis, C. Hipólito, O.R. Oliveira, Environ. Pollut. 120 (2002) 79–86.
[99] R. Gerdol, L. Bragazza, R. Marchesini, A. Medici, P. Pedrini, S. Benedetti, A. Bovolenta, S. Coppi, Atmos. Environ. 36 (2002) 4069–4075.
[100] X. Liu, G. Zhang, K.C. Jones, X. Li, X. Peng, S. Qi, Atmos. Environ. 39 (2005) 5490–5499.

[101] M. Kunert, K. Friese, V. Weckert, B. Markert, Environ. Sci.Technol. 33 (1999) 3502–3505.
[102] E. Ötvös, T. Pázmándi, Z. Tuba, Sci. Total Environ. 309 (2003) 151–160.
[103] J. Poikolainen, E. Kubin, J. Piispanen, J. Karhu, Sci. Total Environ. 318 (2004) 171–185.
[104] T. Berg, O. Røset, E. Steinnes, M. Vadset, Environ. Pollut. 88 (1995) 67–77.
[105] M.V. Frontasyeva, E. Steinnes, Analyst 120 (1995) 1437–1440.
[106] I. Türkan, E. Henden, Ü. Çelik, S. Kivilcim, Sci. Total Environ. 166 (1995) 61–67.
[107] C. Domeño, E. Canellas, P. Alfaro, A. Rodriguez-Lafuente, C. Nerin, J. Chromatogr. A 1252 (2012) 146–154.
[108] D. Tarcau, S. Cucu-Man, J. Boruvkova, J. Klanova, A. Covaci, Sci. Total Environ. 456–457 (2013) 317–324.
[109] J. Bustamante, O. Liñero, I. Arrizabalaga, J.A. Carrero, G. Arana, A. de Diego, Chemosphere 122 (2015) 295–300.
[110] E. Ötvös, I.O. Kozák, J. Fekete, V.K. Sharma, Z. Tuba, Sci. Total Environ. 330 (2004) 89–99.
[111] M.S. McLachlan, K. Welsch-Pausch, J. Tolls, J. Environ. Sci. Technol. 29 (1995) 1998–2004.
[112] J. Schönherr, M. Riederer, Foliar penetration and accumulation of organic chemicals in plant cuticles, in: Reviews of Environmental Contamination and Toxicology, Springer, New York, 1989, pp. 1–70.
[113] R. Gerdol, L. Bragazza, R. Marchesini, R. Alber, L. Bonetti, G. Lorenzoni, M. Achilli, A. Buffoni, N. De Marco, M. Franchi, S. Pison, S. Giaquinta, F. Palmieri, P. Spezzanto, Environ. Pollut. 108 (2000) 201–208.
[114] J. Klánová, P. Čupr, D. Baráková, Z. Šeda, P. Anděl, I. Holoubek, Environ. Pollut. 157 (2009) 3248–3254.
[115] R. Bargagli, M.L. D'Amato, F.P. Losco, Atti. Soc. Tosc. Sci. Nat. 93 (1986) 27–38.
[116] K.D. Wenzel, A. Hubert, M. Manz, L. Weissflog, W. Engewald, G. Schüürmann, Anal. Chem. 70 (1998) 4827–4835.
[117] E. Lehndorff, L. Schwark, Atmos. Environ. 38 (2004) 3793–3808.
[118] H.M. Hwang, T.L. Wade, J.L. Sericano, Atmos. Environ. 37 (2003) 2259–2267.
[119] N. Ratola, S. Lacorte, D. Barceló, A. Alves, Talanta 77 (2009) 1120–1128.
[120] I. Holoubek, P. Kořínek, Z. Šeda, E. Schneiderová, I. Holoubková, A. Pacl, J. Tříska, P. Cudlín, J. Čáslavský, Environ. Pollut. 109 (2000) 283–292.
[121] E. Lehndorff, L. Schwark, Atmos. Environ. 42 (2008) 833–845.
[122] R. Bargagli, Trace Elements in Terrestrial Plants—An Ecophysiological Approach to Biomonitoring and Biorecovery, vol. 325, Springer, London, Heidelberg, 1998.
[123] G. Eriksson, S. Jensen, H. Kylin, W.M.J. Strachan, Nature 341 (1989) 42–44.
[124] W.M.J. Strachan, H. Kylin, G. Eriksson, S. Jensen, Environ. Toxicol. Chem. 13 (1994) 443–451.
[125] J.M. Amigo, N. Ratola, A. Alves, Atmos. Environ. 45 (2011) 5988–5996.
[126] N. Ratola, S. Lacorte, A. Alves, D. Barceló, J. Chromatogr. A 1114 (2006) 198–204.
[127] J.A. Silva, N. Ratola, S. Ramos, V. Homem, L. Santos, A. Alves, Anal. Chim. Acta 858 (2015) 24–31.
[128] N. Ratola, V. Homem, J.A. Silva, R. Araújo, J.M. Amigo, L. Santos, A. Alves, Sci. Total Environ. 476 (2014) 114–124.
[129] A. Hubert, P. Popp, K.D. Wenzel, W. Engewald, G. Schüürmann, Anal. Bioanal. Chem. 376 (1) (2003) 53–60.

[130] N. Hanari, Y. Horii, T. Okazawa, J. Falandysz, I. Bochentin, A. Orlikowska, N. Yamashita, J. Environ. Monit. 6 (2004) 305–312.
[131] N. Ratola, P. Herbert, A. Alves, Anal. Bioanal. Chem. 403 (2012) 1761–1769.
[132] A. Wyttenbach, S. Bajo, L. Tobler, in: G.N. Schrauzer (Ed.), Biological Trace Element Research, Humana Press, Clifton, UK, 1990, pp. 213–221.
[133] R. Bargagli, C. Barghigiani, B.E. Maserti, Chemosphere 15 (1986) 1035–1042.
[134] D. Čeburnis, E. Steinnes, Atmos. Environ. 34 (2000) 4265–4271.
[135] E. Lehndorff, L. Schwark, Atmos. Environ. 44 (2010) 2822–2829.
[136] W. Dmuchowski, A. Bytnerowicz, Environ. Pollut. 87 (1995) 87–104.
[137] M.T.M. Al-Alawi, K.L. Mandiwana, J. Hazard. Mater. 148 (2007) 43–46.
[138] B. Kord, A. Mataji, S. Babaie, Int. J. Environ. Sci. Technol. 7 (2010) 79–84.
[139] A. Hellström, Swed. Univ. Agric. Sci. Rapp. 1 (2004) 2004.
[140] H. Schulz, P. Popp, G. Huhn, H.J. Stärk, G. Schüürmann, Sci. Total Environ. 232 (1999) 49–58.
[141] K.C. Jones, G. Sanders, S.R. Wild, V. Burnett, A.E. Johnston, Nature 356 (1992) 137–140.
[142] F. Guéguen, P. Stille, M. Millet, Chemosphere 85 (2011) 195–202.
[143] K. Steindor, B. Palowski, P. Góras, A. Nadgórska-Socha, Pol. J. Environ. Stud. 20 (2011) 619–622.
[144] A.A. Olajire, E.T. Ayodele, Bull. Chem. Soc. Ethiop. 17 (2003) 11–17.
[145] S.K. Prajapati, B.D. Tripathi, Environ. Pollut. 151 (2008) 543–548.
[146] Y.Q. Wang, S. Tao, X.C. Jiao, R.M. Coveney, S.P. Wu, B.S. Xing, Environ. Pollut. 151 (2008) 158–164.
[147] S. Orecchio, Atmos. Environ. 41 (2007) 8669–8680.
[148] H. Lin, S. Tao, Q. Zuo, R.M. Coveney, Environ. Pollut. 148 (2007) 614–619.
[149] M. Barriada-Pereira, M.J. González-Castro, S. Muniategui-Lorenzo, P. López-Mahía, D. Prada-Rodríguez, E. Fernández-Fernández, Chemosphere 58 (2005) 1571–1578.
[150] B. Viard, F. Pihan, S. Promeyrat, J.C. Pihan, Chemosphere 55 (2004) 1349–1359.
[151] A. Celik, A.A. Kartal, A. Akdoğan, Y. Kaska, Environ. Int. 31 (2005) 105–112.
[152] G.L. Calzoni, F. Antognoni, E. Pari, P. Fonti, A. Gnes, A. Speranza, Environ. Pollut. 149 (2007) 239–245.
[153] G. Lorenzini, A. Panicucci, L. Guidi, Bull. Environ. Contam. 45 (1990) 408–414.
[154] A. Rühling (Ed.), Atmospheric Heavy Metal Deposition in Europe:–Estimation Based on Moss Analysis, Nordic Council of Ministers, 1994.
[155] B. Wolterbeek, Environ. Pollut. 120 (2002) 11–21.
[156] M. Nadal, J. Atmos. Chem. 55 (2006) 241–252.
[157] H.T. Wolterbeek, P. Bode, Sci. Total Environ. 176 (1995) 33–43.
[158] M. Tretiach, A. Carpanelli, Lichenologist 24 (1992) 81–90.
[159] L.A. Di Lella, S. Loppi, G. Protano, F. Riccobono, Atmos. Environ. 40 (2006) 225–237.
[160] N. Ratola, J.M. Amigo, M.S. Oliveira, R. Araújo, J.A. Silva, A. Alves, Environ. Exp. Bot. 72 (2011) 339–347.
[161] F.J. Doucet, J. Carignan, Atmos. Environ. 35 (21) (2001) 3681–3690.
[162] M.V.B. Krishna, J. Arunachalam, Anal. Chim. Acta 522 (2004) 179–187.
[163] P. Pasławski, Z.M. Migaszewski, Pol. J. Environ. Stud. 15 (2006) 154–164.
[164] P. Adamo, S. Giordano, S. Vingiani, R. Castaldo Cobianchi, P. Violante, Environ. Pollut. 122 (2003) 91–103.
[165] D. Kuehnelt, J. Lintschinger, W. Goessler, Appl. Organomet. Chem. 14 (2000) 411–420.
[166] S. Loppi, L. Frati, L. Paoli, V. Bigagli, C. Rossetti, C. Bruscoli, A. Corsini, Sci. Total Environ. 326 (2004) 113–122.

[167] M.D. Luque de Castro, L.E. Garcıa-Ayuso, Anal. Chim. Acta 369 (1998) 1–10.
[168] M.D. Luque de Castro, F. Priego-Capote, J. Chromatogr. A 1217 (2010) 2382–2389.
[169] S. Augusto, M.J. Pereira, C. Máguas, C. Branquinho, J. Toxicol. Environ. Health 75 (2012) 1–10.
[170] S. Orecchio, A. Gianguzza, L. Culotta, Environ. Res. 107 (2008) 371–379.
[171] M. Barriada-Pereira, E. Concha-Grana, M.J. Gonzalez-Castro, S. Muniategui-Lorenzo, P. López-Mahıa, D. Prada-Rodrıguez, E. Fernandez-Fernandez, J. Chromatogr. A 1008 (2003) 115–122.
[172] R. Bajpai, N. Karakoti, D.K. Upreti, Environ. Sci. Pollut. Res. 20 (2013) 5577–5592.
[173] C. Schauer, R. Niessner, U. Pöschl, Environ. Sci. Technol. 37 (2003) 2861–2868.
[174] L. Guo, H.K. Lee, J. Chromatogr. A 1286 (2013) 9–15.
[175] V. Camel, TrAC 19 (2000) 229–248.
[176] N. Fidalgo-Used, E. Blanco-González, A. Sanz-Medel, Anal. Chim. Acta 590 (2007) 1–16.
[177] V. Camel, Analyst 126 (2001) 1182–1193.
[178] J.R. Dean, G. Xiong, TrAC 19 (2000) 553–564.
[179] K. Ganzler, A. Salgo, K. Valko, J. Chromatogr. A 371 (1986) 299–306.
[180] K. Ganzler, A. Salgo, Z. Lebensm. Unters. Forsch. 184 (1987).
[181] L. Xu, H.K. Lee, J. Chromatogr. A 1192 (2008) 203–207.
[182] M. Tomaniova, J. Chromatogr. A 827 (1998) 21–29.
[183] P. Baumard, Mar. Pollut. Bull. 36 (1998) 951–960.
[184] B. Koz, N. Celik, U. Cevik, Ecol. Indic. 10 (2010) 762–765.
[185] H.A. Carreras, M.L. Pignata, Environ. Pollut. 117 (2002) 77–87.
[186] V. Pandey, D.K. Upreti, R. Pathak, A. Pal, Environ. Monit. Assess. 73 (2002) 221–228.
[187] M.A. Wilson, R. Burt, C.W. Lee, Commun. Soil Sci. Plant 37 (2006) 513–524.
[188] M. Montes-Bayón, M.J.D. Molet, E.B. González, A. Sanz-Medel, Talanta 68 (2006) 1287–1293.
[189] T.G. Kazi, N. Jalbani, M.B. Arain, M.K. Jamali, H.I. Afridi, A.Q. Shah, Environ. Monit. Assess. 154 (2009) 155–167.
[190] C. Bendicho, I. Lavilla, Academic Press, 2000, 4424–4426.
[191] R.A. Sánchez-Moreno, M.J. Gismera, M.T. Sevilla, J.R. Procopio, Phytochem. Anal. 21 (2010) 340–347.
[192] J. Namiesnik, P. Szefer, in: J. Namiesnik, P. Szefer (Eds.), Analytical Measurements in Aquatic Environments, LLC: Taylor and Francis Group, 2008, pp. 431–474.
[193] A. Huerga, I. Lavilla, C. Bendicho, Anal. Chim. Acta 534 (2005) 121–128.
[194] I.D. Manariotis, H.K. Karapanagioti, C.V. Chrysikopoulos, Water Res. 45 (2011) 2587–2594.
[195] M. Anastassiades, S.J. Lehotay, D. Stajnbaher, F.J. Schenck, J. AOAC Int. 86 (2003) 412–431.
[196] European Committee for Standardization (CEN). EN 15662:2008. CEN, Brussels, 2009.
[197] A. Wilkowska, M. Biziuk, Food Chem. 125 (2011) 803–812.
[198] A. Albinet, S. Tomaz, F. Lestremau, Sci. Total Environ. 450 (2013) 31–38.
[199] J. Zhao, Chin. J. Anal. Lab. 31 (2012) 33–37.
[200] M. João Ramalhosa, J. Sep. Sci. 32 (2009) 3529–3538.
[201] K. Kalachova, Anal. Chim. Acta 707 (2011) 84–91.
[202] L. Drabova, Talanta 100 (2012) 207–216.
[203] A. Sadowska-Rociek, M. Surma, E. Cieslik, Bull. Environ. Contam. Toxicol. 90 (2013) 508–513.

[204] J. Haib, I. Hofer, J.M. Renaud, J. Chromatogr. A 1020 (2003) 173–187.
[205] K. Ramu, N. Kajiwara, S. Tanabe, P.K. Lam, T.A. Jefferson, Mar. Pollut. Bull. 51 (2005) 669–676.
[206] S. Tao, L.Q. Guo, X.J. Wang, W.X. Liu, T.Z. Ju, R. Dawson, B.G. Li, Sci. Total Environ. 320 (2004) 1–9.
[207] A. Meudec, J. Dussauze, M. Jourdin, E. Deslandes, N. Poupart, J. Chromatogr. A 1108 (2006) 240–247.
[208] M. Barriada-Pereira, M.J. Gonzalez-Castro, S. Muniategui-Lorenzo, P. Lopez-Mahia, D. Prada-Rodriguez, E. Fernandez-Fernandez, J. Chromatogr. A 1061 (2004) 133–139.
[209] A. Sadowska-Rociek, M. Surma, E. Cieslik, Environ. Sci. Pollut. Res. 21 (2014) 1326–1338.
[210] M. Jánská, M. Tomaniová, J. Hajšlová, V. Kocourek, Anal. Chim. Acta 520 (2004) 93–103.
[211] K.E. Smith, G.L. Northcott, K.C. Jones, J. Chromatogr. A 1116 (2006) 20–30.
[212] D.L. Poster, M.M. Schantz, L.C. Sander, S.A. Wise, Anal. Bioanal. Chem. 386 (2006) 859–881.
[213] J.C. Fetzer, Chemical analysis of polycyclic aromatic compounds, in: T. Vo Dinh (Ed.), Chemical Analysis, Wiley, New York, 1989, pp. 59–109.
[214] P. Tremolada, V. Burnett, D. Calamari, K.C. Jones, Environ. Sci. Technol. 30 (1996) 3570–3577.
[215] O. Pani, T. Górecki, Anal. Bioanal. Chem. 386 (2006) 1013–1023.
[216] R.J. Vreuls, J. Dallüge, U.A.T. Brinkman, J. Microcolumn Sep. 11 (1999) 663–675.
[217] J. Schnelle-Kreis, W. Welthagen, M. Sklorz, R. Zimmermann, J. Sep. Sci. 28 (2005) 1648–1657.
[218] M. Nadal, M. Schuhmacher, J.L. Domingo, Environ. Pollut. 132 (2004) 1–11.
[219] Ş. Marin, Ş. Lăcrimioara, R. Cecilia, J. Plant Dev. 18 (2011) 87–93.
[220] A.A. Ammann, J. Mass Spectrom. 42 (2007) 419–427.
[221] B. Michalke, Ecotoxicol. Environ. Saf. 56 (2003) 122–139.
[222] E. Galarneau, Atmos. Environ. 42 (2008) 8139–8149.

Chapter 4

Ambient Air Sampling with Whole-Air, In-Field Concentration and Particulate Matter (PM) Methodologies

Jason S. Herrington
Innovations Group Restek Corporation, Bellefonte, PA, USA
E-mail: jason.herrington@restek.com

Chapter Outline

1. Introduction **109**
2. Very Volatile Organic Compounds/Volatile Organic Compounds **112**
2.1 Whole-Air Sampling 112
2.1.1 Whole-Air Sampling with Bags 112
2.1.2 Whole-Air Sampling with Canisters 120
2.1.3 Sample Preconcentration 130
3. Volatile Organic Compounds/Semi-volatile Organic Compounds **132**
3.1 In-field Concentration 132
3.1.1 Air Sampling with Impingers/Bubblers 133
3.1.2 Air Sampling with Solid Sorbents 136
4. Semi-volatile Organic Compounds/Particulate Matter **143**
4.1 Methodology 144
4.2 Preparation 144
4.3 Fractionators 144
4.4 Applications 147
4.5 SVOCs/Artifacts 147
4.6 Storage 148
5. High Volume Sampling **148**
6. Conclusions **150**
References **151**

1. INTRODUCTION

The field of measuring airborne mixtures may be broken down into the following two very distinct processes: (1) sampling and (2) analysis. Air sampling is the act of collecting a volume of air or filtering target analytes

Comprehensive Analytical Chemistry, Vol. 70. http://dx.doi.org/10.1016/bs.coac.2015.09.004

from a volume of air. Analysis is the qualitative act of identifying compounds and quantitative act of determining their concentrations. Some methods combine sampling and analysis into one (e.g. Fourier transform infrared spectroscopy). This chapter focuses exclusively on sampling as 'real-time' instruments are covered in Chapter 6.

Air samples may be collected in a 'grab' or 'time-integrated' fashion. A grab sample (sometimes referred to as a catch sample) is executed just like the name implies; a sample is grabbed from an atmosphere or source in a relatively short time period (i.e. generally on the order of seconds). One may visualise the following scenario for grab sampling: a stream from which an 8 cm^3 glass of water is collected in approximately 2 s. A time-integrated sample (sometimes referred to as a composite sample) is collected over a period of time (e.g. 24 h to 2 weeks) and represents the average concentration for that sampling time interval. The concentration of a target analyte obtained from a time-integrated sample is often referred to as a time-weighted average, as the peaks and valleys in concentrations for the target analyte have been averaged over the total sampling period. One may visualise the following scenario for time-integrated sampling: a stream from which 1 cm^3 h^{-1} of water is diverted into an 8 cm^3 glass, over an 8 h period.

Both the grab and time-integrated samples produce an 8 cm^3 glass of water. However, if those samples are taken to the next phase and analysed for the drinking water disinfection by-product bromodichloromethane, their results may be drastically different. The grab sample was quick and relatively easy; however, the results only represent the bromodichloromethane concentrations for the drinking water during that 2 s period of time. Therefore, if the bromodichloromethane concentrations were 'abnormally' low or high during the 2 s collection interval relative to a typical 8 h period; then the results may not be representative of the bromodichloromethane concentrations one could expect to encounter. In contrast, the time-integrated sample was more time-consuming and difficult; however, it represents the average bromodichloromethane concentrations for the drinking water over an 8 h period, which may be more representative of the 'true' concentration one might expect to encounter. Both grab and time-integrated sampling approaches have their applications, and the methods covered herein are usually capable of both approaches; however, the field of ambient air sampling is dominated by time-integrated methods.

The field of air sampling may also be partitioned based on sources. For example, point/stationary (e.g. stacks, refineries, manufacturing plants), mobile (e.g. cars, boats, planes, trains), indoor (e.g. office buildings, homes), outdoor (often referred to as ambient and the focus of this book) and personal (often associated with industrial hygiene/occupational health monitoring) sampling. The current chapter will focus on ambient air sampling methods, which in general covers relatively low (e.g. part-per-trillion by volume (pptv) and ng m^{-3} to part-per-billion by volume (ppbv) and μg m^{-3}) concentrations;

however, the methods covered within this chapter are often utilised for sampling the other sources.

Air sampling techniques often overlap sources, technologies and compounds. Consequently, the selection and application of the proper ambient air sampling method can be daunting. In an attempt to provide a natural linear progression across the covered ambient air sampling methods, the current chapter will move across methods in order of volatility, starting with the very volatile organic compounds (VVOCs); followed by the volatile organic compounds (VOCs); then the semi-VOCs (SVOCs); and ending with particulate matter (PM). This conveniently coincides with the three main sections of this chapter: whole-air sampling (i.e. VVOCs and VOCs), in-field concentration (VOCs and SVOCs) and PM sampling [1].

Volatility is defined as the propensity of a substance to vaporise, which is directly related to the substance's vapour pressure. The World Health Organization (WHO) has defined VVOCs as compounds with boiling points from <0 to 50−100 °C [2]. Some examples of VVOCs include the chlorofluorocarbons like dichlorodifluoromethane (R-12), formaldehyde and propylene. The WHO has defined VOCs as any compound whose boiling point is in the range from (50−100 °C) to (240−260 °C) [2]. Some examples of VOCs include benzene, toluene, ethylbenzene, xylenes (o, m and p), methyl tertiary butyl ether, trichloroethylene, methylene chloride and 1,1,1-trichloroethane. The WHO has defined SVOCs as organic compounds with boiling points ranging from (240−260 °C) to (380−400 °C) [2]. Some examples of SVOCs include compound classes such as phthalates (PVC plasticisers), polycyclic aromatic hydrocarbons (PAHs), polychlorinated biphenyls (PCBs), polybrominated diphenyl ethers and polybrominated diphenyl furans. The WHO distinguishes organic compounds associated with PM or particulate organic matter as having a boiling point range >380 °C [2].

It is outside the scope of the current chapter to discuss all of the potential applications of the covered methods, especially for all of the possible individual compounds and/or classes of compounds. However, some example compounds are provided for each method, so as to provide the potential end user with a starting point for appropriate method application. In addition, it is outside the scope of the current chapter to discuss potential environmental and/or human health implications. The justifications for sampling ambient air are unrelated to the goals of this chapter. Lastly, the current chapter will not cover sources, transport, atmospheric reactions, gas-to-particle conversions, fate, risks, etc. The aforementioned details are irrelevant as they pertain to the current chapter. The goal of the current chapter is to give the novice enough information to make an educated decision on which ambient air sampling approach may best fit their application based on volatility; understand the principles of how the method works; understand the materials required to execute it and understand the scope of sample preparation needed for the selected method.

2. VERY VOLATILE ORGANIC COMPOUNDS/VOLATILE ORGANIC COMPOUNDS

2.1 Whole-Air Sampling

One of the most widely used approaches for sampling ambient VVOCs and VOCs is to utilise a whole-air method. Whole-air sampling techniques will collect everything suspended in the sampled atmosphere, hence the name whole air. This means that even though most whole-air sampling strategies are utilised strictly for the collection of VVOCs and VOCs, a whole-air sample will gather SVOCs and PM as well; unless some physical barrier (e.g. a filter) is utilised to prevent their entrance into the whole-air sample collection vessel. Some of the whole-air sampling techniques include gas bulb samplers, bag sampling, glass bottle (often referred to by their trade name Bottle-Vac™) samplers and canisters. Inarguably, canisters are the most popular whole-ambient air sampling method for VVOCs and VOCs. Therefore, the following section will focus primarily on canisters. However, bag sampling remains popular enough to warrant an appropriate discussion. On the contrary, gas bulb samplers are fading in popularity; and glass bottle samplers are relatively new to the whole-ambient air sampling market.

2.1.1 Whole-Air Sampling with Bags

Gas sampling bags (Figure 1) are often referred to as just bags. As the name implies, bags are made of relatively thin (0.051−0.127 mm), pliable, synthetic plastic polymers, which means they have similar physical properties to that of a plastic freezer bag. However, beyond the aforementioned textural similarities bags quickly diverge from the classical 'bag' definition, which conjures images of an open-ended grocery bag. It may be more appropriate to refer to bags as bladders or balloons; however, that is not the industry standard. Bags may best be conceptualised as a bicycle inner tube, in that there is a valve (Figure 2) to control the entrance and egress of gas into and out of the bag. Bag valves are usually made of a relatively inert material, and the two most popular valve materials are polypropylene and stainless steel. Bags are often square or rectangular in shape and they come in a variety of sizes ranging from 0.5 L to 100 L; however, they are easily made to custom sizes.

FIGURE 1 Tedlar® bags with polypropylene valve (left) and stainless-steel valve (right) utilised for collecting, storing and transporting whole-air samples. *Photo courtesy of Restek Corporation.*

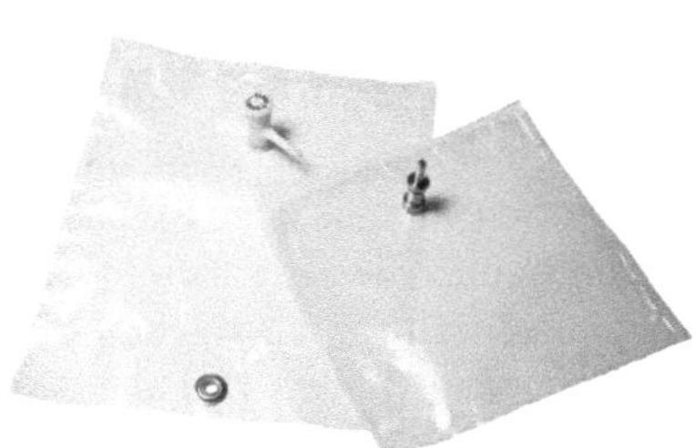

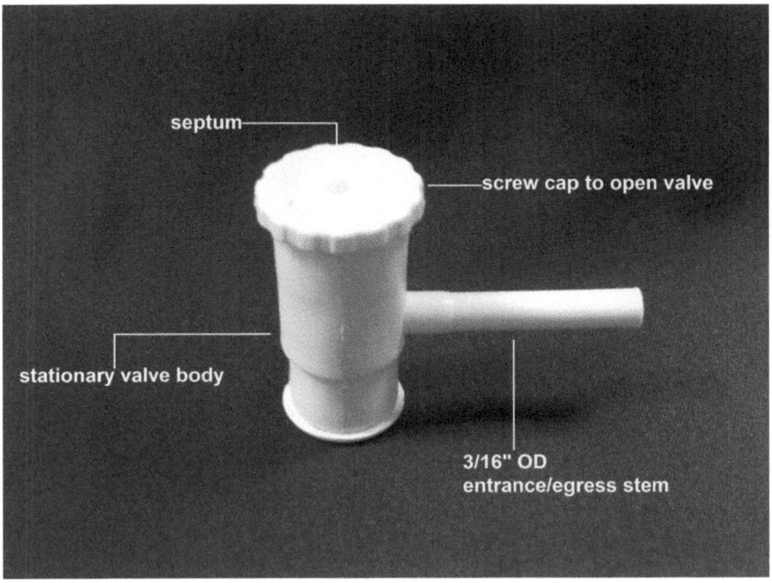

FIGURE 2 Polypropylene valve used on bags for the collection and removal (for analysis) of whole-air samples.

2.1.1.1 Methodology

A new bag is shipped to the field empty (i.e. flat). A field sampling technician then proceeds to fill the bag with a whole-ambient air sample, similar to how someone fills a bicycle inner tube. It is important to note that bags are only intended to reach approximately 80% capacity (to allow for expansion due to environmental changes) and/or pressures of approximately 830–880 torr. Beyond these capacities/pressures there is a risk of creating leaks and/or catastrophic failures (i.e. pop like a balloon). In order to fill a bag with a whole-ambient air sample, the bag is often connected to a positive pressure source. Typically this entails the use of a peristaltic, rotary vane, diaphragm and/or piston pump (Figure 3). A bag needs to be filled at an appropriate flow rate and over the desired sampling duration, which will require the use of a flow meter (Figure 4) to verify/calibrate the pump flow rate. For example, an end user would like to fill a 10 L bag over an 8 h sampling duration. With the 80% fill capacity in mind the 10 L bag should only be filled with 8 L of sample, which means the pump should be flowing at $1.0\ L\ h^{-1}$. So the end user will utilise the flow meter to adjust the pump flow rate to $1.0\ L\ h^{-1}$; temporarily turn the pump off; connect the bag to the pump; open the valve on the bag; begin sampling by turning on the pump; stop sampling 8 h later by turning off the pump; close the valve on the bag and ship the bag to the laboratory for analysis. Regardless of how a bag is filled the main consideration for sampling is the bag volume, the desired duration and needed flow rate.

FIGURE 3 Piston pump used for pressurised flow to fill bags or vacuum flow to draw samples into impingers/bubblers, across thermal desorption tubes and through particulate matter fractionators. *Photo courtesy of Restek Corporation.*

FIGURE 4 Flow meter used for calibrating and verifying pressure and vacuum flows. *Photo courtesy of Restek Corporation.*

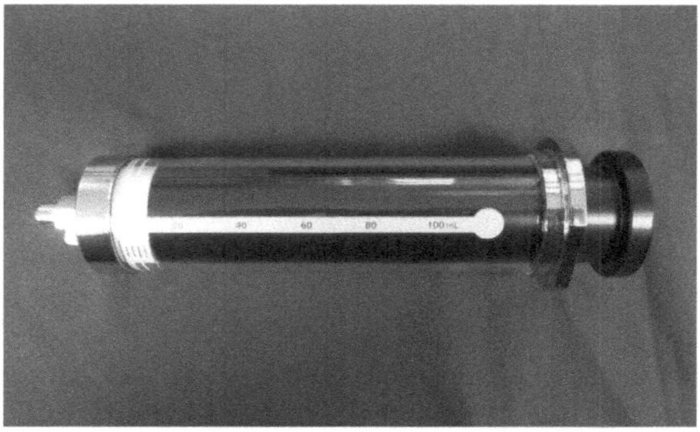

FIGURE 5 Gas-tight syringe utilised for grab samples and/or sample/standard dilutions.

In addition to pumps, gas-tight syringes (Figure 5) may be used to collect a grab air sample by quickly drawing on the plunger. This air sample may then be injected into a bag via a septum, which is generally found in the valve body (Figure 2) on a bag. It is important to note that if high concentrations of any particular compound are collected through a pump or syringe it may possibly contaminate the pump/syringe internals, which may bias results for subsequent air samples utilizing the same sampling device. Due to the aforementioned phenomenon/concern, another popular technique for filling a bag sample is by using a rigid container (often referred to as a 'lung sampler') sampling approach. For this technique a bag is filled by placing it in a rigid airtight container (with only a connection from the bag's valve to the external air). The rigid container (Figure 6) is then placed under vacuum by using an internal or external vacuum pump. The process of evacuating the rigid container results in drawing an air sample into the bag. All of the aforementioned bag volume,

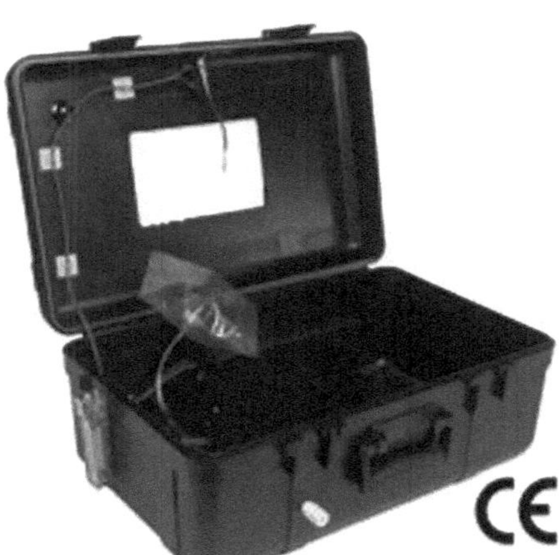

FIGURE 6 Rigid box often referred to as bag or lung sampler used to fill bags under a vacuum source; whilst eliminating the possibility for carry over from sample to sample.

sampling duration and flow rate considerations are applicable to using a lung sampler.

2.1.1.2 Preparation

Bags are manufactured and sold with the intent for single use. Therefore, sample preparation is as simple as securing bags. They are not intended for multiple uses like the other whole-air sampling methods; as studies have demonstrated bags are relatively difficult to clean for reuse [3]. Bags are challenging to clean for the following two reasons: (1) they are usually square or rectangular in shape, which means they have corners. The corners act as sinks for compounds to become trapped and contaminate subsequent samples. (2) They are not rigid and collapse on themselves whilst under vacuum. This means they are not subject to the cleaning procedures carried out on canisters (discussed in section 2.1.2.2.2 of this chapter). Despite the aforementioned information, many end users attempt to cut corners and save money by cleaning them with procedures analogous to the canister cleaning procedures. However, the current chapter will not address bag cleaning, as they are not intended to be used in this manner.

2.1.1.3 Bag Types

Bags are most often made from one of the three following materials: (1) Tedlar® (polyvinyl fluoride), which is ideal for low level (ppbv) analyses of VOCs, due to its low background concentrations and relatively inert properties. (2) Multi-layer foil bags often consisting of polyethylene as the innermost layer, followed by aluminium foil, then polyethylene again, then metalised aluminium and finally an outer nylon layer. These bags are ideal for low molecular weight compounds like carbon monoxide, carbon dioxide, methane and permanent gases, as well as some sulphur-containing compounds. (3) Other proprietary materials such as Altef® or Tecaflon® (polyvinylidene fluoride), which have very similar characteristics to Tedlar®. Bags have also historically been made of Saran® (polyvinylidene chloride) and Teflon®. Table 1 lists the three most popular aforementioned bag materials and their properties.

2.1.1.4 Applications

There are numerous applications for whole-air bag sampling; however, the majority of these methods are targeting relatively high ppbv to ppmv concentrations of non-reactive and non-polar VOCs, with short sample storage durations. Table 2 provides several examples of bag methods/applications for whole-air sampling; however, due to the aforementioned general characteristics, bags have found more applications in the industrial hygiene/occupational health and source sampling fields, than in the ambient air arena; although they maintain a niche in the ambient air sampling arena, potential end users should be aware of the limitations.

TABLE 1 Bag Types and Properties for Whole-Air Sampling

Property	Tedlar®	Multilayer	Altef®/Tecaflon®
Material	Polyvinyl fluoride	**1.** 0.002″ polyethylene **2.** 0.0003″ aluminium foil **3.** Polyethylene **4.** 60 gauge nylon	Polyvinylidene fluoride
Thickness	0.002″	0.005″	0.003″
Tensile strength	8000 psig	24 psig	6100 psig
Continuous operating temperature	107 °C	88 °C	150 °C
Maximum operating temperature	176–204 °C	87–88 °C	150 °C
Oxygen permeability	50 mL $(m^2 \times d)^{-1}$	0.0006 mL $(m^2 \times d)^{-1}$	58 mL $(m^2 \times d)^{-1}$
Water vapor permeability	9–57 g $(m^2 \times d)^{-1}$	0.0006 g 100 square inches day^{-1}	12–15 g $(m^2 \times d)^{-1}$
Carbon dioxide permeability	172 mL $(m^2 \times d)^{-1}$	0.0005 mL $(m^2 \times d)^{-1}$	172 mL $(m^2 \times d)^{-1}$
Specific gravity	1.70 g mL^{-1}	1.09 g mL^{-1}	1.78 g mL^{-1}
Application	Best for low-level (ppbv) VOCs due to low background concentrations and inertness	Best for low molecular weight compounds like carbon monoxide, carbon dioxide, methane and permanent gases Minimises photodegradation 48 h sulphur compounds	Best economical replacement for Tedlar®. Not recommended for ketones, acetates, hydrogen sulfide or permanent gases

TABLE 2 Bag Methods and Application Suitability for Whole-Air Sampling

Method	Target Compounds	Tedlar® Suitability	Multilayer Foil Suitability	Altef®/Tecaflon® Suitability
U.S. EPA TO-3	Volatile organic compounds (VOCs)	Y	N	Y
U.S. EPA 0040	VOCs	Y	N	Y
U.S. EPA 18	VOCs	Y	N	Y
ASTM D5504-01/-08	Sulphur compounds in natural gas and fuels	Y*	Y*	Y*
NIOSH 3704	Perchloroethylene	Y	N	Y
NIOSH 6603	Carbon dioxide	N	Y	N
OSHA ID-172/-210	Carbon dioxide/carbon monoxide	N	Y	N

** Verify with individual manufacturer recommendations/limitations.*

2.1.1.5 Storage

Unused bags may be stored almost indefinitely, so long as they are not exposed to extreme conditions (i.e. high heat, oxidative gases). However, bags are generally not considered to have long sample storage times. For example, bag hold times have been reported to range only from several hours to 72 h [4], whereas canister samples are considered valid up to 30 days [5]. Hsieh et al. [6] evaluated the 7 day stabilities of 56 VOCs at 5–30 ppbv with 30% and 90% relative humidity (RH) and at 24 and 35 °C. They observed that the average half-life of the 56 VOCs was 37 ± 4, 45 ± 6 and 52 ± 6 days for Tedlar® bags, SUMMA® canisters and SilcoCan® canisters, respectively. A significant contributor to bags relatively short-term sample storage durations is the fact that bags are permeable (i.e. sample diffuses through the bag) and/or off-gas VOCs (i.e. the VOCs are generated/emitted from the bags themselves). Most bags are made of a porous polymer, which may also result in sample loss due to adsorption, and/or result in off-gassing of target VVOCs and SVOCs. Methylene chloride, toluene, acetone, ethanol, 2-propanol, phenol and other VOCs have been identified in new Tedlar® bags [7]. In addition, bags are transparent and studies have demonstrated the loss of compounds in bags due to photooxidation from the sun [8]. End users need to be cognizant of the aforementioned when utilizing bags for ambient air sampling.

2.1.1.6 Advantages/Limitations

When evaluating the advantages/limitations of bags, it is important to consider that bags are a whole-air sampling approach intended for compounds in the VVOC to VOC range. Therefore, one could argue that bags should only be compared directly to other whole-air sampling approaches. However, the following discussion will not be limited to a comparison with other whole-air sampling approaches; rather the discussion will cover bag advantages/limitations on a broader scale. For example, a potential end user may be trying to decide between bags and thermal desorption (TD) tubes for the sampling and analysis of benzene. As stated earlier, bags are a whole-air sampling approach and therefore that is the first main advantage of bags when compared to in-field concentration approaches like TD tubes. All whole-air sampling approaches hold the following significant advantage over non-whole-air sampling approaches: the flexibility to sample an unknown atmosphere with the confidence that all/most VVOCs and VOCs were collected. In addition to the aforementioned advantage, bags are selected as a sampling approach for the following three main reasons: bags are lightweight, collapsible and relatively inexpensive. This means end users may purchase many bags for large sampling campaigns and ship large numbers for relatively little money.

Bags have several significant disadvantages, especially when compared to other whole-air sampling approaches. As mentioned earlier, bags are only intended for single use. Another significant disadvantage of bags is the need for a sampling pump to fill the bag, which means they are prone to power failures. In addition, ideally a lung box with pump is required to fill bags and avoid sample contamination due to carry-over. A vacuum bag sampler costs approximately $1500 and weighs roughly 8 kg, so this quickly diminishes the cost and weight advantages initially observed with bags. In addition, vacuum bag samplers possess a relatively large footprint, when compared to canister sampling. Bags are generally not as durable as alternative sampling approaches. Although significantly stronger than a balloon, they may be popped like a balloon. In addition, bags contain seams, which are prone to leaks. Bags generally do not have as long of sample hold times as other sampling methods and do not preconcentrate the analytes in the air sample. In addition, bags are often made of a polymer and off-gassing of target VVOCs and SVOCs is an issue. Lastly, most bags are transparent and studies have demonstrated the loss of compounds in bags due to photooxidation from the sun. Due to the aforementioned limitations, canisters are often preferred for ambient whole-air sampling. All of the aforementioned should be considered when selecting bags for an ambient air sampling application.

2.1.2 Whole-Air Sampling with Canisters

One of the most popular approaches for sampling ambient VVOCs and VOCs is to utilise a whole-air canister sampling method, especially in the United States (US) where canisters dominate the ambient air sampling field. However, in the European Union (EU) TD tubes dominate the ambient air sampling arena. For the purposes of this chapter the discussion of TD tubes will follow in the VOCs/SVOCs section, as TD tubes are generally utilised for sampling in this volatility range. That is not to say TD tubes are not applicable to VVOCs, as they certainly are. To elaborate, canisters are often used for the collection of several SVOCs; however, in general canisters are considered to occupy the VVOCs/VOCs niche. These are some generalisations utilised strictly to help structure the flow of this chapter.

Canisters (Figure 7) are often referred to as cans. As the name implies, canisters are made of stainless-steel (typically 304) metal spheres with a wall thickness typically around 1.07 mm. Canisters come in a variety of shapes and sizes ranging from 400 mL to 15 L and they are outfitted with stainless-steel (typically 316) valves to control the entrance and egress of sample. These valves are typically diaphragm or bellows type valves. Canisters are unique in the fact that they act as both their own power source for sampling (i.e. they are their own sampling pump) and sample storage vessel. Canisters achieve this by the fact that their walls are rigid and therefore the canisters are placed under vacuum and the vacuum within the canister is used to draw in a sample.

2.1.2.1 Methodology

The ‘gold standard’ for whole-ambient air sampling with stainless-steel canisters is the U.S. Environmental Protection Agency’s (EPA) Toxic Organic (TO) Compendium Method TO-15 [5]. For the most part the international

FIGURE 7 Canisters utilised for collecting, storing and transporting whole-air samples. *Photo courtesy of Restek Corporation.*

community mirrors Method TO-15. Therefore, the majority of the following discussion as it pertains to whole-ambient air sampling with canisters will directly relate to Method TO-15. The bulk of the following discussion is also directly applicable to gas bulb sampling and glass bottles; however, the discussion will focus on canisters.

A cleaned and evacuated (i.e. under vacuum typically 50 mtorr) canister is shipped to the field, where it is filled with a whole-ambient air sample. As stated previously, a canister does not need to be filled with a sampling pump. This is made possible by the rigid construction of the canister thereby enabling end users to place the canister under vacuum. The vacuum in the canister is then used to draw in the air sample. Canister samples are typically collected in one of the following four fashions: (1) a grab sample which is executed by opening the canister valve and allowing the canister to fill itself completely or near to completion (i.e. some vacuum remains) in approximately 30 s or less. (2) With the use of a critical orifice to regulate the flow going into the canister, so as to provide a longer sampling duration. This method is generally regarded as providing stable flows until the canister is almost 50% full. (3) With the use of a mechanical flow controller (Figure 8) which is capable of maintaining a near-constant sample flow rate over time, up until the canister is about 5/6 full (approximately 630 torr). (4) With the use of an electronic mass flow controller. Canisters may also be filled with a pump like discussed earlier with bags; however, this is rarely executed. Just like a bag sample, a canister needs to be filled at an appropriate flow rate and over the desired sampling duration, which will require the use of a flow meter (Figure 4) to verify/calibrate the

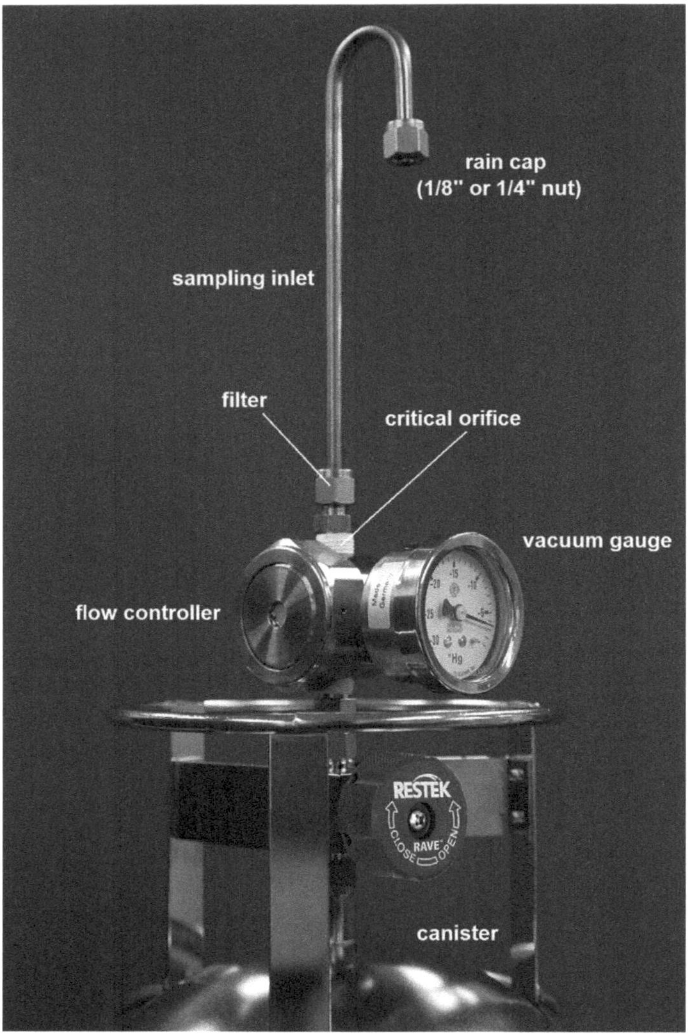

FIGURE 8 Mechanical flow controller utilised to regulate the flow of whole-air samples into canisters. *Photo courtesy of Restek Corporation.*

flow rate. An example canister filling scenario will be discussed later in section 2.1.2.2.3. Regardless of how a canister is filled, after sampling the filled canister is then shipped to a laboratory for analysis.

As a general rule of thumb a canister should always be sampled with a filter in place, so as to prevent particles from entering the canister valve and/or canister sphere. Particle damage to the valve, resulting in a leaking valve, is the most significant cause of canister failure. In addition, particles in the canister may adhere to the canister walls and act as a constant source of contamination in future samples and/or as an active sink for the degradation/loss of future samples.

2.1.2.2 Preparation

2.1.2.2.1 Canister Types There are three main canister types, which are differentiated by the surface treatment applied to the inside walls of the canister. All three treatments are executed to passivate the canister interior in order to reduce the chemical reactivity and protect against contamination. The following three main canister types are available: (1) SUMMA®; (2) electropolished (often referred to as SUMMA® like); (3) silicon-lined (SL). Some of the earliest canisters in widespread use were untreated aluminium and untreated stainless steel. Eventually the first treated/passivated canisters found widespread acceptance. SUMMA® treatment is a nickel chromium oxide layer, which has been applied via liquid deposition. Electropolished, often referred to as electrochemical polishing or electrolytic polishing, is accomplished by immersing the metal into a bath of electrolyte and through which DC power is passed. A reduction reaction occurs, which removes iron from the surface and enriches the chromium and nickel content at the surface. SL canisters are stainless-steel canisters, which have been electropolished; however, they go through an additional passivation step in which a silicon layer is applied to the canister walls via chemical vapour deposition. The implications of the aforementioned canister types will be covered in more detail in section 2.1.2.2.5.

2.1.2.2.2 Canister Cleaning Canisters may be cleaned and reused multiple times. If a canister is not exposed to harsh sampling conditions (i.e. acidic or caustic gases), which is the case for most ambient air sampling, a canister may theoretically be used *ad infinitum*. It is not uncommon to find a 20-year-old canister still in active service. The majority of canisters reach their lifetime due to physical abuse, which results in leaks. In order to reuse a canister, the canister should be subjected to a cleaning procedure. Many end users will purchase a fully assembled canister cleaning oven (which one may conceptualise as a dishwasher) or custom fabricate their own canister cleaning system. The main cleaning system components are a vacuum pump, oven, humidifier and clean gas (ultrahigh purity (UHP) nitrogen, helium or air). The basic cleaning protocol is to have the canister heated in the oven (typically 100 °C) and then draw the canister under vacuum. This step will remove the most recent sample from the canister. The canister is then typically held at this vacuum point for some period of time (typically 1 h). The combination of heat and vacuum is utilised to pull contaminants from the canister walls. Subsequently, the canister is then filled with a 'sweep' gas. This gas is typically UHP nitrogen, helium or air. It is important that the sweep gas is humidified. The water vapour is critical to cleaning as it helps hydrolyse compounds sticking to the walls of the canister. The canister is then evacuated and brought back under vacuum and the cycle repeats itself. Subsequent to cleaning the canister is evaluated for cleanliness. In order to do this, the canister is then filled with humidified UHP gas to 30 psig. The canister is then analysed. A canister is generally considered clean if no contaminate is present at a concentration greater than 0.2 ppbv. Following cleaning, canisters are generally considered to be viable and useable for up to 30 days of storage. Beyond 30 days the canister needs to be re-cleaned and re-certified as clean [5].

2.1.2.2.3 Mechanical Flow Controllers One of the most popular approaches for canister sampling is to use a mechanical flow controller (Figure 8 and Figure 9). Mechanical flow controllers are often referred to as passive samplers, passive flow controllers or flow controllers. A mechanical flow controller maintains a near-constant sample flow over an integrated time period, despite changes in canister vacuum. The critical orifice acts as a flow restrictor, upstream of a constant back pressure (i.e. vacuum in the canister). This constant back pressure is established by the balance between the mechanical spring rate of the diaphragm and the pressure differential across the diaphragm. The latter is established by the pressure difference between the atmospheric pressure (hence the atmospheric reference), the vacuum in the canister and the flow through the critical orifice. The net result is a constant flow up to approximately 630 torr and then the sampling rate drops off. The critical orifice determines the gross flow range. The adjustable piston is used to set a specific flow rate within the flow range. An adjustment to the position of the piston changes the back pressure, which changes the pressure differential across the critical orifice. If the piston is lowered away from the diaphragm, the flow rate will increase. If the piston is raised toward the diaphragm, the flow rate will decrease. As stated previously, it is important to use a filter in the passive sampler, so particles do not clog the critical orifice.

With a basic understanding of how a mechanical flow controller works, the following scenario will help demonstrate the proper critical orifice selection: an end user would like to fill a 6 L canister over an 8 h sampling duration.

The scenario breaks down as follows:

- An evacuated canister starting at 50 mtorr
- 50 mtorr sampled to 630 torr
- 630 torr (~83%) consumed
- 83% of 6 L = 5 L
- 5000 mL/(8 h * 60 min h^{-1}) = 10.3 mL min^{-1}

The desired flow rate is approximately 10.3 mL min^{-1}, which means based on Table 3 the end user would need to install a critical orifice of 0.0020″ in order to fall in the flow range of 8–15 mL min^{-1}. So the end user will install a critical orifice of 0.0020″ into an assembled flow controller; affix the flow controller to a clean/evacuated canister; open the canister valve to begin flow/sampling; utilise a flow meter to adjust the flow rate to 10.3 mL min^{-1}; leave the canister sampling; stop sampling 8 h later by closing the canister valve; cap the canister valve as insurance in case the valve leaks and ship the canister to

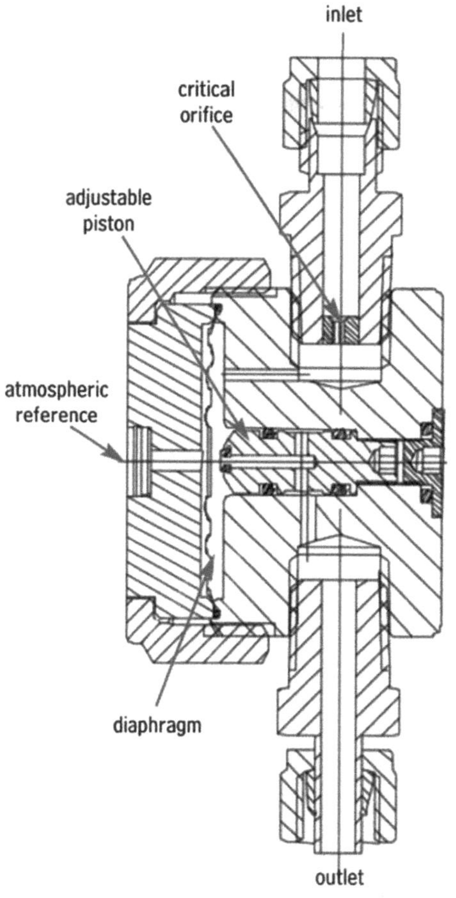

FIGURE 9 Crosscut view of mechanical flow controller used for whole-air sampling into canisters. *Photo courtesy of Veriflo Division of Parker Hannifin Corporation.*

TABLE 3 Mechanical Flow Controller Critical Orifice Sizes and Flow Ranges

Flow Range (mL min^{-1})	Critical Orifice Size (″)
0.5–2	0.0008
2–4	0.0012
4–8	0.0016
8–15	0.0020
15–30	0.0030
30–80	0.0060

the laboratory for analysis. The main considerations for canister sampling are the canister volume, the desired duration and needed flow rate.

It is important to note that because the mechanical flow controllers are within the sample flow path, they will become 'contaminated' just like a canister. Therefore, they will also need to be cleaned. In the event the flow controller does not clean with the aforementioned protocol it may be cleaned with the following more aggressive cleaning procedures: (1) completely disassemble the flow controller; (2) sonicate everything but the critical orifice and O-ring in methanol for 30 min; (3) bake everything but the O-rings at 100 °C for 1 h.

2.1.2.2.4 Applications There are numerous applications for whole-ambient air canister sampling. Stainless-steel canisters have been utilised as a whole-ambient air sampling method for over three decades. In addition to ambient air sampling, canisters have been used in a variety of applications including stack testing, soil vapour testing and automobile emissions testing to name a few. Inarguably, the most popular methods for utilizing canister sampling are the U.S. EPA's Compendium Methods TO-14A [9] and TO-15 [5]. Both of these TO methods are for the 'Determination of Volatile Organic Compounds (VOCs) in Air Collected in Specially-Prepared Canisters'; however, TO-14A is for 'Subsequent Analysis by Gas Chromatography' and TO-15 is 'Analyzed by Gas Chromatography/Mass Spectrometry (GC-MS)'. In addition, a few industrial hygiene/occupational health methods like the American Society for Testing and Materials (ASTM) D5466 and Occupational Safety and Health Administration (OSHA) PV2120 utilise canisters. In general, canisters are utilised for sampling low pptv to ppbv levels of VVOCs and VOCs from $C_{1/2}$ through $C_{10/12}$ (including permanent gases, polar VOCs and reactive compounds); however, this is often extended into the SVOC range with the collection and analysis of SVOCs such as naphthalene, 2-methylnaphthalene and hexachlorobutadiene to name a few.

2.1.2.2.5 Storage It is generally accepted that unused canisters are viable up to 30 days of storage. Congruently, it is widely accepted that most VOCs may be stored in a canister for a period of up to 30 days of storage. Electropolished and SUMMA® passivated, stainless-steel canisters have been utilised for over three decades to collect VOC samples. Numerous studies have evaluated the storage stability of several VOCs in air canisters under various conditions. Westberg et al. [10], Holdren et al. [11], Oliver et al. [12], Gholson et al. [13], Jayanty [14], Holdren et al. [15], Parmar [16], Merrill and Zapkin [17], Oliver [18], Kelly et al. [19], Holdren et al. [20], Brymer et al. [21] and Sin et al. [22] evaluated stabilities in SUMMA® canisters. In addition, Pate et al. [23] evaluated stainless-steel canisters with and without SUMMA® passivation; and Gholson et al. [24] evaluated aluminum canisters. Kelly and Holdren [25] provide a comprehensive review of the majority of these studies; however, the following general conclusions may be deduced from this body of peer-reviewed literature:

The majority of VOCs targeted in Methods TO-14A and TO-15 are stable in electropolished and SUMMA® stainless-steel canisters under routine ambient air conditions after storage of up to 30 days [5]. A minimum amount of water vapour is required in electropolished and SUMMA® stainless-steel canisters in order for particular VOCs (e.g. polar VOCs) to be stable. It is theorised that water occupies active sites on air canister walls, thereby increasing the inertness of the canister.

Over the past decade, SL canisters have grown in popularity. However, despite their increasing popularity, only a few studies have evaluated the effectiveness of SL air canisters. Holdren et al. [20] evaluated the 7-day stabilities of 9 VOCs at 20 ppbv with 21% and 52% RH and at 2 ppbv with 21% RH. They concluded that there were no significant differences between SUMMA® passivated and SL stainless-steel canisters. Hsieh et al. [6] evaluated the 7-day stabilities of 56 VOCs at 5–30 ppbv with 30% and 90% RH and at 24 and 35 °C. They observed that the average half-life of the 56 VOCs was 37 ± 4, 45 ± 6 and 52 ± 6 days for Tedlar® bags; SUMMA® canisters; and SilcoCan® canisters, respectively. Although the observed differences in loss rates between SUMMA® canisters and SilcoCan® canisters were not statistically significant, they concluded that SilcoCan® canisters had fewer active sites and thus adsorbed less; therefore, VOCs were relatively more stable in SL air canisters. Ochiai et al. [26] evaluated the 28-day stabilities of 58 VOCs at 1.7–2.5 ppbv with 1.6, 8.0, 27, 39, 53 and >99% RH. They concluded that SL canisters had increased inertness when compared to SUMMA® canisters, as evidenced by the recoveries of all analytes (>95%) under dry conditions (1.6% RH). Like Hsieh et al. [6], Ochiai et al. [26] concluded that this was the direct result of competitive adsorption of water vapour and VOCs on active sites in the air canister interior surfaces. However, it is important to note that performance differences amongst canister types at low RH levels has very little relevance, as RH conditions that dry (i.e. RH <10%) will seldom be encountered in the majority of ambient air sampling applications [27].

Therefore, the majority of VOCs are stable in electropolished, SUMMA® and SL stainless-steel canisters under routine ambient air sampling conditions after storage of up to 30 days [5]. The only significant consideration for canisters is when sampling sulphur-containing compounds. Several studies like Ochiai have reported the complete loss of thiols in electropolished and SUMMA® passivated canisters [26]; however, these compounds were reported stable in SL canisters under low RH conditions.

Regardless of canister type, the loss mechanisms for VOCs in canisters are as follows:

1. Physical adsorption process, which is from reaction with 'active sites' on canister walls [6] and is generally considered a short-term (i.e. takes place on the order of minutes to hours).

2. Dissolution of soluble compounds in condensed water, which can be exacerbated by the pressurisation of canisters [26].
3. Chemical reactions with other gas-phase compounds (e.g. ozone), which are generally considered short-term (minutes to hours) [5].

2.1.2.2.6 Advantages/Limitations When evaluating the advantages/limitations of canisters it is important to consider that canisters are a whole-air sampling approach intended for compounds in the VVOC to VOC range. Therefore, one could argue that canisters should only be compared directly to other whole-air sampling approaches for VVOCs and VOCs (e.g. bags). However, canisters generally compete with TD tubes in the ambient air sampling arena. Therefore, the following discussion will also include a comparison with TD tubes and other in-field concentration approaches.

Canisters offer several distinct advantages over other whole-ambient air sampling approaches (i.e. bags, gas bulbs and glass bottles). In fact, the only advantages held by the other whole-ambient air sampling methods over canisters are size, weight and cost. Canisters are significantly more durable than bags, gas bulbs and glass bottles; however, this does come with a side limitation that canisters require more space for shipping and have heavier shipping weights, which obviously has higher associated costs. In addition, the canisters and associated sampling equipment generally costs more. Aside from the aforementioned, canisters hold an edge over all other aspects including sample storage stability, compound application, sensitivity and durability.

The remainder of this discussion will focus on advantages and disadvantages of canisters relative to in-field concentration techniques (e.g. TD tubes). First and foremost, canisters are a whole-air sampling approach, which is a significant advantage over in-field concentration methods, as everything in the VVOC, VOC and lower-molecular weight SVOC range may be successfully collected by a canister. Therefore, the potential end user does not need to know their atmosphere like they need to for using TD tubes. Canisters are able to act as their own sampling pump, which is a distinct advantage over virtually every other time-integrated sampling approach, which requires the use of a sampling pump. Being reliant on a sampling pump is a significant disadvantage, because sampling pumps require calibration and are reliant on power, which sometimes may be lost/interrupted and/or drained in the case of batteries, all of which can result in lost samples. Canisters offer the advantage of turn-key operation, whereby field personnel may receive an evacuated canister and calibrated flow controller, attach the flow controller and start and stop sampling by opening and closing the valve. In fact, some laboratories are now shipping canisters with pre-calibrated flow controllers already attached, so field personnel only have to take the setup out of the box and open and close the valve at the beginning and end of the sampling session, respectively. Canisters offer the advantage of being able to utilise automated sampling timers, which are placed between the canister and the flow controller as shown in Figure 10. The sampling timer allows end users to place the canister into

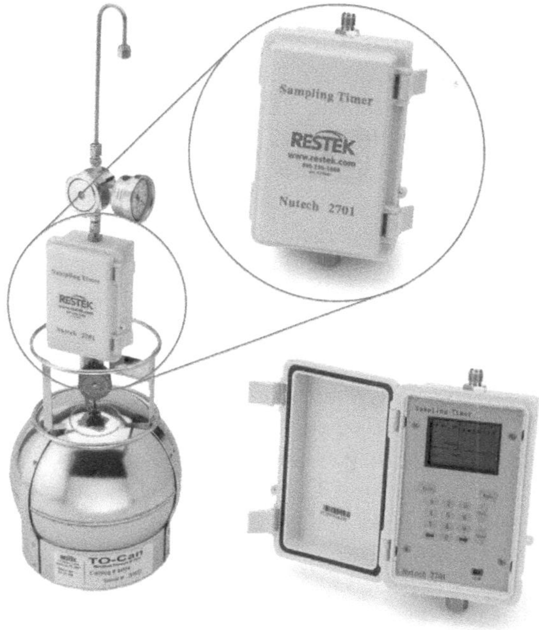

FIGURE 10 Sampling timer utilised for whole-air sampling with canisters. This allows potential end users to place canisters in the field days before the scheduled sampling event. *Photo courtesy of Restek Corporation.*

the field days to weeks before the actual sampling event as the timer opens and closes the valve at prescribed times, and then come back to pick the sampler up from the field. Canisters offer the advantage of virtually limitless uses for whole-ambient air sampling. As discussed in the section 2.1.2.2.2, canisters may be cleaned and reused multiple times, so theoretically their lifetime of use is indefinite. Most in-field concentration techniques have a limited lifetime (e.g. approximately 100 samples on a TD tube). Canisters afford the sampling and analysis of low-level concentrations of reactive compounds, which are often lost in bag and adsorbent samples.

However, canisters are not without their limitations. One of the most significant disadvantages of canister sampling can be cost. There can be an overall higher capital expenditure associated with canisters, as they require the items like a canister cleaning system. In addition, the shipping costs associated with canisters can be higher than the other time-integrated sampling approaches. However, this is largely dependent on the field sampling design. For example, sending out four canisters with four mechanical flow controllers for a one-time field investigation vs four bags with four lung samplers, actually costs significantly more for the bag field sampling approach (i.e. this is not accounting for any laboratory costs). However, if you take the scenario and change it to four canisters and four bags are collected at the same site weekly over a year period, the canisters will cost significantly more, as each week four canisters (approximately 20 kg) are being shipped vs four bags (<1 kg). Various sampling scenarios versus TD tubes and the associated sampling pumps could calculate out differently. So ultimately the potential end user needs to consider their short- and long-term sampling plan and the associated costs. All of the aforementioned should be considered when selecting canisters for an ambient air sampling application.

2.1.3 Sample Preconcentration

Following field sampling, shipping and storage; ambient air samples need to be analysed. Although liquid chromatography is used in several popular air analysis methods (e.g. U.S. EPA Method TO-11A (carbonyls collected on 2,4-dinitrophenylhydrazine (DNPH)-coated cartridges)) [28], the following discussion will focus on sample introduction into a GC, as the majority of ambient air sample analysis is conducted on GC. The analytical instrumentation including GCs, capillary columns, liners and detectors (e.g. flame ionisation detector (FID), MS) is covered in Chapters 9 and 10. The following section will cover how whole-air samples collected in bags, gas bulbs, glass bottles and canisters are transferred and introduced into a GC. The three main approaches for this include (1) direct injection via a gas-tight syringe into the GC inlet; (2) via a sample loop and rotary valve configuration; (3) via a sample preconcentrator.

The concentrations of target analytes encountered in ambient air sampling generally range in the single-digit pptv to double-digit ppbv levels. It is not possible to take a typical 1–10 μL aliquot of an ambient air sample with pptv to ppbv concentrations and inject this directly onto a GC-FID or GC-MS, as these concentrations are far too low for detection. It is also not possible to increase the volume of sample injected into the GC inlet, as that would result in significant band-broadening. The same holds true for sample loops. Direct injection via a gas-tight syringe and sample loop injection requires relatively high concentrations (i.e. ppmv to % level). So these two methods are generally not selected for ambient air sample analysis. In order to inject a sufficient volume of ambient air sample and reach reliably detectable levels, whilst maintaining reliable chromatography, a sample preconcentrator is often employed.

When using a sample preconcentrator the sample flow is as follows (see Figure 11): from canister to preconcentrator, then preconcentrator to GC. Within the preconcentrator there are several steps which will be addressed in the following discussion. A preconcentrator serves the following two primary purposes for sample introduction to the GC: (1) a preconcentrator will concentrate a volume (generally 200–500 mL) of sample, so as to provide a sufficient mass of target analytes to the analytical instrumentation. (2) When a sample is concentrated, everything in the sample is concentrated, so therefore unwanted species such as water vapor, nitrogen, oxygen, carbon dioxide, etc. are also concentrated in the sample. Some of these compounds will interfere

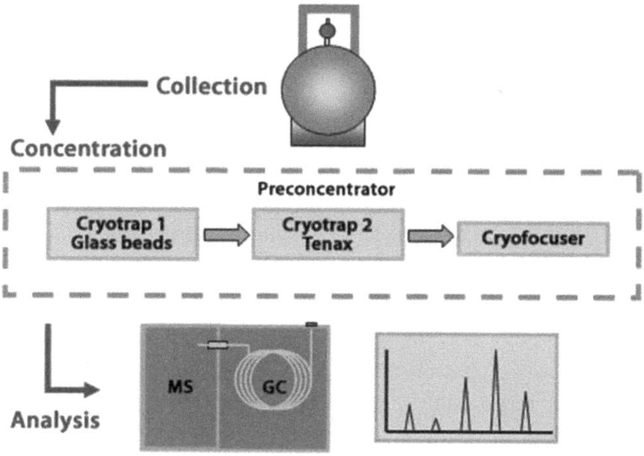

FIGURE 11 General sample flow for utilising a preconcentrator to introduce whole-air samples from bags, gas bulbs, glass bottles and canisters to a GC-MS.

with the analytical instrumentation (e.g. water vapour in an MS). Therefore, the second main function of the concentrator is to remove/reduce the concentrations of the aforementioned compounds.

Most sample preconcentrators accomplish the aforementioned tasks via one of the two following approaches: (1) the sample is concentrated with the use of a single trap (e.g. Markes CIA Advantage™). With these concentrators the volume of sample is carried through an adsorbent trap, thereby trapping both target analytes and contaminants. The sample is then ballistically heated to drive off the target analytes and contaminants, which are then carried via a transfer line directly to the head of the GC capillary column or the GC sample inlet. The removal of contaminants is accomplished by executing a dry purge after sample collection and/or a sample split. This method of sample concentration is analogous to purge and trap. In some preconcentrators the sample is passed through a Nafion® (a copolymer of tetrafluoroethylene and fluorosulphonyl monomer) dryer prior to reaching the adsorbent trap. The Nafion® permeable membrane dryer is used to remove water vapour from the sample stream. However, polar organic compounds permeate this membrane in a manner similar to water vapour, and rearrangements can occur in some hydrocarbons due to the acid nature of the dryer. Therefore, Nation® dryers have fallen out of favour, and multi-trap preconcentrators are the dominant preconcentration technique. (2) The sample is concentrated with the use of a multi-trap (e.g. Nutech 8900DS) (as shown in Figure 11). The sample is concentrated on a series of sorbent beds which concentrate/focus target analytes (often referred to as 'microscale purge and trap'). A very common configuration consists of three traps. In order of sample flow: trap 1 consists of glass beads used to remove water vapour, nitrogen, oxygen and carbon dioxide from the sample stream; trap 2 consists of Tenax® (poly(2,6-dephenylphenylene oxide)) material used to focus the VOCs and remove any residual water vapour and CO_2 and trap 3 is a proprietary trap used for final analyte focusing and transfer of analytes to the GC column. The single-trap preconcentrators are reliable and proven systems; however, the most popular preconcentrators utilise a multi-trap system.

The aforementioned multi-bed trap systems execute their two primary goals with the following four stages: (1) trap 1 with glass beads is cryogenically cooled to −150 to −180 °C. −150 to −160 °C is the most popular; however, the colder trapping temperatures are utilised for VVOCs in the $C_{1/2}$ range. In general, 200–500 mL of the sample is passed across this trap and most everything freezes out on trap 1, whilst N_2 and O_2 pass on. At this point in time a relatively large volume of water vapour has also been frozen out. For example, an ambient air sample containing 80% RH has approximately 16 μL of water per every 1 L of air. This means a 500 mL sample would result in around 8 μL of water frozen on trap 1. This volume of water would result in poor chromatography, reduce column life and interfere with detection (especially MS). (2) Trap 1 is gradually heated and helium is gently passed across trap 1 so as to allow the VVOCs and VOCs to volatilise, whilst leaving the water behind. At this point in time the analytes continue on to trap 2 containing Tenax® which is cryogenically cooled to −40 to 10 °C. At this stage the CO_2 is allowed to pass through the Tenax®. CO_2 will often coelute with early eluting VVOCs like propylene. (3) Trap 2 is heated and helium is passed across trap 2 so as to sweep the analytes on to trap 3 which is a proprietary trap cooled to the same temperature as trap 1 was for final analyte focusing. This stage will minimise any band-broadening. (4) Trap 3 is ballistically heated and all of the analytes are transferred to the GC directly on column (i.e. a transfer line from the preconcentrator is coupled directly to the GC column).

3. VOLATILE ORGANIC COMPOUNDS/SEMI-VOLATILE ORGANIC COMPOUNDS

3.1 In-field Concentration

One of the most widely used techniques for ambient air sampling of VOCs and SVOCs is to utilise a solid or liquid sorbent to trap target analytes out of an atmosphere. The target analytes are retained through adsorption and/or absorption, which are then desorbed or extracted in the laboratory for analysis. Such sampling approaches have been referred to as 'in-field concentration' sampling. This terminology is appropriate when one considers that with time the concentration of any target analyte held constant in the atmosphere would increase in concentration on the sorbent. It is also very appropriate when one considers how a whole-ambient air sample is eventually concentrated in the laboratory via a preconcentrator (discussed in section 2.1.3 of this chapter).

With the current approach, this concentration step already takes place 'in-field'. This sampling approach generally relies on the physical surface strength of the sorbent to effectively trap a target analyte and then efficiently release the analyte later for analysis. It is important to note that some in-field concentration techniques rely on absorption. Historically, in-field concentration methods have occupied the niche of VOCs/SVOCs. That is not to say they cannot be applicated to VVOCs, as they certainly are. There are a variety of in-field concentration approaches, and the following techniques will be covered in this section: impingers, solid sorbents and coated solid sorbents.

3.1.1 Air Sampling with Impingers/Bubblers

Impingers/bubblers (Figure 12) are often Pyrex® glass tubes designed for the collection of VOCs/SVOCs and sometimes PM into a liquid medium. Impingers and bubblers are almost identical, with the exception that bubblers utilise a sintered glass filter to reduce the size of bubbles and concurrently increase the number of bubbles. The smaller bubbles have an increased surface area to volume ratio; combined with the increased number of bubbles, a bubbler may increase the exchange rate between gas-phase compounds and the trapping medium, thereby increasing overall collection efficiency.

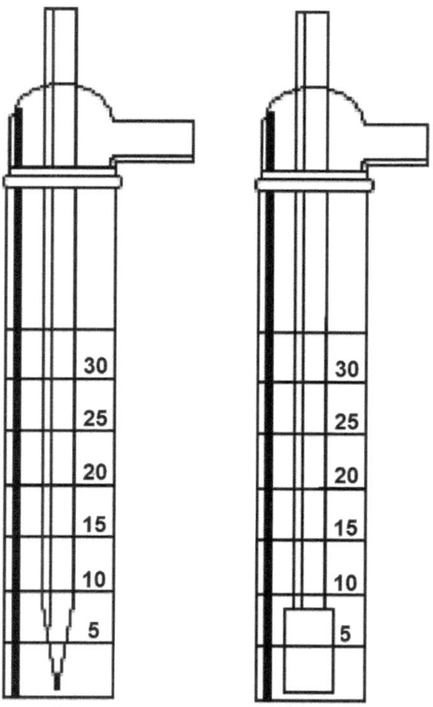

FIGURE 12 Impinger (left) and bubbler (right) used for trapping volatile organic compounds/semi-volatile organic compounds in a liquid media.

3.1.1.1 Methodology

A conditioned impinger/bubbler (referred to as impinger herein) is shipped to the field empty with the liquid medium (usually in separate sealed bottle). A field sampling technician then proceeds to fill the impinger with a prescribed volume of the liquid medium, and attaches the impinger to a sampling pump (Figure 3). A volume of air is passed through the glass tube, which results in the gas stream bubbling through the liquid. The liquid retains the compounds of interest and in some cases dissolves the compounds of interest. The liquid is then analysed in the laboratory to determine airborne concentrations.

As with other sampling approaches utilizing pumps, end users need to be cognizant of the sampling duration and desired flow rates, which will require the use of a flow meter (Figure 4) to verify/calibrate flows. In addition, to protect the pump from splashed or spilled impinger liquid, it is important to install a liquid trap between the impinger and the pump. Approximately 1 inch of solid sorbent may be added to act as a trap in front of the pump. It is important to note that some methods specify the use of a particulate filter pre-impinger, so as to prevent particles from being sampled and biasing results. The following three considerations are vital to setting the sampling flow rate and duration on an impinger: (1) a sufficient flow to adsorb enough material to meet detection limits. (2) Not too high of a flow to cause a breakthrough of compounds. (3) Not too high of a flow to cause excessive volatilisation of the impinger/bubbler solution.

3.1.1.2 Preparation

The following two main steps need to be executed in preparation for impinger/bubbler sampling: (1) the impinger needs to be cleaned; (2) the impinger solution needs to be prepared. The first of these two preparation steps generally calls for the impinger to be thoroughly rinsed with a non-target solvent (e.g. methanol) and then dried in an oven. As for the latter of the two preparation steps there are numerous impinger solutions. In many cases the solution is water. However, depending on the target analytes, a solvent such as methanol may be employed. Many impinger methods take advantage of acidic or basic buffer solutions. In other cases a water solution containing a derivatisation agent (e.g. 2,4-dinitrophenylhydrazine for carbonyls) is employed. In addition, some of the impinger methods require the impinger solution to be chilled in an ice bath during sampling, so as to increase the trap efficiency and reduce re-volatilisation of VOCs from the solution. It is outside the scope of this chapter to discuss every solution preparation method; however, it is central for the potential end user to know that said procedures are outlined in each method.

3.1.1.3 Applications

Numerous applications utilise impinger/bubbler samplers for the collection of VOCs/SVOCs. However, the majority of impinger applications are for the sampling of stationary sources (e.g. smoke stacks, kilns). Impingers are selected for high temperature applications/moisture levels, because the impinger solution helps buffer the effects of the high temperatures and moisture, which otherwise may cause significant damage to other sampling approaches (e.g. canisters and TD tubes). In addition, impinger samplers have been historically utilised for industrial hygiene/occupational health; however, study participants are often very mobile/active and impinger solution spills were very common. Therefore, impingers have largely been replaced by TD tubes and passive samplers (i.e. badges). With all of the aforementioned, impinger applications do exist for ambient air sampling. The following three considerations are vital to successful method selection and application: (1) it is imperative that the end user know what their target analytes are. (2) The proper impinger solution must be chosen to effectively trap/release those target analytes. (3) The concentrations to be encountered need to be anticipated so as to determine a proper flow rate that meets detection limits and avoids breakthrough (to be discussed more in section 3.1.2). Table 4 includes a list of some appropriate ambient air sampling applications, as well as some stationary source methods, only to help elucidate the flexibility/versatility of impinger methods.

3.1.1.4 Storage

Storage stability for impinger/bubbler solutions is of concern regarding the following two aspects: (1) before sampling, as some impinger solutions are to be used within a day or two of preparation [29]. In addition, some solutions

TABLE 4 Impinger/Bubbler Methods and Application Suitability

Method	Target Compounds	Solution	Type
U.S. EPA TO-5	Aldehydes and ketones	2,4-Dinitrophenylhydrazine	Ambient air
U.S. EPA TO-8	Phenols/cresols	Chilled NaOH	Ambient air
U.S. EPA TO-11	Aldehydes and ketones	2,4-Dinitrophenylhydrazine	Ambient air
U.S. EPA 26/26A	Hydrogen halide/halogens	Sulphuric acid/sodium hydroxide	Source sampling
U.S. EPA 29	Metals	Hydrogen peroxide and potassium permanganate	Source sampling
U.S. EPA 12	Inorganic lead	Nitric acid	Source sampling

must be kept on ice or refrigerated at all times [30]. (2) After sampling most impinger solutions should be analysed within a day or two; however, some (e.g. oxime solutions for carbonyls) may be stable over several weeks of storage. Each impinger method will usually state the appropriate blank and sample solution storage stabilities, which the end user should be aware of.

3.1.1.5 Advantages/Limitations

One of the most significant advantages impingers/bubblers offer over other competing methods (e.g. TD tubes) is the ability to sample relatively high-temperature gas streams and gas streams with relatively high-water vapour concentrations. Hence, their primary application to source sampling. Impingers offer the advantage of flexibility (i.e. the number of applications is only limited by the solution utilised) over other competing methods. Like canisters, impingers may be emptied, cleaned and reused repeatedly; so theoretically their lifetime of use is indefinite.

One of the most significant limitations of impingers/bubblers is the fact that most often they are made of borosilicate glass, which means they are prone to breaking. This, coupled with the fact that solution spills are very common, means impingers result in lost samples and exposure hazards. Therefore, impingers/bubblers have lost a significant market share in the field of air sampling, to alternative methods like TD tubes and badges. Although the impinger/bubbler lifetime may be indefinite, the solution may only have a shelf life of up to 30 days. Impingers and sorbents may be over sampled, especially if concentrations of target compounds are higher than expected. Therefore, a second or backup impinger may be used in series in the event of compound breakthrough from the first impinger. The phenomenon of breakthrough is not unique to impingers and is a limitation almost all in-field concentration techniques suffer from. In addition, the potential end user must be cognizant of the atmosphere they plan to sample and target analytes, so as to make the proper impinger solution selection and sample flow rate settings. All of the aforementioned should be considered when selecting impingers for an ambient air sampling application.

3.1.2 Air Sampling with Solid Sorbents

Solid sorbent materials are often referred to as solid sorbents or sorbent tubes. It is important to note that the term solid is used as this is a significant differentiator over impingers/bubblers, which take advantage of liquid sorbents. It is also imperative to highlight that the term sorbent is used as this covers both absorption and adsorption, either of which may take place depending on the sorbent and target analytes. However, the majority of solid sorbents are referring to adsorption in which target analytes are physically attracted to the surface of the sorbent. There are countless solid sorbent materials and tube configurations; however, in general a solid sorbent consists of a tube-shaped container, which is usually made of a relatively inert material like glass, stainless steel or polypropylene. The tube-type container is packed with the solid sorbent material and porous end caps/plugs like quartz wool, Teflon® (polytetrafluoroethylene (PTFE)) frits or stainless-steel mesh are placed on either end of the tube to keep the sorbent material from falling out.

A successful solid sorbent will accomplish the following three goals: (1) preferentially trap and retain the compounds of interest; (2) leave the compounds unaltered; (3) release the compounds of interest during thermal or solvent extraction. To best illustrate the solid sorbent sampling methodology, TD (Figure 13) tubes will be used as an example throughout this section. In particular, U.S. EPA Method TO-17 [31] will be used as a focal point for the discussion. As noted earlier, TD tubes dominate the ambient air sampling arena in the EU. In addition, TD tubes have been increasing in popularity in the US in recent years.

3.1.2.1 Methodology

A conditioned solid sorbent is shipped to the field. A field sampling technician then proceeds to uncover/uncap the solid sorbent and begin sampling. Sampling is usually accomplished by drawing an air sample through the container, which is filled with the sorbent material. For example, air is drawn through the ¼ mm OD × 3.5 mm long stainless-steel tube of a TD tube, which is packed with Tenax®. The compounds of interest in the air sample are sorbed onto the Tenax®. The air sample is usually drawn through a TD tube with the vacuum

FIGURE 13 Stainless-steel (left) and glass (right) solid sorbent tubes used for sampling volatile organic compounds/semi-volatile organic compounds in ambient air.

end of a sampling pump (Figure 3), which means the entire sampling train (i.e. with sample media in line) must be calibrated with a flow meter (Figure 4). Solid sorbents may also be used by utilizing a gas-tight syringe (Figure 5) to draw a volume of gas across the solid sorbent for a grab sample. In addition, there are numerous solid sorbent sampling methods, which utilise diffusion to accomplish the sampling (i.e. passive samplers). These methods are discussed in Chapter 2. Following sampling, the solid sorbent is returned to the laboratory for analysis where the analytes are either solvent extracted or thermally extracted (i.e. TD) from the tubes.

The 8 h sampling examples provided in the previous bag and canister sampling sections highlight that the main considerations for those approaches are choosing and setting an appropriate flow rate for the desired sampling duration, which results in a set sample volume. The same target volume approach holds true for solid sorbents; however, only as it directly relates to target analyte mass loading. A given mass of solid sorbent is only able to retain a given mass of a target analyte. Once a solid adsorbent has been loaded to maximum capacity, all of the available physical sites are occupied and/or inaccessible. Therefore, any further sampled target analyte will pass through the solid sorbent resulting in a phenomenon referred to as 'breakthrough'. Most solid sorbent sampling methods have breakthrough definitions, which are all variations on the aforementioned theme. Manufacturers provide breakthrough volumes (BVs) and methods like the U.S. EPA Method TO-17 [31] lists safe sampling volumes (SSVs) for some of the more popular sorbents and target analytes. The BVs are often expressed as litres per gram (i.e. litres of gas per gram of adsorbent), and SSVs are typically 2/3 the BV value. It is important to note that the majority of these BVs are determined at room temperatures (i.e. 20 °C) with little to no RH; therefore, BVs will go down with increasing temperature and/or RH. In the event that a BV or SSV is not available for the selected sorbent and target analyte, the following three general TD sampling guidelines exist: (1) some researchers have shown 50 mL min^{-1} is the theoretical optimum flow rate [31] for ¼ mm OD TD tubes. (2) Pump flow rates above 10 mL min^{-1} are generally required in order to minimise errors, as flow rates as low as 1–4 mL min^{-1} may be significantly biased by diffusion [32]. (3) Flow rates in excess of 200 mL min^{-1} are not recommended for ¼ mm TD tubes, unless for short-term (e.g. 10 min) monitoring [31].

As with bag and canister sampling, the main consideration for sorbent sampling is the SSV, the desired duration and the needed flow rate. So if a potential end user would like to sample benzene for an 8 h sampling duration with a Tenax® tube they would know that U.S. EPA Method TO-17 recommends an SSV of 6 L. This means a sampling rate of 12.5 mL min^{-1} is needed to stay below the SSV, which is also conveniently above the recommended minimum sampling flow rate of 10 mL min^{-1}. So now the end user will connect a Tenax® TD tube to the vacuum end of a pump; turn the pump on; connect a flow meter to the other end of the TD tube; read and adjust the flow to 12.5 mL min^{-1}; sample the TD tube for 8 h; come back and stop the pump; cap the ends of the TD tube and ship the TD tube to the laboratory for analysis.

Although TD tubes were referenced for some of the aforementioned solid sorbent considerations, sample breakthrough is applicable to all solid sorbent methods, whether it is for low- or high-volume sampling (to be discussed later). For example, it is well known that naphthalene will break through polyurethane foam (PUF) cartridges (Figure 14), so XAD-2® is a suggested substitute sorbent to minimise naphthalene breakthrough. In the event that an alternative suggested solid sorbent is not available; BVs are not available for the target analytes and solid sorbent; and general sampling guidelines, like the aforementioned for TD tubes are not available; end users may use a 'backup' tube/cartridge in series to investigate the presence of breakthrough. In general, if more than 5% of the target analyte is observed on the backup cartridge, breakthrough has occurred. There are more standardised laboratory testing procedures for determining BVs [33]; however, those are outside the scope of this chapter.

In addition to breakthrough, end users need to be aware of potential matrix effects. As stated earlier, the ideal solid sorbent will leave the compounds of interest intact; however, sometimes the sampled atmosphere itself will have oxidative/reactive species, which will result in a matrix effect on the solid sorbent. Therefore, some solid sorbent sampling methods will recommend surrogate compounds be added to each cartridge prior to field sampling. In general, surrogate compounds are inert compounds with similar chemical properties to the target analytes. In many cases the surrogate compounds are deuterated analogues of the target analytes. Each solid sorbent sampling method will make the appropriate surrogate compound recommendations and

FIGURE 14 Polyurethane foam plugs with glass holders utilised for sampling semi-volatile organic compounds. The smaller cartridges are utilised for low volume sampling and the larger cartridges are utilised for high volume sampling. *Photo courtesy of Restek Corporation.*

procedures. The principle behind surrogate standards is that the recovery of the surrogate compounds can help monitor for matrix effects and/or breakthrough.

3.1.2.2 Preparation

In preparation for field sampling, the solid sorbents need to be conditioned. This procedure is analogous to canister cleaning. In general, solid sorbents are cleaned either with solvents and/or thermally. For example, a new or used Tenax® TD tube will require conditioning for at least 2 h at 320 °C whilst passing at least 50 mL min^{-1} of pure helium carrier gas through it, whilst the PUF cartridges used as part of U.S. EPA Method TO-13A [34] need to be cleaned in a Soxhlet extractor with solvents and then vacuum dried. In general, a reasonable criterion for a clean solid sorbent is to reduce artifacts to 10% or less of individual analyte masses; however, individual methods/applications may specify certain criteria.

3.1.2.3 Applications

Solid sorbent sampling is most often applied to the sampling of VOCs and SVOCs, including PAHs, PCBs, pesticides, etc. There are numerous solid sorbents, each best at trapping particular chemicals or chemical classes. Solid sorbents may be applied to the sampling of ambient air, automobile testing, industrial hygiene/occupational health, etc. There is no one single sorbent capable of collecting all VOCs and SVOCs; however, the advent of multi-sorbent (multi-bed) traps has greatly increased the applicable range of compounds collected. In addition to the aforementioned classical solid sorbent paradigm, there are solid sorbent methods based on absorption, such as multi-channel tubular traps containing polydimethylsiloxane, which have been used for the sampling ambient PAHs [35,36] and airborne dichlorodiphenyltrichloroethane (DDT) [37]. These traps have been employed in the denuder configuration as discussed in Chapter 5. Due to the countless iterations of sorbents/configurations/applications it is not possible to capture all of the available solid sorbent options. The following three considerations are vital to successful method selection and application: (1) it is imperative that the end user know what their target analytes are. (2) The proper sorbent must be chosen to effectively trap/release those target analytes. (3) The concentrations to be encountered need to be anticipated so as to determine a proper flow rate that meets detection limits and avoids breakthrough. Table 5 provides a list of some of the more popular solid sorbent methods/applications.

3.1.2.4 Coated Solid Sorbents

As mentioned in section 3.1.1.5 of the impingers/bubblers discussion, impingers have lost favour to solid sorbent methods. For example, impinger-based methods such as U.S. EPA TO-5 [29] are available as U.S. EPA TO-11A 'Determination of Formaldehyde in Ambient Air Using Adsorbent Cartridge

TABLE 5 Solid Sorbent Methods and Application Suitability

Method	Compounds	Sorbent
U.S. EPA TO-1	Volatile organic compounds (VOCs)	Tenax®
U.S. EPA TO-2	VOCs	Carbon molecular sieve
U.S. EPA TO-9A	Polychlorinated dibenzo-*p*-dioxins and dibenzofurans and polybrominated dibenzo-*p*-dioxins and dibenzofurans	Polyurethane foam (PUF)
U.S. EPA TO-10A	Pesticides, polychlorinated biphenyls (PCBs)	PUF
U.S. EPA TO-13A	PCBs/polycyclic aromatic hydrocarbons (PAHs)	XAD-2®
U.S. EPA TO-17	VOCs	Carbotrap 300®

Followed by High Performance Liquid Chromatography (HPLC) (Active Sampling Methodology)' [28]. As discussed earlier, uncoated solid sorbents generally rely on adsorption being the mode of action for trapping the compounds of interest. In the case of coated solid sorbents there is also absorption taking place. However, the sorbent terminology covers both aspects. It is outside the scope of the current chapter to cover the entire subject of carbonyl sampling with DNPH-coated solid sorbents. However, the potential end user should understand the general methodology and preparations remain relatively unchanged from uncoated solid sorbents. The potential end user should refer to U.S. EPA Method TO-11A for more information on the method [28] and be aware of the considerations outlined by Herrington and Hays [38].

3.1.2.5 Storage

It is difficult to make any general claims regarding solid sorbent sample storage, as there are countless solid sorbent sampling applications. Each solid sorbent sampling method/application will specify recommended sample storage durations and conditions. For example, it is recommended that TD tubes be stored in a refrigerator at <4 °C and analysed within 30 days of sample collection for most VOCs [31]. PUF cartridges used for collecting PAHs should also be refrigerated at 4 °C; however, they need to be analysed within 24 h of collection. It is important to note that glass TD tubes and glass PUF cartridges are subject to analyte loss due to photooxidation. Another significant consideration is the storage of clean and unused solid sorbents, because of the ability of VOCs to diffuse into solid sorbents. Therefore, most solid sorbent sampling methods make recommendations on the storage of blanks/unsampled sorbents. For example, it is suggested that TD tubes be sealed with compression fittings outfitted with PTFE ferrules, and then stored within a glass storage vessel. In addition, adding some sacrificial bulk sorbent to the container is another strategy employed to minimise the contamination of blank samples. Regardless of the recommended blank/sample storage protocols, potential end users need to be aware of what the method stipulates and that unfit storage conditions may significantly bias and/or contaminate their results, especially when dealing with relatively low ambient air concentrations.

3.1.2.6 Advantages/Limitations

Solid sorbent sampling methods have several significant advantages when compared to other ambient air sampling methods. Arguably, the most significant advantage solid sorbents offer is their diverse and wide-ranging applications. Potential end users may select an appropriate solid sorbent capable of reliably sampling from $C_{2/3}$ to $C_{30/32}$, whereas whole-air sampling methods may only reliably sample from $C_{1/2}$ to $C_{10/12}$. For example, professionals looking to sample compounds like pentachlorophenol or lindane may reliably utilise a solid sorbent like PUF. Compounds like lindane may easily be sampled into bags and canisters; however, due to their relatively low vapour pressures the compounds will adhere to the bag/canister walls and cannot be extracted for analysis. When compared to the whole-air canister approach (TO-15 vs TO-17), one of the advantages solid sorbents offer is the fact that one may ship hundreds of TD tubes for the same cost it takes to ship several canisters.

Despite their advantages, solid sorbent sampling methods are not without their limitations. Most notably, solid sorbents generally have a prescribed number of thermal/solvent extraction cycles. For example, in general TD tubes may be thermally cycled approximately 50–100 times. Solid sorbents are often very difficult to clean once they are heavily contaminated and require special care and consideration for storage of samples and blanks; and some sorbents are subject to forming artefacts from the sorbent themselves. For example, researchers have detected benzaldehyde, phenol and acetophenone artifacts when sampling high-concentration (100–500 ppbv) ozone atmospheres with Tenax® [39]. In addition, solid sorbents are generally not considered suitable for inorganic gases like oxides of carbon, nitrogen, sulphur, ozone and other permanent gases. Like many other air sampling methods, solid sorbents are dependent on sampling pumps, which means they are prone to sample loss due to power/pump failures.

Single sample extraction is another significant limitation of solid sorbents. Most solid sorbent methods rely on a single thermal/solvent extraction of the sample. In the event something goes wrong during the extraction process the sample is lost. Whereas a whole-air sampling approach affords multiple analyses of the original unadulterated sample. It is important to note that in recent years manufacturers have designed TD units capable of splitting some of the sample extract onto a separate TD tube during the thermal extraction process, thereby affording a duplicate sample for a secondary analysis. However, this duplicate TD tube does not contain the original sample, as the sample was split. Therefore, any concentrations which were already close to the analytical detection limits may not be detectable on the secondary analysis. So while the advancement in TD tube splitting during analysis is an improvement, it remains a limitation, especially when compared to whole-air sampling approaches.

Lastly, the potential solid sorbent end user must be cognizant of the atmosphere they plan to sample, the target analytes and anticipated concentrations; so as to make the proper sorbent selection and sample flow rate settings, whilst avoiding sample breakthrough and matrix interferences. This requires the end user to exercise a certain degree of forethought and consideration, which is not necessary for the whole-air sampling approach. All of the aforementioned should be considered when selecting solid sorbents for an ambient air sampling application.

4. SEMI-VOLATILE ORGANIC COMPOUNDS/PARTICULATE MATTER

This section focuses on the sampling of PM. However, SVOCs are listed in the section heading for the following three reasons: (1) having just covered solid sorbents for sampling compounds in the VOC/SVOC volatility range, the SVOC/PM range is the natural progression across the volatility scale. (2) The PM sampling methodologies discussed herein are not utilised with the specific intent of collecting SVOCs; however, SVOCs are often inadvertently collected which requires consideration. The implications and remediation strategies of SVOC co-sampling will be discussed later. (3) Some ambient air sampling methods like U.S. EPA Method TO-13A call for the sampling and analysis of PAHs collected as PM on filters and as SVOCs on PUF cartridges.

PM is a subset of ambient aerosols, which are a suspension of solid and/or liquid particles in air. Examples of aerosols include dusts (e.g. sawdust), mists (e.g. sea spray, spray paint), bioaerosols (e.g. mould spores, pollen), smoke (e.g. diesel exhaust, wood fire) and the list goes on *ad infinitum*. Ambient aerosols are ubiquitous in indoor, outdoor and personal environments. Over the last several decades epidemiological, toxicological and depositional studies have linked adverse human health effects to exposure to airborne particles, which are capable of entering the thoracic region of the respiratory system. For example, exposure to silica and asbestos particles has been directly linked to the adverse health outcomes silicosis and asbestosis, respectively. The ubiquitous nature and such adverse human health effects associated with PM drives the significant research efforts on ambient PM. As indicated in the introduction to this chapter, >500 page books are written on the subject of PM alone [1] and potential end users should refer to these. Therefore, the following section only attempts to provide a cursory overview of ambient PM sampling.

4.1 Methodology

Particles are actively sampled with pumps (Figure 3) by drawing the air stream over a filter. There are passive particle sampling methodologies; however, they are outside the scope of this chapter. Filter media include nylon, polypropylene, polyvinylidene fluoride, mixed cellulose ester, PTFE and quartz fiber filter; with the latter two being the most popular. Filters are often in the shape of a circle and range in size from 25 to 102 mm in diameter. The particles are deposited on the filter, which is then taken back to a laboratory for analysis. PM analysis is most often a gravimetric analysis (i.e. the mass of the filter was weighed pre-sampling and post-sampling) to determine the total aerosol mass or total suspended particulate (TSP).

4.2 Preparation

The main preparation required for PM sampling is the conditioning of the filters. Most PM sampling methods will stipulate a filter conditioning protocol like the following example: filters are to be desiccated under a vacuum (−15 mm Hg) at 60 °C with a dry air bleed for 36 h. The filters are then to be desiccated again under vacuum at 40 °C using dry air for 15 h. Subsequent to the aforementioned desiccations, pre-sampling weighing requires the filters to be equilibrated for at least 24 h at 21 °C ± 1 °C and RH of 35% ± 5%. Finally, after conditioning and equilibration, the filters are weighed with an electronic balance capable of weighing to the nearest 0.001 mg. Again, each PM method will specifically outline the proper conditioning, equilibration and weighing protocol for pre- and post-sampling. In addition to gravimetric PM analysis there are PM methods which utilise solvent and/or thermal extraction for the analysis of metals, elemental carbon, organic carbon (OC) and individual organic compound species collected on filters [40]. The potential end user should be aware of any extra filter conditioning requirements for these methods.

4.3 Fractionators

Devices often called 'fractionators' are used to sample aerosols of a particular aerodynamic equivalent diameter (AED). The AED of a particle is the diameter of a unit density sphere that would have the identical settling velocity as the particle. It is a function of particle diameter, density, shape and surface characteristics. AED often determines the site of deposition in the lung and effects air sampling characteristics. Particles are often classified as inhalable, thoracic and high-risk respirable. Particles with an AED less than 100 μm are considered the inhalable fraction and may be breathed into the nose and/or mouth. Particles with an AED less than 25 μm are considered the thoracic fraction and may penetrate head airways (i.e. nose and mouth) and enter the lung airways. Particles with an AED less than 10 μm are considered the respirable fraction and may penetrate beyond the bronchioles and into the gas exchange region of the lung. Aerosols in natural environments are bimodal in size and the relative modal concentrations will be dependent on location, activity, weather, seasons, etc. Modes are lognormal in shape and a divided into 'coarse' mode and 'fine' mode. In general, the coarse fraction of thoracic particles is the mass fraction of particles between 10 and 2.5 μm AED. And the fine fraction of respirable particles is particles less than 2.5 μm AED. It is these distinctions which dictate the use of fractionators designed to sample for either the coarse (PM_{10}) or fine ($PM_{2.5}$) fraction [9].

For example, the Mesa Labs BGI Very Sharp Cut Cyclone (VSCC™) (Figure 15) will allow end users to collect the 'fraction' of aerosols within a given air sample, whose aerodynamic diameter is 2.5 μm and smaller. Particles whose AED is greater than 2.5 μm are not collected on the filter. In general, these devices rely on inertia to conduct the fractionation. For example, the Mesa Labs BGC VSCC™ is a cyclonic fractionator, which means the sampled

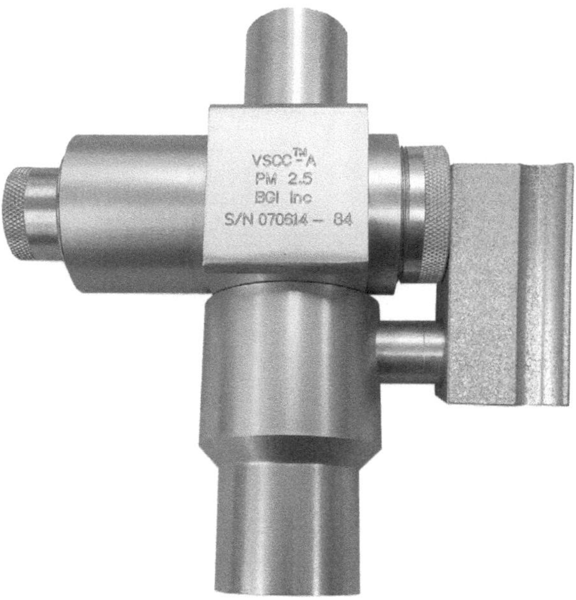

FIGURE 15 Mesa Labs BGI $PM_{2.5}$ Very Sharp Cut Cyclone (VSCC™) used for fractionating particulate matter (PM). *Photo courtesy of Mesa Labs.*

air stream is passed through a cyclone (envision a tornado) and particles with an AED greater than 2.5 μm have enough inertia at a set flow rate ($16.67\ L\ min^{-1}$ for the Mesa Labs BGI VSCC™) that they hit the walls of the cyclone and drop out of the air stream. Whereas the particles with an AED of 2.5 μm and smaller do not hit the walls, so they are able to travel onto the filter for collection (Figure 16). It is important to note that fractionators are not 100% efficient at the specified 'cut point', but are designed with 50% collection efficiency at the particular AED. Sometimes this is referred to as the mean AED. This means larger particles are not excluded altogether, but are collected with substantially decreasing efficiency and smaller particles are collected with increasing efficiency [41]. To further elaborate, it means if one were to collect a sample with said fractionator and characterise the AED of every particle on that filter, the average AED would be the 'cut point'. There

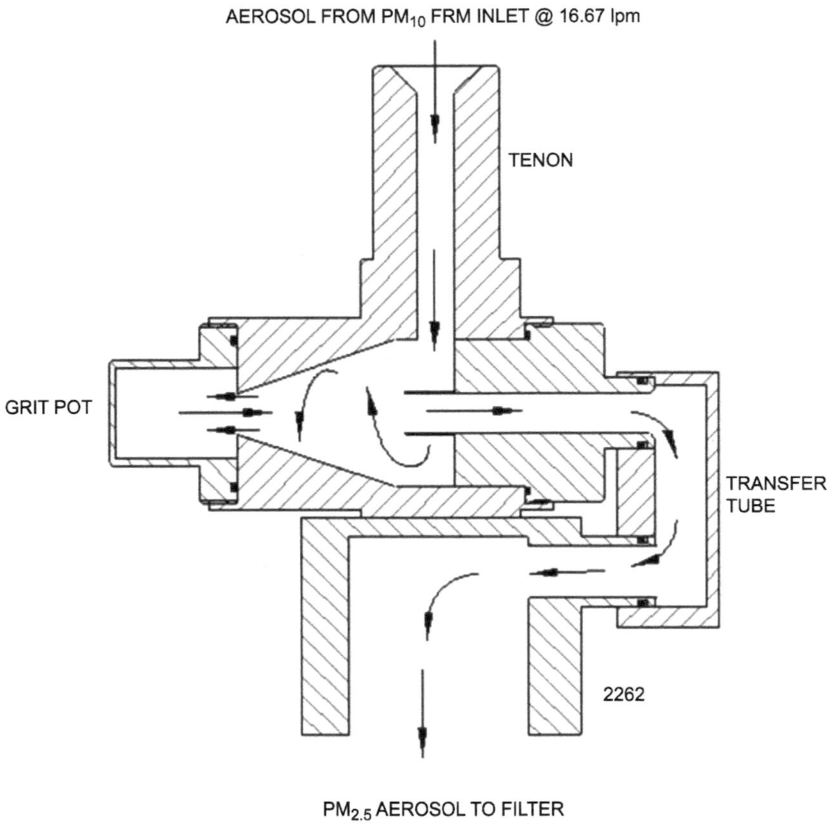

FIGURE 16 Mesa Labs BGI $PM_{2.5}$ Very Sharp Cut Cyclone (VSCC™) sample flow path. Particulate matter (PM) > 2.5 μm is removed from sample stream due to inertial impaction created by the cyclone and are deposited in a 'grit pot', whilst $PM_{2.5}$ proceeds on to the filter for collection. *Photo courtesy of Mesa Labs.*

are a variety of fractionators available; however, the majority of the fractionators rely on inertia, whilst impactors are another example. A variety of PM samplers exist in PM_1, $PM_{2.5}$, PM_4 and PM_{10} ranges.

4.4 Applications

PM in ambient air with an AED of less than 10 μm (PM_{10}) and 2.5 μm ($PM_{2.5}$), and 25−45 μm (TSP) have been regulated by the National Ambient Air Quality Standards [41]. The concentrations of PM in ambient air are measured using the EPA-approved Federal Reference Method (FRM) samplers. These methods are based on 24 h sampling. Size selective PM concentration measurements are made using a two-stage process. The first stage consists of a pre-separator designed to remove the larger particles. The second stage consists of a filter used to accumulate the PM mass after fractionation. For example, for a PM_{10} sampler, the first stage theoretically removes particles larger than 10 μm AED. The net mass of PM collected on the filter from the second stage divided by the volume of air sampled provides a measure of the concentration in units of mass per unit volume (μg m^{-3}). Table 6 lists the U.S. EPA FRM methods for PM sampling as well as several source and industrial hygiene/occupational health PM sampling methods.

4.5 SVOCs/Artifacts

As mentioned previously, there are SVOC co-sampling considerations for PM sampling. There are the two following possible SVOC biases, which may occur when sampling PM with quartz fibre filters: (1) the positive/adsorption; (2) the negative/evaporation. The first of which may be a positive bias resulting from the adsorption of gas-phase OC on the quartz fibre filter. If these samples are analysed for PM OC and/or TC (EC + OC), the aforementioned adsorption artefact will result in an artificially high estimation of the true OC. It is

TABLE 6 Particulate Matter (PM) Sampling Methods and Application Suitability

Agency	Method	Application
ASTM	E2515-11	Solid fuel burning appliances
U.S. EPA	5	Stationary sources
OSHA	PV2121	Gravimetric determination
U.S. EPA 40 CFR	Appendix B to Part 50	Total suspended particulate
U.S. EPA 40 CFR	Appendix J to Part 50	PM_{10}
U.S. EPA 40 CFR	Appendix L to Part 50	$PM_{2.5}$

important to note that EC remains unaffected. The second bias is a negative bias in the measurement of PM OC and/or TC resulting from the evaporation of SVOCs from the particles. Several studies have indicated the adsorption/positive artefact to be the dominant artefact.

The positive/adsorption artefact is often accounted for by using tandem (i.e. in series) quartz fibre filters or by sampling with a Teflon® filter followed by a quartz fibre filter. For the tandem quartz fiber filter (QFF) approach, PM is collected only by the first filter, whilst both the first and second filters are exposed to and adsorb gas-phase OC. The OC on the second filter may then be subtracted from the OC on the first filter to account for the adsorbed gas OC. However, in order for this approach to be correct, both filters must be in equilibrium with the sampled airstream, adsorb the same amount of gas-phase OC and not have a significant amount of OC loss through evaporation. For the tandem Teflon® and QFF correction, the Teflon® filter collects PM, but OC gas adsorption is negligible, so only the QFF adsorbs gas-phase OC. For this correction to be effective the Teflon® filter cannot adsorb any gas-phase OC and the Teflon® filter cannot lose particulate OC. Several studies have shown no difference when using either type of correction. Some researchers contend that the use of two quartz filters is preferable to Teflon® and quartz because both the collection and blank filters are in the same sampler [42]. Regardless, potential end users should be aware of the aforementioned artefacts; the possible remediation strategies and make an appropriate informed decision for their application. Some methods/applications are not concerned with SVOC artefacts. For example, sampling for PAHs with U.S. EPA Method TO-13A utilises a filter to collect PM in front of a PUF cartridge to collect SVOCs. In this case it does not matter what artefact is present or not, as both the PM filter and PUF cartridge are extracted together and analysed as total PAHs. Further discussion on gas/particle partitioning artefacts can be found in Chapter 5.

4.6 Storage

The majority of ambient PM sampling applications will have pre- and post-sampling storage requirements. In general, the same pre-sampling weighing requirements are applied to post-sampling requirements. For example, the filters are to be equilibrated for at least 24 h at 21 °C ± 1 °C and RH of 35% ± 5%.

5. HIGH VOLUME SAMPLING

Some of the methods referenced in the previous VOCs/SVOCs and SVOCs/PM sections of this chapter are termed ‘high volume’ methods. High volume methods exist for SVOCs and PM, as larger sampling volumes are needed in order to meet the detection limits for the concentrations encountered during ambient air sampling. In general, high volume methods do not exist for

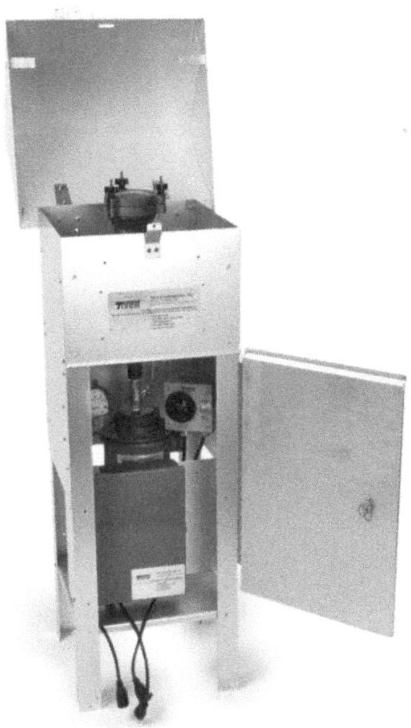

FIGURE 17 Tisch Environmental, Inc. TE-1000PUF high volume sampler used for collecting ambient levels of particulate matter and semi-volatile organic compounds. *Photo courtesy of Tisch Environmental, Inc.*

VVOCs and VOCs. The major difference between high-volume sampling methods and low-volume sampling methods is the amount of air sampled (and therefore the sampling pump specifications required) and the size of the sampling media. High volume samplers usually collect 1500 m^3 of air over a 24-h period and low volume samplers collect 24 m^3 or less. High volume samplers operate at a flow rate of 1.1 $m^3\ min^{-1}$. In general, low volume sample flow rates are 1 $m^3\ h^{-1}$ or less. In addition, the filters and sorbents are generally larger for high volume applications. Figure 14 provides an example of PUF plugs with glass holders, which may be utilised for sampling SVOCs. As shown in the figure, the smaller cartridges are utilised for low volume sampling and the larger cartridges are utilised for high volume sampling. A variety of high volume samplers are available from several different manufacturers; they usually consist of an aluminium housing approximately 1.2 m(W) × 0.6 m(L) × 0.6 m(H), a size selective inlet for PM sampling, a mechanical timer, a brush or brushless electrical motor, a mass flow controller, a stainless-steel filter holder, elapsed time indicator and pressure recorder (Figure 17). It is important to note that in recent years the U.S. EPA has created low volume methods as alternatives to high volume methods. For example, Compendium Method IO-2.3 is for the sampling of PM_{10} at

TABLE 7 High Volume Sampling Methods and Application Suitability for SVOCs and PM

Method	Compounds	Sorbent
U.S. EPA TO-4A	Pesticides, polychlorinated biphenyls	Polyurethane foam (PUF)
U.S. EPA TO-9A	Polychlorinated dibenzo-*p*-dioxins and dibenzofurans and polybrominated dibenzo-*p*-dioxins and dibenzofurans	PUF
U.S. EPA TO-13A	Polycyclic aromatic hydrocarbons	XAD
U.S. EPA CFR, Appendix B to Part 50	Particulate matter (PM)	Total suspended particulate (TSP)
Appendix J to Part 50	PM	PM_{10}
Appendix L to Part 50	PM	$PM_{2.5}$

16.7 L min^{-1} as a substitute to U.S. EPA CFR Appendix J to Part 50, which relates to sampling PM_{10} with high volume samplers at 1100–1700 L min^{-1}. In general, the same sample methodology, preparation, sample storage and considerations are applied to low- and high volume sampling. Table 7 provides some examples of the more popular high-volume sampling methods.

6. CONCLUSIONS

There are numerous ambient air sampling methods, which often makes it difficult for the potential end user to determine what technique is the best overall approach for their given scenario. This chapter was written to provide the potential end user with a practical overview of whole-air, in-field concentration and PM methods; by sampling with bags, canisters, impingers/bubblers, solid sorbents (coated and uncoated) and PM fractionators. After reading this chapter, the potential end user should be able to now appropriately choose which method best suits their application; understand how sampling is conducted with the method; know what sample preparation is associated with it; recognise how samples are stored and be able to consider what the advantages/limitations of the method are.

This chapter was written in order of increasing volatility; starting with VVOCs and ending with PM. This is important to recognise, because in an ideal world the potential end user should select the sampling approach based on the target analyte(s), as opposed to the sampling approach dictating the target analyte(s). In addition to all of the aforementioned considerations, the potential end user should be cognizant of the sampling volume, the sampling duration and sampling flow rate. These three variables are interrelated and can have significant implications with regards to analytical detection limits and/or analyte capacity of the sampling method, especially for ambient air sampling, which generally has relatively low concentrations of target analytes.

Lastly, it is important for the potential end user to remember that this chapter was only written as a perfunctory overview of the covered topics. Therefore, when it comes to method selection and/or application, the potential end user is strongly encouraged to refer to the specific methods referenced which contain a wealth of helpful information beyond the pages of this chapter.

REFERENCES

[1] J.H. Vincent, in: J.W. Sons (Ed.), Aerosol Sampling: Science, Standards, Instrumentation and Applications, John Wiley & Sons, England, 2007.
[2] World Health Organization, in: Regional Office for Europe (Ed.), Indoor Air Quality: Organic Pollutants, World Health Organization, Copenhagen, 1987.
[3] S. Ghosh, K.H. Kim, J.R. Sohn, Sci. World J. 11 (2011) 2160–2177.
[4] Y. Wang, et al., Environ. Sci. Technol. 30 (10) (1996) 3115–3117.
[5] US Environmental Protection Agency, Compendium of methods for the determination of toxic organic compounds in ambient air. Compendium method TO-15. Determination of volatile organic compounds (VOCs), in: O.O.R.A. Development (Ed.), Air Collected in Specially-Prepared Canisters and Analyzed by Gas Chromatography/Mass Spectrometry (GC/MS), U.S. EPA, Cincinnati, OH, 1999.
[6] C.C. Hsieh, S.H. Horng, P.N. Liao, Aerosol Air Qual. Res. 3 (2003) 17.
[7] S.L. Trabue, J.C. Anhalt, J.A. Zahn, J. Environ. Qual. 35 (5) (2006) 1668–1677.
[8] W.A. Lonneman, R.L. Seila, J.J. Bufalini, Environ. Sci. Technol. 12 (1978) 459–463.
[9] US Environmental Protection Agency, Compendium of methods for the determination of toxic organic compounds in ambient air. Compendium method TO-14A. Determination of volatile organic compounds (VOCs), in: O.O.R.A. Development (Ed.), Air Collected in Specially-prepared Canisters with Subsequent Analysis by Gas Chromatography, U.S. EPA, Cincinnati, OH, 1999.
[10] H.H. Westberg, M.W. Holdren, H.H. Hill Jr., in: C.R. Council (Ed.), Analytical Methodology for the Identification and Quantification of Vapor Phase Organic Pollutants, Washington State University, 1981.
[11] M. Holdren, et al., Evaluation of cryogenic trapping as a means for collecting organic compounds in ambient air, in: Final Report to U.S. Environmental Protection Agency, 1984.
[12] K.D. Oliver, J.D. Pleil, W.A. McClenny, Atmos. Environ. 20 (7) (1986) 1403–1411.
[13] A.R. Gholson, et al., Evaluation of canisters for measuring emissions of volatile organic air pollutants from hazardous waste incineration, JAPCA 39 (9) (1989) 1210–1217.
[14] R.K.M. Jayanty, Atmos. Environ. 23 (4) (1989) 777–782.
[15] M.W. Holdren, et al., Stability of volatile organic compounds in SUMMA polished canisters, in: F.R.R.U.S.E.P. Agency (Ed.), Contract No. 68-D0-0007, 1991. Battelle, Columbus, OH.

[16] S.S. Parmar, A study of holding times for H2S, COS, and CH3SH samples (gas phase) in different containers, in: I.M.O.T.A.R.A. Pollutants (Ed.), Proc. 1991 EPA/AWMA Int. Symp., Air and Waste Management Association, Pittsburgh, Pennsylvania, 1991.
[17] R.G. Merrill Jr., M.A. Zapkin, Application of canister sampling and analysis methodology to voltile polar organic compounds, in: Masurement of Toxic and Related Air Pollutants, 1991, pp. 602–606. Publication number VIP-21.
[18] K.D. Oliver, Sample integrity of trace level polar VOCs in ambient air stored in summa-polished canisters, in: Technical Note TN-4420-93-03, I. ManTech Environmental Technology, 1993. Submitted to U.S. EPA Research Triangle Park, NC.
[19] T.J. Kelly, et al., American Chemical Society (1993) 1146–1153.
[20] M.W. Holdren, et al., Effectiveness of Silcosteel Treatment of Canisters, Unpublished Final Report, 1994.
[21] D.A. Brymer, et al., Environ. Sci. Technol. 30 (1) (1995) 188–195.
[22] D.W. Sin, et al., Analyst 126 (3) (2001) 310–321.
[23] B. Pate, et al., J. Air Waste Manage. Assoc. 42 (4) (1992) 460–462.
[24] A.R. Gholson, R.K.M. Jayanty, J.F. Storm, Anal. Chem. 62 (17) (1990) 1899–1902.
[25] T.J. Kelly, M.W. Holdren, Atmos. Environ. 29 (19) (1995) 2595–2608.
[26] N. Ochiai, et al., J. Environ. Monit. 4 (6) (2002) 879–889.
[27] J.S. Herrington, Storage Stability of 66 Volatile Organic Compounds (VOCs) in Silicon-Lined Air Canisters for 30 Days, 2014. EVAN2066-UNV.
[28] US Environmental Protection Agency, Compendium of methods for the determination of toxic organic compounds in ambient air, in: O.O.R.A. Development (Ed.), Compendium Method TO-11A. Determination of Formaldehyde in Ambient Air Using Adsorbent Cartridge Followed by High Performance Liquid Chromatography (HPLC) (Active Sampling Methodology), second ed., U.S. EPA, Cincinnati, OH, 1999.
[29] US Environmental Protection Agency, Method for the Determination of Aldehydes and Ketones in Ambient Air Using High Performance Liquid Chromatography (HPLC), U.S. EPA, 1984.
[30] US Environmental Protection Agency, Method for the Determination of Phenol and Methylphenols (Cresols) in Ambient Air Using High Performance Liquid Chromatography, U.S. EPA, 1986.
[31] US Environmental Protection Agency, Compendium of methods for the determination of toxic organic compounds in ambient air, in: O.O.R.A. Development (Ed.), Compendium Method TO-17. Determination of Volatile Organic Compounds in Ambient Air Using Active Sampling Onto Sorbent Tubes, second ed., U.S. EPA, Cincinnati, OH, 1999.
[32] S. Batterman, T. Metts, P. Kalliokoski, J. Environ. Monit. 4 (6) (2002) 870–878.
[33] E. Woolfenden, J. Air Waste Manage. Assoc. 47 (1) (1997) 20–36.
[34] US Environmental Protection Agency, Compendium of methods for the determination of toxic organic compounds in ambient air, in: O.O.R.A. Development (Ed.), Compendium Method TO-13A. Determination of Polycyclic Aromatic Hydrocarbons (PAHs) in Ambient Air Using Gas Chromatography/Mass Spectrometry (GC/MS), second ed., U.S. EPA, Cincinnati, OH, 1999.
[35] P.B. Forbes, et al., Anal. Chim. Acta 730 (2012) 71–79.
[36] P.B. Forbes, E.R. Rohwer, Environ. Pollut. 157 (8–9) (2009) 2529–2535.
[37] Y. Naudé, E.R. Rohwer, Anal. Chim. Acta 730 (2012) 112–119.
[38] J.S. Herrington, M.D. Hays, Atmos. Environ. 55 (2012) 179–184.

[39] E.D. Pellizzari, B. Demian, K.J. Krost, Anal. Chem. 56 (1984) 793–798.
[40] J.S. Herrington, et al., Atmos. Environ. 54 (2012) 53–59.
[41] EPA, U.S., in: OAQPS (Ed.), Review of the National Ambient Air Quality Standards for Particulate Matter: Policy Assessment of Scientific and Technical Information, 2005. RTP, NC.
[42] J. Noll, M.E. Birch, Environ. Sci. Technol. 42 (14) (2008) 5223–5228.

Chapter 5

Denuders

Patricia B.C. Forbes* **and Egmont R. Rohwer**
Department of Chemistry, Faculty of Natural and Agricultural Sciences, University of Pretoria, Pretoria, South Africa
**Corresponding author: E-mail: patricia.forbes@up.ac.za*

Chapter Outline

1. Introduction **156**
- 1.1 Definitions 156
- 1.2 The Need for Denuder-Based Sampling 156
- 1.3 Reasons for Determining Gas and Particle-Associated Analyte Concentrations Separately 159
- 1.4 Disadvantages of Conventional Filter/Adsorbent-Based Sampling Systems 159

2. Types of Denuders **162**

3. Theory of Denudation **165**
- 3.1 Air Flow Considerations 165
- 3.2 Particle Transmission Efficiency 166
- 3.3 Gas Phase Collection Efficiency 168
  - 3.3.1 Cylindrical Denuders 168
  - 3.3.2 Annular Denuders 170

4. Denuder Sampling Applications **171**
- 4.1 Denuders Used for the Sampling of Inorganic Analytes 171
- 4.2 Denuders Used for the Sampling of Organic Analytes 173
  - 4.2.1 Adsorbant-Based Denuders 173
  - 4.2.2 Sorbent-Based Denuders 174

5. Denuder Sampling Considerations **175**

6. Analyte Extraction Methods and Analytical Techniques Employed **179**
- 6.1 Extraction of Analytes, Pre-concentration and Use of Internal Standards 179
- 6.2 Analytical Techniques Generally Employed 179

7. Challenges and Future Outlook with Respect to Denuder Sampling of Air Pollutants **179**

References **180**

Comprehensive Analytical Chemistry, Vol. 70. http://dx.doi.org/10.1016/bs.coac.2015.09.005

1. INTRODUCTION

1.1 Definitions

The word 'denuder' comes from the verb 'denude' which means to strip or lay bare [1], as a denuder serves to remove an analyte from an air sample mixture. Specifically, in denuder samplers the gas phase analyte is stripped from an air sample in the denuder, whilst particle-associated analyte remains in the airstream and may be collected on a downstream device, such as a filter. This process is known as 'denudation' and it is the result of two mechanisms governing the movement of gas molecules and particles through the denuder (Figure 1); firstly movement in the direction of the gas stream as a result of forced gas flow through the system (active pumping) and secondly movement perpendicular to the longitudinal gas flow through the denuder due to radial diffusion [2].

1.2 The Need for Denuder-Based Sampling

Denuders are useful in a number of air monitoring applications, which can be broadly classified as either (1) removal of a non-target analyte from the air sample during sampling or (2) simultaneous sampling of the gas and particle phases of an airborne analyte.

In the former case, the removal of a non-target analyte from the air sample during sampling is necessary if it may react with the analyte of interest or other compounds present in the air sample, thereby potentially causing analyte loss or generation and sampling artefact formation. Examples are the removal of HNO_3 in nitro-polycyclic aromatic hydrocarbon monitoring [3] and the removal of oxidising compounds such as ozone in polycyclic aromatic hydrocarbon (PAH) determinations [4,5]. A denuder system has also been used to remove both organic gaseous compounds and oxidants (NO_2 and O_3) during carbonaceous aerosol sampling, which would otherwise lead to sampling artefacts and interferences [6]. The relative importance of these sampling artefacts depends on a number of parameters, primarily the ambient temperature and atmospheric concentrations of the compounds concerned (analyte and oxidants, for example) [5]. Alternatively, the nontarget analyte concerned may interfere with the analyte of interest during subsequent analysis, and removal at the sampling stage is an alternative to separation during sample preparation. An example is in the measurement of particle-bound organic carbon (OC), where gas phase OC is removed by denuder-based sampling methods [7].

Denuders were initially used to denude air particulate samples of gaseous components [8] until in 1979 when Ferm used an oxalic-acid-coated denuder as a gas sampling device to sample atmospheric ammonia, whilst interfering particle-borne ammonium ions passed through [9]. The use of denuders for air sampling was then primarily for the separation of gas phase inorganic air pollutants, such as SO_2, HNO_3 and NH_3 from aerosols, where filter-based sampling

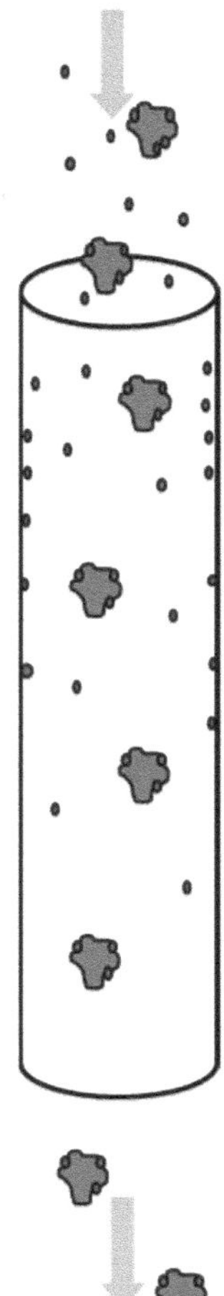

FIGURE 1 Schematic diagram of the denudation process for sampling air pollutants, where gas phase analytes (depicted by small circles) diffuse to the walls of the denuder and are collected, whilst particle-associated analytes pass through the denuder and may be collected separately downstream on a filter.

had previously been employed. The problems which were thus overcome were possible retention of the gases on the filter and reaction of the gases with the collected particulate matter [10].

More recently, denuders have been found to be useful for the simultaneous sampling of gas and particle phase semi-volatile organic compounds (SVOCs), as this class of air pollutants can exist in both phases in the atmosphere as a result of their vapour pressure range of 10^{-8}–10^{-2} Torr or $10^{-5.9}$–$10^{0.1}$ Pa [11]. In terms of air pollutant monitoring, SVOCs typically of interest include PAHs, polychlorinated biphenyls (PCBs) and organochlorine pesticides, although the partitioning of polychlorinated naphthalenes [12] and nitronaphthalenes [3] has also been investigated. The smaller, more volatile species, such as phenanthrene and hexachlorocyclohexane exist primarily in the vapour phase, whilst the larger, less volatile compounds, such as benzo[*a*]pyrene, PCBs and DDT are primarily particle associated [13]. The partitioning process is also governed by the ambient temperature and the concentration of particles present in the atmosphere onto which partitioning can occur, therefore equilibrium partitioning can be expressed as [14,15]:

$$K_{vp} = \frac{(F/TSP)}{A} \tag{1}$$

where F is the particulate phase concentration of the compound (ng m^{-3}); TSP is the total suspended particulate loading in the atmosphere (μg m^{-3}); and A is the vapour phase concentration of the compound in the atmosphere (ng m^{-3}).

The partition coefficient is a function of atmospheric temperature, T [14];

$$\frac{d(\ln K_{vp})}{d(1/T)} = \frac{H_d}{R} \tag{2}$$

where H_d is the energy required to desorb and vaporise the compound from the particle (in kJ mol^{-1}), which is of similar magnitude to the heat of vaporisation of the analyte, and R is the gas constant.

The octanol-air partition coefficient has also been used as a predictor of SVOC partitioning to aerosols [16,12].

In denuder samplers, the gas phase analyte is removed from the air sample flowing laminarly through the device by diffusion onto an adsorbent or into a sorbent coated on the walls of the denuder, prior to removal of particulates (and thereby particle phase analyte) by filtration. This is possible because the diffusion coefficients of the gas phase analyte are typically three to six orders of magnitude higher than for submicron particles [17]. The incorporation of the second adsorbent or sorbent downstream of the filter serves to collect any analytes which have volatilised from the collected particulate matter [18] or which have broken through the gas phase sorbent bed under higher volume sampling conditions [19]. Additional potential sampling artefacts which may occur with diffusion denuders are:

- Loss of fine particulate matter to the gas phase sampling medium or the walls of the denuder, as a consequence of sedimentation or Brownian diffusion [10], which would lead to an overestimation of the gas phase analyte concentration and an underestimation of the particle-associated concentration. This effect may be elevated under dry conditions, which enhance electrostatic effects, whilst vertical placement of the denuder reduces sedimentation [5]. Laminar flow is also crucial [9].
- Desorption of analytes from particles as they travel through the denuder would lead to an underestimation of particle phase and an overestimation

of gas phase concentrations. This effect arises as the partitioning equilibrium is perturbed during sampling and it may be minimised by reducing the transit time for particles through the denuder.

Means by which to theoretically estimate the extent of these sampling artefacts are discussed in Section 3.

1.3 Reasons for Determining Gas and Particle-Associated Analyte Concentrations Separately

Partitioning of SVOCs in the atmosphere between the gas and particle phases affects the air–water exchange processes as well as the rates of both wet and dry deposition of these compounds [16]. Partitioning therefore affects the atmospheric removal of SVOCs [20], where in terms of wet deposition, vapour phase pollutants may dissolve into raindrops whilst particle-associated compounds may be washed out of the atmosphere [13]. The determination of the concentrations of air pollutants in both phases is therefore important from an environmental fate perspective. Partitioning studies also assist in elucidating atmospheric physicochemical processes, where gas phase SVOCs may condense to form aerosols or be physically adsorbed onto primary particles [21]. Such studies also improve our understanding of long-range atmospheric transport processes [22].

Partitioning influences analyte degradation including photolytic reactions, therefore denuder-based sampling has proven valuable in elucidating photochemical smog mechanisms [3]. Denuder systems have been employed in understanding formation and decay of SVOCs, such as PAHs and nitro-PAHs, in the atmosphere [23,24].

The impact of SVOCs, such as PAHs, on human health has been shown to be heavily dependent on their atmospheric physical state [5], as lung deposition patterns vary according to whether the SVOC is in the gas or particle phase [25].

1.4 Disadvantages of Conventional Filter/Adsorbent-Based Sampling Systems

It is important that the partitioning of analytes between the gas and particle phases not be disturbed during sampling and analysis, so that results reflect the partitioning thereof in the atmosphere at the time of sampling.

In order to determine the concentrations of SVOC analytes in both the gas and particle phases, sampling trains consisting of a filter to collect particles followed by downstream gas phase collection by means of an adsorbent, such as polyurethane foam (PUF), have been employed. This high volume (hi-vol) air sampler methodology is recommended by the U.S. Environmental Protection Agency (EPA) for the sampling of SVOCs from air [26], and it has found favour due to its robustness and ease of field use [18]. This type of sampling system has inherent limitations, however, due to the sampling configuration and the high volumetric flow rates which are typically employed, in the order of 1 m^3 min^{-1} [11].

These sampling artefacts include adsorption of a portion of the target SVOC gas phase analytes onto the filter medium or particles collected on the filter, which would result in an overestimation of the particle phase concentration and an underestimation of the gas phase concentration of the analyte (Figure 2). Studies have therefore been conducted to estimate the extent of this effect and to provide a means of correcting for it, which have typically employed the second filter behind the first one [11]. It has been found that the filter medium affects the extent of this sampling artefact, with less adsorption of gas phase analytes onto Teflon filters as compared to quartz fibre filters [11] and glass fibre filters (GFF) [27]. The increase in back pressure across the sampling system upon incorporating an additional filter for such experiments, however, may perturb the equilibrium and lead to variations in sampling artefacts observed. Some studies have indicated that artefacts arising from adsorption onto collected particles are not significant under typical filter loading [28].

In a denuder setup, this sampling artefact is avoided by removal of the gas phase analyte prior to collection of the particle phase analyte downstream.

The second potential sampling artefact which can occur with filter/adsorbent sampling relates to volatilisation of particle phase analytes from the filter, commonly referred to as 'blowoff' (Figure 3), which is a consequence of a disturbance in the equilibrium conditions upon sampling [21]. This leads to an underestimation of particle phase concentrations and an overestimation of gas phase concentrations of target SVOCs, especially those of intermediate vapour pressure, which in terms of PAHs was found to be from phenanthrene to chrysene, as the higher volatility compounds are primarily in the gas phase and therefore blowoff effects are negligible and the heavier compounds do not easily partition into the gas phase [21]. The collection of particles on the filter

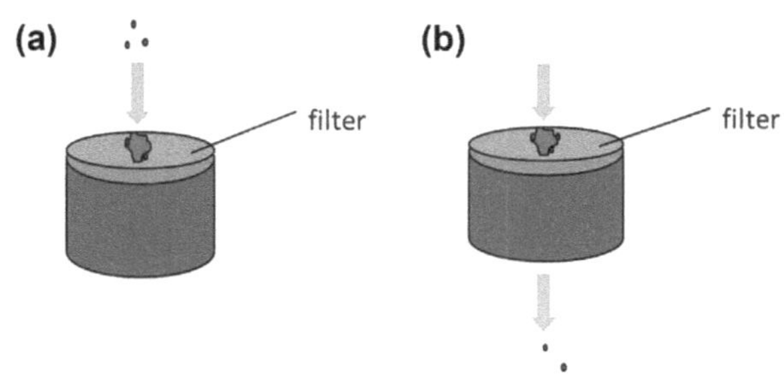

FIGURE 2 Schematic diagram of filter/adsorbent system adsorption effects (a) gas phase analyte is sampled through a particle-loaded filter and (b) a portion of the gas phase analytes are adsorbed onto the particles present on the filter (or on the filter itself), leading to lower gas phase concentration of the analyte in the air downstream of the filter.

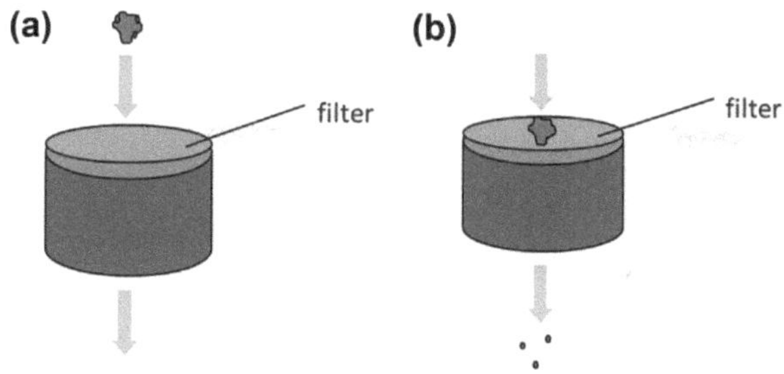

FIGURE 3 Schematic diagram of filter/adsorbent system blowoff effects (a) particle phase analyte is sampled through a filter and (b) a portion of the analyte which was adsorbed onto the particles present on the filter volatilises, leading to higher gas phase concentration of the analyte in the air downstream of the filter.

surface leads to an increase in pressure drop across it, which promotes volatilisation [21], and atmospheric particulate PAH concentrations have been reported to be underestimated by a factor of two using conventional hi-vol samplers [5] as a result of this effect. Preferential depletion of the submicrometre particle fraction has been noted, due to the large surface area associated with this fraction [21].

With denuder sampling, this sampling artefact may be addressed by the incorporation of a second gas phase sampling device downstream of the filter, which serves to collect any analyte blown off the filter, allowing for the correction of particle phase concentrations if necessary. Blowoff can also be minimised by employing denuder sampling devices which have low pressure drops [29] and by sampling smaller volumes at low flow rates [21].

For both denuder and filter/adsorbent sampling systems, sampling artefacts can be minimised by ensuring sampling times are short to reduce the possibility of temperature and concentration fluctuations during sampling [11], although short sampling times consequently require sensitive analytical methods which can detect trace levels of these air pollutants.

Studies have been conducted where hi-vol air samplers have been compared to denuder samplers in the determination of gas and particle phase analyte concentrations, where the two systems are operated concurrently. The inherent differences in the sampler geometries and designs make drawing conclusions about sampling efficiencies and partitioning difficult, however. These include differences in the particle size range collected as well as variations in sample volume and sampling flow rates [18]. It has been noted, however, that breakthrough of more volatile SVOCs (including 2- and 3-ring PAHs) from the PUF cartridges may occur in hi-vol samplers at ambient temperatures, whilst denuder sampling employing XAD-4 gas phase adsorbent medium was more efficient [18].

Although hi-vol systems have the advantage of collecting larger air volumes than denuders, which contributes to sensitivity, denuder systems have been shown to exhibit superior recoveries and lower blank values, resulting in similar or even lower limits of detection (LODs) [22].

2. TYPES OF DENUDERS

In designing a denuder air sampling system for a particular application, a number of parameters need to be considered, including:

- The choice of sorbent/adsorbent

  It is important that the material employed serves as an effective sink or collection medium for the analyte(s) concerned. Adsorbent-based denuders (such as XAD-4 resin systems) require coating before use, whilst many sorbent-based denuders (such as PDMS systems) can be re-used.
- The denuder dimensions

  The internal diameter of the denuder will influence the flow regime (refer to Section 3), and the dimensions will also influence portability and ease of use. Increasing the length of the denuder and/or decreasing the channel diameter will increase the pressure drop across the denuder (i.e. the difference in pressure at the inlet and outlet) [2], which affects the specifications of the pump required to draw air through the system.
- The denuder tube material (outer housing) should be robust and allow for portability.
- The air sampling flow rate which should be employed

  Laminar flow conditions must be maintained and the analyte should have sufficient residence time in the denuder to allow for efficient trapping of the gas phase in the adsorbent/sorption medium. Again, the flow rate determines the specifications of the pump required.

Initially, cylindrical denuders were employed which consist of an open tube, where the inner surface is coated with a material which strips the gaseous analyte from the sample airstream (Figure 4). These denuders were used primarily for inorganic aerosol monitoring in terms of acidic gases (such as separately sampling SO_2 and sulphate aerosols); however, these denuders need to be very long in order to achieve acceptable sorption efficiencies at flow rates of 1–2 L min^{-1}, as there is a relatively small absorptive surface area [10]. As a consequence, annular type denuders were developed, which have enhanced sorption surface areas, as they consist of a number of concentric tubes housed inside each other coaxially (Figure 5). An equivalent sampling efficiency is thus obtained for a shorter annular system, operating at higher sampling flow rates [8]. Alternative geometries, such as composite cylindrical denuders (also called multi-channel systems) consisting of a cluster of parallel tubes (such as Ref. [19]), as well as parallel-plate denuders (such as Ref. [30]), honeycomb denuders [31] and coil denuders (such as Ref. [32]) have also been developed.

Denuder systems are available commercially, such as the University Research Glassware (URG) Corporation systems, which are designed for use

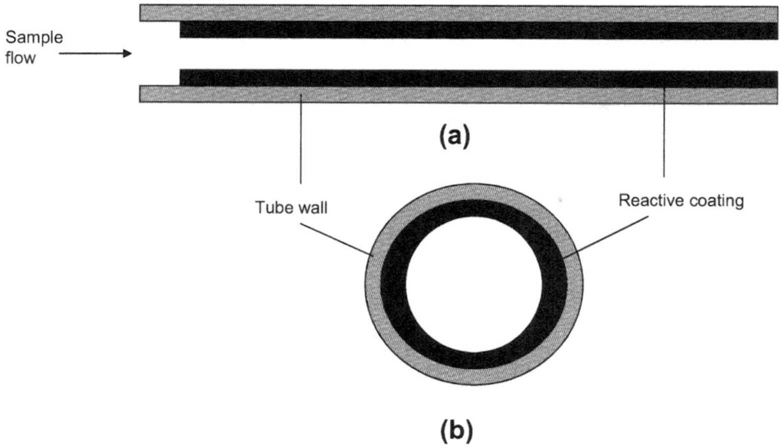

FIGURE 4 Schematic diagram of a cylindrical denuder (a) longitudinal view and (b) in cross-section.

in the monitoring of a wide range of both organic and inorganic air pollutants, depending on the coating employed. Three, four or eight channel annular URG denuders, for example, are 0.1−0.5 m long, and the interior glass is etched to provide an increased area for coating and thereby gas collection [33]. Coiled and parallel-plate denuders (Figures 6 and 7) are also commercially available [33].

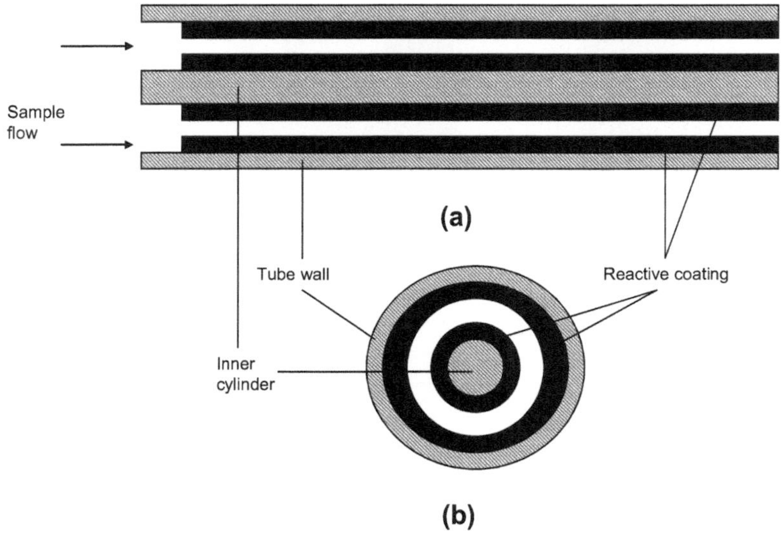

FIGURE 5 Schematic diagram of an annular denuder (a) longitudinal view and (b) in cross-section.

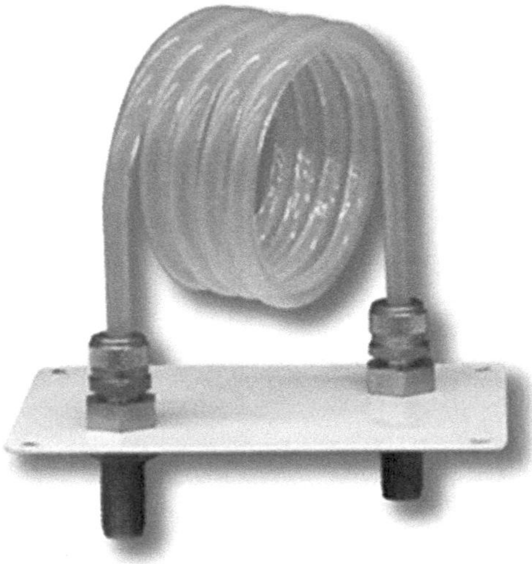

FIGURE 6 A coiled denuder. *Courtesy of URG Corporation, USA.*

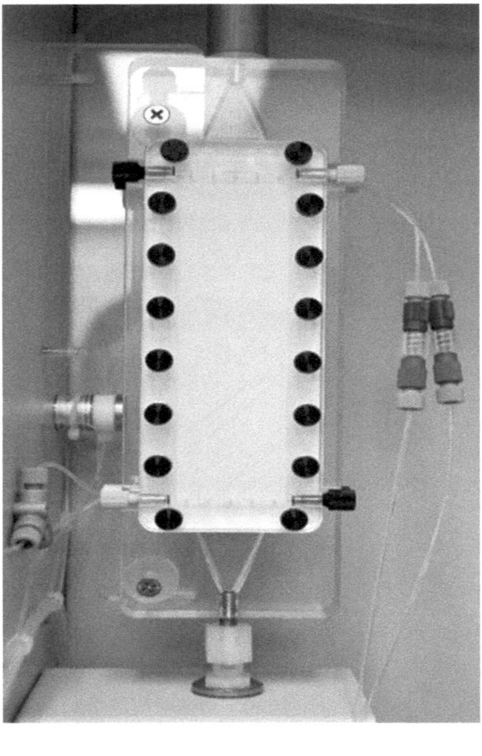

FIGURE 7 A parallel-plate denuder. *Courtesy of URG Corporation, USA.*

Some denuders require large volumes of solvent to extract the target analytes after sampling with pre-concentration and clean-up steps often needed prior to analysis, therefore denuders have been developed which allow for solvent-free thermal desorption, introducing the target analytes directly into the analytical instrument (such as a GC-MS), as cost and time efficient alternatives. In some cases a custom-built thermal desorber is needed (such as Ref. [34]), however, it is also possible with some denuder designs to employ commercial thermal desorber systems (such as Ref. [19]).

3. THEORY OF DENUDATION

In terms of denuder sampling of atmospheric SVOCs, it is important to note that neither 100% gas phase collection nor 100% particle transmission through the denuder to the downstream filter is practically achievable, however, efficiencies should be as close to this as possible. In terms of designing novel denuders and in comparing denuder performance, it is useful to employ theoretical and empirical equations to evaluate potential denudation efficiencies. These are summarised in this section; the reader is referred to Ali et al. [8] and Katsanos and Roubani-Kalantzopoulou [35] for more details on the development of the mathematical expressions concerned.

3.1 Air Flow Considerations

Laminar air flow through the denuder device is a prerequisite for denudation, as turbulence will result in losses of particles (and therefore particle bound SVOC analytes) to the walls of the denuder by impaction, therefore, gas and particle phases will not be separately sampled.

In order for laminar flow conditions to be satisfied, conventional theory for flow in pipes suggests that the Reynolds number (R_e) needs to be <2300, although the transitional R_e at which the flow changes from the laminar to turbulent flow regime would decrease with an increase in the surface roughness of the inner surface of the denuder [36], as in the case of resin coated denuders.

The R_e can be calculated for a specific cylindrical denuder configuration from Eqn (3):

$$R_e = \frac{VD}{\upsilon} \tag{3}$$

where V is the linear velocity (m s^{-1}); D is the internal diameter of the denuder tube (m); and υ is the kinematic viscosity (m^2 s^{-1}).

Further, υ can be determined from

$$\upsilon = \frac{\mu}{\rho} \tag{4}$$

where μ is the dynamic viscosity (kg m^{-1} s^{-1}) and ρ is the gas density (kg m^{-3}).

Using typical values of $\mu = 1.87 \times 10^{-5}$ kg m^{-1} s^{-1} and $\rho = 1.168$ kg m^{-3} for air at 25 °C and 1 atm, a kinematic viscosity of 1.60×10^{-5} m^2 s^{-1} is obtained.

It is evident that the R_e increases with an increase in linear flow velocity, therefore high flow rates will result in turbulent conditions and poor denudation. It should be further noted that even when the R_e indicates laminar flow conditions, this is only achieved after a short distance l, from the denuder inlet of diameter D [8]:

$$l = 0.07DR_e \tag{5}$$

In the case of a denuder tube where $D = 0.3$ mm, as in the case of the multi-channel silicone rubber trap denuder (refer to Section 4.2.2) and $R_e = 100$, laminar conditions would thus be established 2.1 mm from the inlet. As this denuder is 55 mm long, the non-laminar portion accounts for <5% of the total length of the denuder [19]. For denuders which have wider internal diameters, the non-laminar region would account for a more significant portion of the denuder length for the same R_e, unless the denuder is proportionally longer.

In addition to laminar flow conditions, it is also important that longitudinal diffusion of the air moving through the denuder is negligible compared to the linear velocity arising from actively pumping air through the denuder, which ensures that the gaseous analyte moves through the system in a defined manner. In order to verify that this condition is met in a denuder, the Peclet number (P_e) is calculated, which must be greater than 10 [2]:

$$P_e = \frac{2Rv}{D_A} > 10 \tag{6}$$

where R is the internal radius of the denuder (cm); v is the linear velocity of the air sample flowing through the denuder (cm s^{-1}) at a given distance from the centre; and D_A is the diffusion coefficient of gaseous analyte A in the sample (cm^2 s^{-1}).

Under laminar flow conditions, the linear flow profile through an open tube (denuder) can be determined from Eqn (7) [2]:

$$v = 2v_{av}\left(1 - \frac{r^2}{R^2}\right) \tag{7}$$

where v_{av} is the linear velocity of the sample gas flow (cm s^{-1}); and r is the distance from the longitudinal axis of the denuder tube (cm).

3.2 Particle Transmission Efficiency

Denuders are operated under laminar flow conditions, as discussed in Section 3.1, in order to minimise particle loss to the walls of the denuder due to turbulence. In addition, denuders are usually positioned vertically during operation to prevent sedimentation within the denuder due to gravitational settling. Even when these criteria for best practice are met, loss of particles in the denuder may occur due to electrostatic effects (especially for small particles under low humidity conditions), where charged particles are not transmitted due to localised electrostatic fields on the walls of the denuder. Experiments to determine such electrostatic losses are difficult to repeat [37], due to variations in environmental and localised conditions. Another means of particle transmission losses through denuders arises from diffusion of the particles to the walls of the denuder, where higher particle transmission losses are experienced for smaller particles, due to their higher diffusion coefficients.

Experimental results relating to the diffusional loss of neutral particles in an annular denuder compared favourably to theoretical calculations based on equations developed for a channel of rectangular cross-section [37]. Here the annular denuder channel was considered as a parallel-plate configuration in an 'unrolled and flattened' state which allowed for the approximation of the channel geometry as a rectangle.

For cylindrical denuders, diffusive loss theories for particles traversing tubes of circular cross-section are appropriate, as applied in Hermann et al. [38] and Ferm [9]. The transport (or transmission) efficiency, E_{diff}, is given by:

$$E_{diff} = 1 - 2.564\mu^{2/3} + 1.2\mu + 0.1767\mu^{4/3} \quad (8)$$

for $\mu \leq 0.02$, and

$$\begin{aligned} E_{diff} = & \ 0.819 \exp(-3.657\mu) + 0.0975 \exp(-22.305\mu) \\ & + 0.0325 \exp(-56.961\mu) + 0.0154 \exp(-107.62\mu) \end{aligned} \quad (9)$$

for $\mu > 0.02$, where

$$\mu = \frac{\pi D_p L}{Q} \quad (10)$$

and D_p is the diffusion coefficient of a particle of a specific diameter ($m^2\ s^{-1}$); L is the length of the denuder (m); and Q is the volumetric flow rate ($m^3\ s^{-1}$).

The particle diffusion coefficients may be calculated from Eqn (11):

$$D_p = \frac{C_c kT}{3\pi d_p \mu_{abs}} \quad (11)$$

where C_c is the Cunningham slip correction factor; k is the Boltzmann constant ($1.3806503 \times 10^{-23}\ m^2\ kg\ s^{-2}\ K^{-1}$); T is the temperature (K); d_p is the diameter of the particle (m); and μ_{abs} is the absolute (or dynamic) gas viscosity ($kg\ m^{-1}\ s^{-1}$).

Particle transport efficiencies, E_{diff}, can therefore be calculated for a range of particle sizes for a specific denuder configuration. It is important that the transmission efficiency of particles in the size range of interest in the particular application be considered, such as combustion-derived particles or diesel vehicle particulate emissions.

The theoretical particle transmission efficiencies can also be verified by laboratory measurements, where particles of a particular size range are generated and passed through the denuder to be tested. Scanning mobility particle sizer and condensation particle counter instrumentation may be employed in such experiments to compare inlet and outlet particle concentrations and sizes [19]. Jet nebulisers [19] and diesel vehicles [39] have been used as sources of particles in such particle transmission studies.

3.3 Gas Phase Collection Efficiency

In any denuder system, it is important that the gas phase analyte is efficiently sampled and pre-concentrated by the adsorbent/sorbent on the walls of the denuder. This critical gas collection efficiency can be determined theoretically for cylindrical and annular denuders, as described in this section, as they are most commonly employed. For theoretical approaches to other denuder geometries, the reader is referred to Katsanos and Roubani-Kalantzopoulou [35].

3.3.1 Cylindrical Denuders

In a cylindrical, sorption-based denuder (where analyte removal is permanent), the rate of change of gaseous concentration of analyte, A, is usually determined from the general mass balance equation [35]:

$$\frac{\partial c_A}{\partial t} = -\upsilon \frac{\partial c_A}{\partial z} + D_A \frac{\partial^2 c_A}{\partial z^2} + D_A \left(\frac{\partial^2 c_A}{\partial r^2} + \frac{1}{r} \frac{\partial c_A}{\partial r} \right) - r_A \tag{12}$$

where $c_A = c_A(t,\ z,\ r)$ is the concentration of analyte A in the gas phase (mol cm^{-3}), as a function of time, t (s); length along the denuder, z (cm); and distance from the cylinder axis, r (cm); υ relates to the axial air flow velocity (refer to Eqn (7)); D_A is the diffusion coefficient of A in air (cm s^{-1}) and r_A is the rate of a gaseous chemical reaction of A (mol cm^{-3} s^{-1}).

The first term on the right-hand side of Eqn (12) describes convective movement along the denuder, whilst the second term is due to longitudinal diffusion along the cylinder axis of the denuder. The third term arises from radial diffusion and the fourth term relates to homogeneous chemical reactions in the denuder.

A number of assumptions can be made in order to simplify this equation and allow for a solution to be found [35]. The first of these is to assume steady-state conditions, where the composition of the air sample does not change, i.e.:

$$\frac{\partial c_A}{\partial t} = 0 \tag{13}$$

Therefore, the left-hand side of Eqn (12) is set to zero. Further, the second term relating to longitudinal diffusion may be ignored if it is assumed that the linear velocity of analyte A resulting from longitudinal diffusion is negligible compared with the linear velocity resulting from actively pumped sample air flow. This assumption holds if the P_e is greater than 10, as discussed in Section 3.1.

If it is also assumed that no homogeneous reactions occur in the denuder which create or destroy analyte A, Eqn (12) may be simplified to:

$$v\frac{\partial c_A}{\partial z} = D_A\left(\frac{\partial^2 c_A}{\partial r^2} + \frac{1}{r}\frac{\partial c_A}{\partial r}\right) \tag{14}$$

Further assuming that pollutant A is a trace gas, that laminar flow is developed in the denuder with constant viscosity, that the temperature is constant and that the amount of analyte A collected on the denuder wall is small compared to its capacity, allows for the Gormley–Kennedy solution of Eqn (14) [40] to be employed:

$$\frac{c_{av}}{c_0} = 0.8191\exp(-7.314z^*) + 0.0975\exp(-44.61z^*) + 0.0325\exp(-113.9z^*) + \ldots \tag{15}$$

where c_0 is the gas concentration of analyte A entering the denuder; c_{av} is the average gas concentration of analyte A leaving the tube; and $z*$ is a dimensionless factor given by:

$$z^* = \frac{\pi}{2}\frac{D_A L}{V} \tag{16}$$

where L is the length of the denuder; and V is the volumetric flow rate [35].

The fraction of analyte retained on the denuder (collection efficiency) can be determined from:

$$\left(1 - \frac{c_{av}}{c_0}\right) \tag{17}$$

The collection efficiency may be increased by increasing L, and by decreasing the sampling flow rate (V). Collection efficiencies are also higher for analytes with higher diffusion coefficients.

A number of alterations to the Gormley–Kennedy solution have been suggested. One of these takes into account an *effective length* of the denuder, which decreases with sampling time as a result of the depletion of the active surfaces of the denuder with increasing sample volume, leading to a decreased gas collection efficiency [8]. This alteration is significant only when the depletion rate is high as a result of high analyte concentrations, high moisture levels or low surface capacities, and therefore would be of minor significance in trace gas analysis, but would be of relevance under breakthrough conditions, which is important when adsorption mechanisms are employed for analyte retention, as the active surface may become saturated.

Another adaption to the estimation of gas collection efficiency is based on the collection surface not performing as a perfect sink. Here a reaction probability is incorporated, which requires knowledge of the reaction probability factor and the diffusional behaviour of the analyte [8]. The differences between approaches were however found to be insignificant under normal sampling conditions [8].

For sorption-based denuder systems, such as silicone-based denuders, the Gormley–Kennedy approximation is valid for SVOC analytes of lower volatility (with higher retention volumes), such as pyrene, but would not be valid for the more volatile SVOCs (with lower retention volumes), such as naphthalene [19]. In this case a chromatographic model is more appropriate due to continuous partitioning behaviour, where the PDMS-air partition coefficients are expected to be proportional to the octanol-air partition coefficients [41]. This is because PDMS may be treated as a liquid, as chromatographic retention is dominated by dissolution of the analyte into the bulk PDMS phase rather than by surface interactions [31]. The reader is referred to Rowe and Perlinger [31] for details on the calculations related to this. Experimentally derived breakthrough volumes for each analyte provide an indication of gas collection efficiencies.

3.3.2 Annular Denuders

The efficiency (E) of an annular denuder to quantitatively retain gas phase analytes from the air sample can similarly be determined as a percentage from the expression [10]:

$$\%E = \left(1 - \frac{C_{av}}{C_0}\right)100 \tag{18}$$

where C_{av} is the concentration of analyte exiting the denuder and C_0 is the gas phase concentration of analyte A entering the denuder (i.e. the concentration in the air sample being monitored), but in the case of annular denuders an empirically modified form of the Gormley–Kennedy equation applies, which incorporates the equivalent diameter of the annular channel [10]:

$$\frac{C_{av}}{C_0} = 0.82\exp[-22.53\Delta_a] \tag{19}$$

where

$$\Delta_a = \frac{\pi D_A L}{4F}\frac{d_1 + d_2}{d_2 - d_1} \tag{20}$$

and D_A is the gas diffusion coefficient of analyte A ($cm^2\ s^{-1}$); L is the denuder length (cm); F is the sampling volumetric flow rate ($cm^3\ s^{-1}$); d_1 is the outer diameter of the denuder annulus (cm) and d_2 is the inner diameter of the annulus (cm).

A gas phase collection efficiency of 99.999999%, for example, corresponds to a potential loss of 10 ppb of the original analyte concentration [18].

Gas collection efficiencies can also be determined or verified experimentally, such as that by Temime-Roussel et al. [25], which focused on gas phase PAH collection by XAD-4-coated commercial annular denuders utilising a dynamic vapour generator. Gas collection efficiencies of over 90% were found for a range of temperatures, humidity and sample volumes. Typically, denuder gas collection efficiencies decrease with increasing volatility of the analyte, therefore, studies need to consider the full volatility range of analytes when determining such efficiencies, where the more volatile analytes may represent a minimum efficiency [25]. The experimental breakthrough fraction (bt) of an analyte can be calculated from:

$$bt = \frac{M_b}{M_f + M_b} \tag{21}$$

where M_f and M_b are the mass of the analyte on the front (primary) and backup (secondary) denuders, respectively [31].

Gas phase collection efficiencies for non-chromatographic-based denuder systems are enhanced by an increase in temperature and an increase in gas diffusion coefficients, although the former can also lead to enhanced vaporisation losses of analytes from the particle phase collected on the filter.

4. DENUDER SAMPLING APPLICATIONS

4.1 Denuders Used for the Sampling of Inorganic Analytes

A 'denuder difference method' has been used for HNO_3 determinations, where two samplers are used in parallel; one being a filter and the other a denuder in series with a filter. The difference in the two filter nitrate concentrations provides the HNO_3 concentration. Denuder sampling is useful in inorganic air pollutant determinations where reaction of gaseous species on the filter surface causes sampling artefacts, such as the reaction of ammonia with sulphuric acid which would lead to an underestimation of ammonia and an overestimation of ammonium ion concentrations [42].

In many cases cylindrical denuders have been used for inorganic air pollutant monitoring. Sulphur dioxide, for example, has been removed from sulphate aerosols by means of denuders employing tetrachloromercurate/maleic acid buffer coatings [10]; whilst oxalic acid coatings have been used for sampling of ammonia when ammonium-based aerosol particles are present [9] and bases have been used for HNO_3 removal (such as by Harrison and Kitto [42]). Examples of denuder applications for inorganic air pollutants are shown in Table 1.

In comparing denuder and filter pack sampling for inorganic analytes, it has been found that the results are site-specific as they are dependent on particulate acidity, humidity and temperature. In addition, apparatus-specific variables including housing and siting of the equipment, sampling flow rates and denuder methods can influence results [42].

TABLE 1 Examples of Sampling, Extraction and Analytical Methods, Where Denuders Were Employed for the Monitoring of Inorganic Air Pollutants

Target Analytes	Location	Type of Denuder	Length (m)	Gas Sampling Medium	Particle Sampling Medium	Flow Rate (L min^{-1})	Sampling Interval	Extraction of Gas Phase	Extraction of Particle Phase	Analysis	LOD	References
SO_2 (standards); sulphate aerosol	N/A	Annular	0.2	TCM-maleic buffer solution						Fluorescent SO_2 monitor		[10]
NH_3	N/A	Cylindrical	0.5	Oxalic acid	N/A	3	24 h	2 mL NaOH	N/A	Ion-specific electrode	0.01 ppb	[9]
Sulphate, nitrate, chloride and ammonium	England	Cylindrical	0.5	Na_2CO_3 or H_3PO_3	PTFE	10	24 h	Deionised water	Isopropanol/ deionised water	Ion chromatography		[42]
Gaseous acids and NH_3	Italy	Annular	0.22	Na_2CO_3 and citric acid	PTFE and Nylon	15	12 h	Dilute H_2O_2	Ultrasonic (H_2O or carbonate buffer)	Ion chromatography (acids) spectrophotometry (NH_3)	0.1 μg m^{-3}	[17]

TCM, tetrachloromercurate; N/A, not applicable.

4.2 Denuders Used for the Sampling of Organic Analytes

4.2.1 Adsorbant-Based Denuders

The Integrated Organic Gas and Particle Sampler (IOGAPS) is an eight-channel annular diffusion denuder coated with ground XAD-4 resin adsorbent. It was originally called the Integrated Organic Vapor/Particle Sampler (IOVPS) and a number of these annular denuders may be employed in series with a cyclone inlet [43]. The IOGAPS denuder has been found to have 5−10% particle diffusion losses for particles less than 0.05 μm in diameter [27]. It has been employed in SVOC determinations, including organochlorine compounds and PAHs in semi-rural air in England [18]. The IOGAPS has also been used for seasonal and diurnal organic and elemental carbon determinations in particulate matter in Canada [7,44,45], which found that particle losses through the denuder were negligible. In these applications, a denuder system was employed to prevent sampling artefacts.

The IOGAPS has also been used in sampling atmospheric per- and polyfluoroalkyl substances, as overestimation of particle phase concentrations collected onto GFF in hi-vol samplers had been reported [22]. In this comparative study, it was found that the total concentrations for the denuder and hi-vol systems were in agreement within a factor of 4, but vapour phase perfluoroalkane sulphonic acids and perfluoroalkyl carboxylic acids adsorbed onto the GFF in the hi-vol system, leading to elevated particle-associated concentrations.

Feilberg et al. [3] utilised an XAD-4-based denuder for nitro-substituted PAH determinations. Here two Teflon impregnated GFF followed by a 5-channel annular gas phase denuder was employed, as losses of 80−90% for the particle-associated nitronaphthalenes were found when the denuder was positioned upstream of the filter. These losses were ascribed to particle losses through the denuder and volatilisation of particle-associated analytes whilst traversing the denuder. The study employed a backup filter to determine gas phase adsorption losses onto the filter and a 0.05−0.1% loss was found, however, blowoff from the filter could impact on gas phase concentrations in such a device. This group has also used this denuder system for PAH determinations (for example Refs. [24] and [23]). A multi-parallel-plate denuder design utilising XAD-4 resin has also been employed in PAH vapour sampling [46] and an XAD-4-based annular denuder has been utilised in the removal of ozone simultaneously with gas phase PAHs in air samples, where the oxidant could degrade target PAHs [5].

Although XAD-4-based denuder systems have been extensively used in air pollution studies, benzene, styrene, naphthalene and fluorene are known to contaminate the resin, which necessitates several solvent washing cycles prior to use in order to achieve clean blanks [25]. Re-coating of the denuder with adsorbent is periodically required and solvent extraction is needed prior to analysis. The coating, drying and washing steps have been associated with problems regarding reproducibility and contamination [47].

An activated-carbon-based parallel-plate denuder was employed by Kavouras et al. [30] in the sampling of both aliphatic hydrocarbons and PAHs in Chile. A concern with the use of this adsorbent is the possibility of irreversible adsorption of target analytes, but in this application the gas phase was not analysed. The parallel-plate denuder simply served to strip the gas phase analytes from the particle phase, which was sampled downstream and analysed. Examples of adsorbent-based denuder studies used for organic air pollutant monitoring are included in Table 2.

4.2.2 Sorbent-Based Denuders

A high-volume compound annular denuder (consisting of a series of 12 nested denuders, 0.20 m long with an overall outer diameter of 0.08 m) was developed for the collection of PAH vapour, which utilised high-vacuum silicone grease as a sorptive medium (coating thickness ranged from 4 to 100 μm) [49]. SE-30 and silicone pump oil had also been tested by these authors, but the SE-30 was found to gradually oxidise upon exposure to air, hindering collection efficiency; and the silicone oil coating was too thin, which decreased the capacity of the device. Contamination from impurities in the silicone grease could be of concern.

A composite cylindrical denuder, composed of a collection of methyl-silicone stationary phase capillary gas chromatography columns (0.15−0.5 m long, 0.53 mm i.d., 8 μm d_f) as the sampling tubes, was developed for use in atmospheric PAH and PCBs sampling [14]. Comparative studies showed that the results obtained from the denuder were not statistically different to those obtained with a hi-vol PUF sampler. These devices required liquid extraction and sample cleanup prior to analysis, similar to the XAD-4-based denuders, as the denuder (specifically the epoxy resin used to bond the capillary tubes together) was not resilient to thermal desorption conditions, which would allow for direct sample introduction without preparative steps [14].

Rowe and Perlinger [31] also used a multi-capillary denuder consisting of 289 sections of capillary GC columns (0.25 m long, 0.53 mm i.d., 5 μm d_f) for SVOC sampling (hexachlorobenzene, PCBs and polybrominated diphenyl ethers (PBDEs)). Here the stationary phase was polydimethylsiloxane (PDMS). Dudek et al. [50] compared the sorption properties of different types of capillary columns for denudation applications. A high-flow denuder has also been constructed from PDMS-coated stainless-steel honeycomb (0.102 m long) [31].

PDMS offers a number of advantages for sorptive sampling: it is reuseable and can be thermally desorbed or solvent extracted; absorptive retention reduces the possibility of irreversible adsorption or catalytic surface reactions; it is hydrophobic therefore humidity effects are minimised during sampling and analysis; diffusion is high in PDMS resulting in rapid uptake kinetics; and finally PDMS provides a smooth surface (as compared to XAD resin coatings), which therefore minimises the potential of the surface to trap particles present in the air sample [31].

A miniature, portable multi-channel silicone rubber trap based denuder system has been employed by Forbes et al. [19] for atmospheric PAH determinations in air. The system consists of two of these traps in series (each 178 mm long, containing 22 parallel 55 mm long PDMS tubes of 0.3 mm i.d.), separated by a quartz fibre filter for particle phase collection (Figure 8). This denuder operates at ~0.5 L min^{-1} therefore a portable battery-operated personal sampling pump can be used. It has been applied in air pollution studies relating to sugar cane burning and diesel exhaust emissions [19,51], as well as pesticides [52].

Sorbent-based denuder systems are generally reuseable, and thermal desorption is an attractive option with these denuders, as it is solvent-free, it is more time efficient and provides the option for automation. As the entire sample is analysed for the analyte, lower limits of detection are possible for these methods, enhancing their sensitivity. Examples of the use of sorbent-based denuder systems for organic air pollutant monitoring are included in Table 2.

5. DENUDER SAMPLING CONSIDERATIONS

It is important to note that the distribution of SVOC air pollutants between the gas and particle phases is a dynamic process, which is dependent on a number of parameters including vapour pressure and concentration of the SVOC concerned, as well as the ambient temperature [21]. Therefore temperature changes and dilution effects will perturb the equilibrium and thus the gas to particle partitioning ratio and denuders should be operated under steady-state temperature and pressure conditions [2]. Ageing of air masses during long-term transport also results in a variation in partitioning of SVOCs [21].

Depletion mechanisms such as ozonolysis and nitration of PAHs may be a source of analyte loss during sampling [21], which applies to both denuder and filter/adsorbent sampling systems, although Goriaux et al. [5] found that the XAD-4 denuder adsorbent also removed ozone from the airstream, which minimised degradation of particle-bound PAHs.

High-humidity sampling conditions may reduce the gas phase sampling efficiency of some denuder systems, as water vapour may compete for adsorption sites of hydrophilic sampling media. This is not considered to be a problem for hydrophobic media such as XAD-4 [18] and PDMS [19]. It should also be noted that the partitioning of hydrophobic air pollutants, such as alkanes and PAHs, onto polar aerosols (wood soot and secondary aerosols) may itself be significantly affected by ambient humidity [53].

Particle losses due to sedimentation can be minimised by positioning the denuder vertically during sampling [5]. Cyclones and impactors may also be employed upstream of the denuder in order to remove large particles and airborne detritus from the sample, which may improve effectiveness of the method [8]. This adapted setup can, however, result in a high pressure drop across the sampling device, which may enhance evaporative losses from particles [29].

TABLE 2 Examples of Sampling, Extraction and Analytical Methods, Where Denuders Were Employed for the Monitoring of Organic Air Pollutants

Target Analytes	Location	Type of Denuder	Length (m)	Gas Sampling Medium	Particle Sampling Medium	Flow rate ($L\ min^{-1}$)	Sampling Interval	Extraction of Gas Phase	Extraction of Particle Phase	Internal Standards	Cleanup and Fraction-ation	Analysis	LODs	References
Organo-chlorines and PAHs	England	8-channel annular (IOGAPS)	0.6	Ground XAD-4	GFF	16.7	48 h	3 × 250 mL hexane extraction, rotary evaporation pre-concentration	Soxhlet using 250 mL hexane for 12 h, rotary evaporation pre-concentration	Gas phase: spike pre-extraction; particle phase: spike post extraction; and pre-analysis	Column chroma-tography, deactivated silica/alumina	Organo-chlorine compounds: GC-ECD; PAHs: GC-MSD	2.0 pg m^{-3} (OCs) and 5.0 pg m^{-3} (PAHs)	[18]
PAHs	UK	8-channel annular (IOGAPS)	0.6	Ground XAD-4	GFF and Teflon-coated GFF	16.7	48 h	3 × 250 mL hexane extraction, rotary evaporation pre-concentration	Soxhlet using 250 mL hexane for 12 h, rotary evaporation pre-concentration	d-PAHs for extraction efficiency, surrogates for sample processing efficiency	Column chroma-tography, deactivated silica/alumina	GC-MS		[27]
OC and EC	Canada	8-channel annular (IOGAPS)	0.6	Ground XAD-4	QF	16.7	9 and 14 h		Filter punch directly analysed			Thermal optical transmission aerosol carbon analyser	0.4 μgC m^{-3}	[7,44,45]
Nitro-PAHs and PAHs	Laboratory chamber	5-channel annular	0.4	Ground XAD-4	Two Teflon-impreg-nated GFF in series	20	20 min	Hexane: acetone: methylene chloride (5:3:2) or hexane: methylene chloride (1:4)	Soxhlet using hexane: acetone: methylene chloride (5:3:2) or hexane: methylene chloride (1:4) for 3 h	2-nitro-biphenyl	As per Ref. [24]	GC-MS		[3,23,24]

PAHs with O_3 removal	France	8-channel annular (URG 2000-30-CF)		Ground XAD-4	4-stage filter pack (Teflon-coated GFF and XAD-2 resin)	31.7–58.3	12 h	Ultrasonic extraction in dichloro-methane	Ultrasonic extraction in dichloro-methane			HPLC-fluorescence detection	LOQ: 0.1–24.5 ng mL^{-1}	[5]
PAHs	France	8-channel annular (URG)		Ground XAD-4	4-stage filter pack (Teflon-coated GFF and XAD-2 resin)	16.71	6–10 h	Ultrasonic extraction in cyclohexane	Not applicable		Filtration	HPLC-fluorescence detection		[25]
PAHs	Rome	1-channel annular	0.18	Ground XAD-4	QFF	6	6 h	Ultrasonic extraction in cyclohexane	Soxhlet using 120 mL dichloro-methane: methanol (9:1)	d-PAHs prior to extraction	Filtration and column chroma-tography (alumina)	GC-MS	0.3–15.3 ng	[48]
PFASs	Canada	8-channel annular (IOGAPS)	0.285 each	Ground XAD-4	GFF	85	24 h	Petroleum ether and methanol	Ultrasonic extraction in dichloro-methane and methanol	Mass-labelled internal standard spike prior to extraction	Filtration	GC-MS & LC-MS	0.001–1.87 pg m^{-3}	[22]
Aliphatic HCs and PAHs	Chile	60 parallel plates	20.3 × 25.4 cm sheets	Activated carbon	Teflon	88.8	12 and 24 h	N/A	Reflux in methylene chloride		Column chroma-tography, silica	GC-MS		[30]
PCBs and PAHs	USA	Composite cylindrical (50–120 tubes)	0.15–0.5	Methyl-silicone	QFF	1–3	24 h	Liquid extraction with dichloro-methane	SFE	PCB and d-PAH internal standards prior to evaporation of extracts	Column chroma-tography, $MgSO_4$/alumina	GC-MS or GC-ECD		[14]

N/A, not applicable; d-PAHs, perdeuterated PAH standards; IOGAPS, Integrated Organic Gas and Particle Sampler; GFF, Glass Fibre Filter; QF, Quartz filter; SFE, Supercritical Fluid Extraction; PFAS, per- and polyfluoroalkyl substances; PDMS, polydimethylsiloxane.

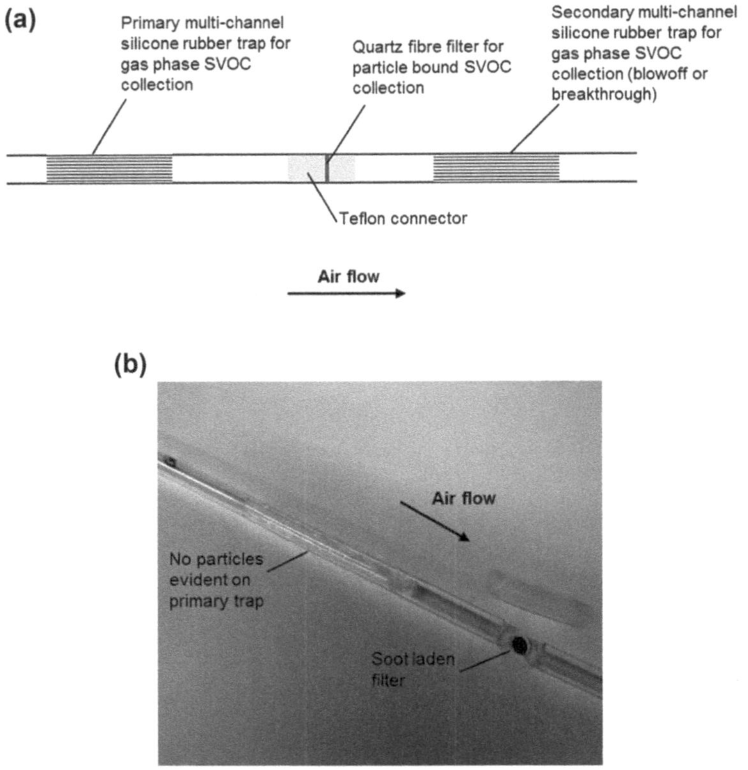

FIGURE 8 Multi-channel silicone rubber trap denuder (a) schematic diagram showing the primary and secondary traps consisting of 22 parallel PDMS tubes each housed in a 6-mm i.d. glass tube and separated by a quartz fibre filter (QFF) held in place by a Teflon connector (b) Primary trap and QFF after sampling emissions from a sugar cane burn, providing visual evidence of good particle transmission through the primary trap. *Adapted from Ref. [19], with permission from Elsevier.*

It is also important that the transit time of particles through the denuder is sufficiently short to prevent re-equilibration of particle-bound analyte to the gas phase during transit, as this would produce a positive bias in terms of the gas phase concentration. For SVOCs, a transit time of $\leq$0.3 s is therefore recommended [54,55]. This can be achieved by increasing the flow rate or reducing the denuder length [39].

Sample storage is critical in maintaining the phase distribution of the analytes as they were when sampled, therefore it is preferable to separate the sampling components (denuder(s) and filter) immediately after sampling to prevent any re-equilibration between them during transport and storage. Components should be wrapped in aluminium foil to minimise photodegradation and stored in well-sealed containers under refrigeration prior to analysis to prevent volatilisation and degradation.

6. ANALYTE EXTRACTION METHODS AND ANALYTICAL TECHNIQUES EMPLOYED

6.1 Extraction of Analytes, Pre-concentration and Use of Internal Standards

Surrogate compounds, such as tetrabromobenzene and mirex for organochlorine analyses [18], as well as deuterated internal SVOC standards, are employed to monitor analytical efficiencies in denuder-based studies. These standards may be spiked onto the denuder (or filter) matrices at different stages of the monitoring process to gain insight into analyte recoveries, namely:

- Prior to sampling to determine any losses during sampling and exposure to transportation environments right through to analysis
- Before analyte extraction, which will indicate overall analyte losses from extraction through to final analysis
- Post-analyte extraction to determine analyte recoveries relating to pre-concentration and clean-up and fractionation steps as well as analysis
- Pre-analysis, where the internal standards are added to the final sample extract just prior to injection in order to evaluate analytical efficiencies and for analyte quantitation

Examples of extraction as well as clean-up techniques typically employed in denuder sampling are provided in Tables 1 and 2.

6.2 Analytical Techniques Generally Employed

A range of analytical methods are employed in denuder studies and depend on the analyte concerned. Examples are included in Tables 1 and 2 and range from GC-MS and LC-MS for the organic air pollutants to ion chromatography and fluorescence spectroscopy for the inorganic species. More detail on analytical methods is provided in Section III of this book.

7. CHALLENGES AND FUTURE OUTLOOK WITH RESPECT TO DENUDER SAMPLING OF AIR POLLUTANTS

Denuder-based sampling has numerous advantages over hi-vol sampling methods in terms of reduced probabilities of artefact formation, whilst simultaneously pre-concentrating the gas phase analyte(s).

Some denuders require labour-intensive coating and extraction procedures and are bulky, therefore choice of an appropriate device is important for the application. Thermal desorption is possible with some denuders, which avoids the coating and extraction steps and allows for automation. Sensitive analytical techniques are necessary when low sampling flow rates and short sampling times are employed, in order to ensure that analytes are present at levels above the limits of detection. Denuders may offer a simple, rugged and portable alternative to other sampling methods and allow for gas phase sampling of analytes even in particle-laden atmospheres. Denudation provides important information in terms of human and environmental health assessments, where different phases of an analyte can significantly alter the environmental fate and health outcomes concerned. Denudation studies also provide insight into atmospheric processes and seasonal variations in air pollutant partitioning, therefore it has an important role to play in the monitoring of ambient air pollutants.

REFERENCES

[1] C. Soanes, A. Stevenson (Eds.), Concise Oxford English Dictionary, eleventh ed. (revised), Oxford University Press, USA, 2006.
[2] A. Kloskowski, M. Pilarczyk, J. Namieśnik, Crit. Rev. Anal. Chem. 32 (4) (2002) 301–335.
[3] A. Feilberg, R.M. Kamens, M.R. Strommen, T. Nielsen, Atmos. Environ. 33 (1999) 1231–1243.
[4] M. Tsapakis, E.G. Stephanou, Environ. Pollut. 133 (2005) 147–156.
[5] M. Goriaux, B. Jourdain, B. Temime, J.-L. Besombes, N. Marchand, A. Albinet, E. Leoz-Garziandia, H. Wortham, Environ. Sci. Technol. 40 (2006) 6398–6404.
[6] P. Mikuška, Z. Večeřa, A. Bartošiková, W. Maenhaut, Anal. Chim. Acta 714 (2012) 68–75.
[7] X. Fan, J.R. Brook, S.A. Mabury, Environ. Sci. Technol. 37 (2003) 3145–3151.
[8] Z. Ali, C.L.P. Thomas, J.F. Alder, Analyst 114 (1989) 759–769.
[9] M. Ferm, Atmos. Environ. 13 (1979) 1385–1393.
[10] M. Possanzini, A. Febo, A. Liberti, Atmos. Environ. 17 (12) (1983) 2605–2610.
[11] K.M. Hart, J.F. Pankow, Environ. Sci. Technol. 28 (1994) 655–661.
[12] T. Harner, T.F. Bidleman, Environ. Sci. Technol. 32 (1998) 1494–1502.
[13] T.F. Bidleman, W.N. Billings, W.T. Foreman, Environ. Sci. Technol. 20 (1986) 1038–1043.
[14] M.S. Krieger, R.A. Hites, Environ. Sci. Technol. 28 (1994) 1129–1133.
[15] H. Yamasaki, K. Kuwata, H. Miyamoto, Environ. Sci. Technol. 16 (4) (1982) 189–194.
[16] A. Finizio, D. Mackay, T. Bidleman, T. Harner, Atmos. Environ. 31 (15) (1997) 2289–2296.
[17] I. Allegrini, F. de Santis, V.D. Palo, A. Febo, C. Perrino, M. Possanzini, A. Liberti, Sci. Total Environ. 67 (1987) 1–16.
[18] A.J. Peters, D.A. Lane, L.A. Gundel, G.L. Northcott, K.C. Jones, Environ. Sci. Technol. 34 (2000) 5001–5006.
[19] P.B.C. Forbes, E.W. Karg, R. Zimmermann, E.R. Rohwer, Anal. Chim. Acta 730 (2012) 71–79.
[20] T.F. Bidleman, Environ. Sci. Technol. 22 (4) (1988) 361–367.
[21] L. Van Vaeck, K. Van Cauwenberghe, J. Janssens, Atmos. Environ. 18 (2) (1984) 417–430.
[22] L. Ahrens, M. Shoeib, T. Harner, D.A. Lane, R. Guo, E.J. Reiner, Anal. Chem. 83 (2011) 9622–9628.
[23] R.M. Kamens, F. Zhi-Hua, Y. Yao, D. Chen, S. Chen, M. Vartiainen, Chemosphere 28 (9) (1994) 1623–1632.
[24] Z. Fan, D. Chen, P. Birla, R.M. Kamens, Atmos. Environ. 29 (10) (1995) 1171–1181.
[25] B. Temime-Roussel, A. Monod, C. Massiani, H. Wortham, Atmos. Environ. 38 (2004a) 1913–1924.
[26] U.S. EPA, Compendium of Methods for the Determination of Toxic Organic Compounds in Ambient Air, EPA/625/R-96/010b, second ed., United States Environmental Protection Agency, Cincinnati, OH, 1999.

[27] D.A. Lane, A.J. Peters, L.A. Gundel, K.C. Jones, G.L. Northcott, Polycyclic Aromat. Compd. 20 (2000) 225–234.

[28] B.J. Turpin, J.J. Huntzicker, S.V. Hering, Atmos. Environ. 28 (19) (1994) 3061–3071.

[29] X. Zhang, P.H. McMurry, Environ. Sci. Technol. 25 (1991) 456–459.

[30] I.G. Kavouras, J. Lawrence, P. Koutrakis, E.G. Stephanou, P. Oyola, Atmos. Environ. 33 (1999) 4977–4986.

[31] M.D. Rowe, J.A. Perlinger, J. Chromatogr. A 1217 (2010) 256–263.

[32] D.Y.H. Pui, C.W. Lewis, C.-J. Tsai, B.Y.H. Liu, Environ. Sci. Technol. 24 (3) (1990) 307–312.

[33] URG (University Research Glassware) Corporation, USA. www.urgcorp.com (accessed on 24.04.15).

[34] D.E. Tobias, J.A. Perlinger, P.S. Morrow, P.V. Doskey, D.L. Perram, J. Chromatogr. A 1140 (2007) 1–12.

[35] N.A. Katsanos, F. Roubani-Kalantzopoulou, J. Chromatogr. A 710 (1995) 191–228.

[36] Gh. M. Mala, D. Li, Int. J. Heat Fluid Flow 20 (1999) 142–148.

[37] Y. Ye, C.-J. Tsai, D.Y.H. Pui, C.W. Lewis, Aerosol Sci. Technol. 14 (1991) 102–111.

[38] M. Hermann, F. Stratmann, M. Wilck, A. Wiedensohler, J. Atmos. Oceanic Technol. 18 (2001) 7–19.

[39] B. Temime-Roussel, A. Monod, C. Massiani, H. Wortham, Atmos. Environ. 38 (2004b) 1925–1932.

[40] P.G. Gormley, M. Kennedy, Proc. R. Ir. Acad. 52 (1949) 163–169.

[41] N. De Coensel, K. Desmet, T. Górecki, P. Sandra, J. Chromatogr. A 1150 (2007) 183–189.

[42] R.M. Harrison, A.M.N. Kitto, Atmos. Environ. 24A (10) (1990) 2633–2640.

[43] L.A. Gundel, V.C. Lee, K.R.R. Mahanama, R.K. Stevens, J.M. Daisey, Atmos. Environ. 29 (14) (1995) 1719–1733.

[44] X. Fan, P.K.H. Lee, J.R. Brook, S.A. Mabury, Aerosol Sci. Technol. 38 (S2) (2004a) 63–69.

[45] X. Fan, J.R. Brook, S.A. Mabury, Atmos. Environ. 38 (2004b) 5801–5810.

[46] J.M. Delgado-Saborit, C. Stark, R.M. Harrison, Environ. Sci. Technol. 48 (2014) 499–507.

[47] J. Namieśnik, M. Pilarczyk, Toxicol. Environ. Chem. 64 (1997) 203–229.

[48] M. Possanzini, V. Di Palo, P. Gigliucci, M.C.T. Scianò, A. Cecinato, Atmos. Environ. 38 (2004) 1727–1734.

[49] R.W. Coutant, P.J. Callahan, M.R. Kuhlman, R.G. Lewis, Atmos. Environ. 23 (10) (1989) 2205–2211.

[50] M. Dudek, A. Kloskowski, L. Wolska, M. Pilarczyk, J. Namieśnik, J. Chromatogr. A 977 (2002) 115–123.

[51] G. Geldenhuys, E.R. Rohwer, Y. Naudé, P.B.C. Forbes, J. Chromatogr. A 1380 (2015) 17–28.

[52] Y. Naudé, E.R. Rohwer, Anal. Chim. Acta 730 (2012) 112–119.

[53] M. Jang, R.M. Kamens, Environ. Sci. Technol. 32 (1998) 1237–1243.

[54] R.M. Kamens, D.L. Coe, Environ. Sci. Technol. 31 (6) (1997) 1830–1833.

[55] L.A. Gundel, D.A. Lane, Sorbent-coated Diffusion Denuders for Direct Measurement of Gas/particle Partitioning by Semi-volatile Organic Compounds, Ernest Orlando Lawrence Berkeley National Laboratory, January 1998. Report LBNL-41278.

Chapter 6

Automated Continuous Air Monitoring

Johan Paul Beukes,[1,*] Andrew Derick Venter,[1] Miroslav Josipovic,[1] Pieter Gideon Van Zyl,[1] Ville Vakkari,[2] Kerneels Jaars,[1] Marcell Dunn[1] and Lauri Laakso[1,2]

[1]*Unit for Environmental Sciences and Management, North-West University, Potchefstroom Campus, Potchefstroom, South Africa;* [2]*Finnish Meteorological Institute, Helsinki, Finland*

**Corresponding author: E-mail: paul.beukes@nwu.ac.za*

Chapter Outline

1. Introduction 183
2. Measurement Motives and the Consequences Thereof 184
2.1 Air Quality versus Scientific Research Interest 184
2.2 Selection of Monitoring Devices 190
2.3 Measurement Accreditation 193
2.4 Site Selection 195
2.5 Data Logging Requirements 197
3. Data Handling, Quality and Processing 200
3.1 Working with Large Data Sets 200
3.2 Data Quality 201
3.2.1 Importance of Regular Checks and Data Transfer 201
3.2.2 Maintaining an Electronic Diary 202
3.3 Data Cleaning 203
4. Conclusions 204
References 206

1. INTRODUCTION

Anthropogenic and anthropogenically induced emissions (such as emissions due to land use change) of various gaseous and particulate atmospheric species have substantially increased since the onset of the industrial revolution. This can be illustrated by considering the atmospheric concentrations of greenhouse gases (GHGs), e.g. methane (CH_4), nitrous oxide (N_2O) and carbon dioxide (CO_2). The global average concentration of CH_4 has increased by a factor of 2.5 from approximately 722 ppb in 1750 to 1803 ppb in 2011. Over the same time period the concentration of N_2O in the atmosphere has increased by a factor of

Comprehensive Analytical Chemistry, Vol. 70. http://dx.doi.org/10.1016/bs.coac.2015.09.006

1.2. Ambient CO_2, which is considered to be the species with the highest overall positive radiative forcing effect (increasing temperature), has increased from approximately 320 ppm in 1960 to just below 400 ppm in 2010 alone [1].

In order to have a clear understanding of the sources (both natural and anthropogenic), transformations (e.g. physical, chemical and biogeochemical), transport, fates and impacts of atmospheric species, automated continuous monitoring is essential on extended spatial and temporal scales. Detailed descriptions of the fundamental principles of instrumentation applied for continuous atmospheric monitoring are readily available for both air quality (e.g. in operational manuals of the instruments and/or from the instrument suppliers) and research orientated instruments (e.g. in research publications). Therefore, rather than reviewing information that is already generally available in the public domain, this chapter will attempt to address issues that are not commonly discussed including the impact of the operational motives on the actual measurements and associated requirements, as well as aspects related to data quality and handling of large data sets.

The perspectives shared by the authors were derived mainly from the development and operation of the Welgegund atmospheric research station in South Africa (http://www.welgegund.org), which is one of the most comprehensively equipped long-term continuously operating air quality and climate change relevant atmospheric research stations in Africa [2–5], as well as from participation in international projects requiring long-term automated continuous atmospheric measurements, e.g. the European Integrated project on Aerosol, Cloud, Climate, and Air Quality Interactions (EUCAARI, European Commission 6th Framework programme project number 036833-2) [6–10], and Academy of Finland supported projects [11–19].

2. MEASUREMENT MOTIVES AND THE CONSEQUENCES THEREOF

Automated continuous atmospheric measurements can be conducted for various reasons, which require different operational philosophies. The two most common reasons are legislative requirements (e.g. measurements conducted to indicate air quality) and scientific research interest. In order to contextualise the different motives and how these influence the actual measurement philosophies applied, the different motives are briefly discussed. Additionally, the impact on issues such as the selection of monitoring devices, whether to acquire measurement accreditation or not, site selection and data logging requirements are also considered.

2.1 Air Quality versus Scientific Research Interest

Most countries have well-defined legislation for ambient air quality. The species included and the limits defined for these species in the legislation mainly depend on the balance between acceptable risk (e.g. to human health and the environment) and economic developmental goals. Although not a set rule, developed countries with strong economies would typically target lower risk and hence include a wider range of atmospheric species with stringent limits in ambient air quality legislation. In contrast, developing countries would typically preferentially target economic development and to a certain degree accept higher risk, implying fewer species and/or less stringent limits being incorporated in ambient air quality legislation and/or less stringent implementation of legislation.

The United States (US) Environmental Protection Agency (EPA) National Ambient Air Quality Standards (NAAQS) can be considered as a typical example of national ambient air quality legislation. The US EPA standards for the six principal pollutants, which are referred to as 'criteria' pollutants, are presented in Table 1 [20]. The afore-mentioned legislation differentiates between two types of standards, i.e. primary standards that provide public health protection, including protecting the health of sensitive populations such as asthmatics, children and the elderly, as well as secondary standards that provide public welfare protection, including protection against decreased visibility and damage to animals, crops, vegetation and buildings.

If ambient air quality monitoring as stipulated by legislation is the main reason to conduct automated continuous monitoring, the atmospheric species to be included in such a monitoring station or campaign are obvious, since these are dictated by the legal requirements associated with the ambient air quality legislation (e.g. as indicated in Table 1), legal requirements associated with emissions monitoring, or climate change relevant legislative standard(s). In contrast, if scientific research interest is the main purpose for conducting automated continuous monitoring the species included will be determined by the specific scientific question(s) considered. However, there are no set boundaries between motives associated with legislatively required measurements and research interest. It is actually quite common for many long-term continuously operating atmospheric research stations to also include monitoring of species related to air quality and/or climate change that could be considered as routine measurements. The Welgegund atmospheric research station in South Africa could be considered as a case study, for which the automated continuous monitoring equipment are presented in Table 2.

Almost all the species that are relevant from a South African air quality legislative perspective (i.e. SO_2, NO_2, PM_{10}, O_3, benzene as part of the VOCs, Pb as part of the aerosol trace metals and CO) are being monitored continuously or during campaigns at the Welgegund atmospheric research station. Additionally, numerous other atmospheric species and parameters are also being monitored (Table 2). The measurement suite at Welgegund was not specifically compiled to enable legislative air quality monitoring but rather to address specific South African and/or international atmospheric research needs/questions.

TABLE 1 The US EPA National Ambient Air Quality Standards for Criteria Pollutants that Differentiate between Primary and Secondary Standards. The Applicable Averaging Period, the Level and the Form of Application is also Indicated [20]. Units of Measure for the Standards are parts per million (ppm) by Volume, parts per billion (ppb) by Volume and micrograms per cubic metre of Air ($\mu g\ m^{-3}$)

Primary/ Secondary	Averaging Time	Level	Form
Carbon Monoxide (CO)			
Primary	8 h	9 ppm	Not exceeded more than once per year
	1 h	35 ppm	
Lead (Pb)			
Primary and secondary	Rolling 3 month average	0.15 $\mu g\ m^{-3}$	Not to be exceeded
Nitrogen Dioxide (NO_2)			
Primary	1 h	100 ppb	98th percentile of 1 h daily maximum concentrations, averaged over 3 years
Primary and secondary	Annual	53 ppb	Annual mean
Ozone (O_3)			
Primary and secondary	8 h	0.075 ppm	Annual fourth-highest daily maximum 8-hr concentration, averaged over 3 years
Particulate Matter ($PM_{2.5}$)			
Primary	Annual	12 $\mu g\ m^{-3}$	Annual mean, averaged over 3 years
Secondary	Annual	15 $\mu g\ m^{-3}$	Annual mean, averaged over 3 years
Primary and secondary	24 h	35 $\mu g\ m^{-3}$	98th percentile, averaged over 3 years
Particulate Matter (PM_{10})			
Primary and secondary	24 h	150 $\mu g\ m^{-3}$	Not to be exceeded more than once per year on average over 3 years
Sulphur Dioxide (SO_2)			
Primary	1 h	75 ppb	99th percentile of 1 h daily maximum concentration, averaged over 3 years
Secondary	3 h	0.5 ppm	Not to be exceeded more than once per year

TABLE 2 Continuously Operating Instruments and Associated Measurement Resolutions for Species Currently Measured at the Welgegund Atmospheric Research Station in South Africa

Parameter		Time Resolution	Analytical Method or Operating Principle of Installed Instrument
Automated Continuous Online Sampling and Analyses			
Aerosol number size distribution (12–840 nm)		7.5 min	Charging of aerosol particles, followed by size classification according to electrical mobility using a differential mobility particle sizer and counting with condensation particle counter
Ion number size distribution (0.8–42 nm)		5 min	Air ion spectrometer classifying ions according to their mobility transverse to air flow and electric field
PM_{10} aerosol mass		1 min	Synchronised hybrid ambient real-time particulate monitor that is a hybrid nephelometric (light scattering)/radiometric aerosol mass monitor
Aerosol light absorption		1 min	Multi-angle absorption photometry to determine aerosol absorption coefficient at a single wavelength (637 nm), from which black carbon (BC) concentration is determined
Aerosol light scattering		1 min	Three wavelength (450, 525 and 635 nm) simultaneous light scattering
Trace gasses	SO_2	1 min	SO_2 molecules excited by UV, which emit UV radiation of a specific wavelengths upon decay (returning to the lower energy state), which is proportional to concentration
	NO_x/NO	1 min	• NO: Measuring the light intensity produced from the chemiluminescence gas phase reaction between NO and O_3 • NO_x: Heated molybdenum converter that reduces NO_x to NO, which is detected as stated
	O_3	1 min	UV light absorption at specific wavelengths, which is directly related to O_3 concentration
	CO	1 min	Infrared light absorption at specific wavelengths, which is directly related to CO concentration

Continued

TABLE 2 Continuously Operating Instruments and Associated Measurement Resolutions for Species Currently Measured at the Welgegund Atmospheric Research Station in South Africa—cont'd

Parameter		Time Resolution	Analytical Method or Operating Principle of Installed Instrument
Solar radiation	Direct and reflected photosynthetic photon flux density (PPFD), direct and reflected global radiation and net radiation	10 s	Up and downward pointing sensors, i.e.: • PPFD in the spectral range of 400–700 nm • Pyranometer measuring broadband solar radiation flux density • Net radiation based on a thermopile sensor for which voltage is proportional to the net radiation
Meteorology	Temperature and relative humidity	1 min	Combined temperature and humidity sensor
	Wind speed and direction	1 min	Wind speed: Anemometer that rotates at speed proportional to the wind speed Wind direction: Potentiometer as a shaft angle transducer, enabling wind direction sensing
	Ambient pressure	1 min	Barometric pressure sensor
	Precipitation	1 min	Tip bucket (0.2 mm) mechanism that measures rain volume and intensity
Flux measurements	H_2O, CO_2 and sensible and latent heat fluxes	0.1 s raw data and calculated fluxes 30 min	Closed path infrared gas analyser. Micrometeorological data collected with a three-dimensional ultrasonic anemometer
	SO_2, NO_2, CH_4 and H_2O fluxes		Dual mid-infrared quantum cascade laser that passes multiple times through a cell to obtain long optical path, with infrared detector. Micrometeorological data collected with a three-dimensional ultrasonic anemometer
Soil measurements	Soil temperature and moisture at different depths, soil heat flux	1 min	Temperature and moisture probes
Vertical atmospheric profile		15 s	Ground mounted ceilometer lidar that measures vertical profile by laser illumination and analysing the reflected light

Automated Continuous Online Sampling and Offline Analyses			
Rain chemistry		Capturing all events	Collection with wet-only sampler and analysis by ion chromatography
Campaign-Based Automated Continuous Online Sampling and Analyses			
Aerosol chemical composition	Nonrefractory PM_1 ammonium (NH_4^+), organics, sulphate (SO_4^{2-}), nitrate (NO_3^-) and chloride (Cl^-)	30 min continuously for 1 year	Aerosol chemical specification monitor that applies flash vaporisation at 600 °C and ionisation by electron impact, followed by ion detection with mass spectrometry
Vertical column chemical composition	Column NO_2 and O_3	2 min continuously for 5 months	Sun-only (280–500 nm) spectrometer, mounted on a computer controlled sun-tracker, to retrieve total column O_3 and NO_2
Campaign-Based Automated Online Sampling, with Offline Analyses			
Volatile organic compounds		2 h 4 times per week (in duplicate) for 2.5 years	Sample collection onto adsorbent tubes, analysis by thermal desorption – gas chromatography–mass spectrometry
Aerosol organic composition	PM_1 (<1 μm), $PM_{1-2.5}$ (1–2.5 μm) and $PM_{2.5-10}$ (2.5–10 μm)	24 h once a week for 1 year	Collection on quartz filters with cascade impactor. Analysis by comprehensive gas chromatography–time of flight mass spectrometry
Aerosol organic (OC) and BC composition	$PM_{2.5}$ and $PM_{2.5-10}$	24 h every day for 2 years	Collection on quartz filters with dichotomous Partisol sampler and analysis of OC and BC aerosol fractions by thermal/optical protocol
Aerosol trace metal composition	$PM_{2.5}$ μm (<2.5μm) and $PM_{2.5-10}$	24 h every day for 1 year	Dichotomous Partisol sampling. Extraction with acidic dissolution and analysis by inductively coupled plasma-mass spectrometry

Most of the large fossil fuel consuming industries in South Africa, e.g. coal-fired power stations, coal-to-liquid fuel petrochemical operations and pyrometallurgical smelters [21,22], do not yet apply off-gas sulphur and nitrogen removal technologies, i.e. de-SO_X and de-NO_X. Therefore, the measurement of ambient SO_2 and NO_2 concentrations and fluxes, rain chemistry, as well as the aerosol chemical composition will assist in quantifying and understanding acid rain and deposition issues. The comprehensive aerosol measurements are aimed at understanding the formation of secondary aerosols that grow to cloud condensation nuclei size on almost all sunny days in the interior of South Africa [11,13,14,16–19]. Carbon fluxes, as well as associated H_2O fluxes and soil moistures indicate in what manner the ecosystem on the Highveld of South Africa, which is at least partially moisture limited, release and absorb CO_2 and H_2O. Vertical measurements were also included, since it is well known that the atmosphere over the South African Highveld is very layered, especially in the colder winter months [8,10]. Many of the measurements also focus on issues related to savannah and grassland fires [4], for which the region is well known. Due to relatively high emissions of O_3 precursor species in many source regions in South Africa and the abundance of sunshine, ambient air quality exceedances of O_3 are common. Therefore, O_3 formation is also important from a research perspective.

From the above-mentioned comparison of legislative and research motives, it is clear that these motives have a significant impact on the species included in automated continuous monitoring at a specific site or during a campaign.

2.2 Selection of Monitoring Devices

If legislative ambient air quality monitoring is the main motive to conduct automated continuous monitoring the measurement techniques applied are usually set. Other measurement techniques could be used, but results obtained with these methods must first be validated as similar to the reference method(s). Alternatively, if scientific research interest is the main motive, the choice of instrumentation is much less restrictive and is dictated to a large degree by the scientific questions/needs, as indicated in Table 2. Obviously measurements conducted for scientific interest must adhere to best practices and peer review to ensure data quality. However, the choice of instrumentation goes beyond the quality of results alone and due consideration should be given to often overlooked issues such as ease of use, availability of calibration standards and equipment, reliability and availability of backup and/or spare parts. These issues are especially important for measurement stations and/or campaigns conducted in countries or regions that are far from the suppliers and/or service agents, since instrument breakdown or maintenance that requires servicing by experts usually results in extensive gaps in data. Therefore, local service backup, availability of spare parts and the technical expertise of the station's own staff should play a significant role in instrument selection. In an effort to illustrate the criteria that must be taken into consideration when selecting monitoring instruments the measurement of NO_2 will be used as an example, since most large industries in South Africa do not yet apply de-NO_x processes and because a prominent NO_2 hotspot is evident from satellite retrievals over the South African Highveld. The tropospheric NO_2 column density of this area is comparable to that observed for central and northern Europe, eastern North America and south-east Asia [23].

NO_2 can be measured by a number of techniques. One of the most commonly internationally applied techniques utilises chemiluminescence. In these instruments, NO_2 is catalytically converted to NO on a heated molybdenum surface and then measured as NO by means of chemiluminescence during the reaction with O_3. However, it has been proven that this method not only converts NO_2 to NO but also partially converts other oxidised nitrogen compounds, e.g. nitric acid (HNO_3) and peroxyacetyl nitrate (PAN), which can cause overestimations of NO_2 of between 17% and 30% (e.g. Ref. [24]). Therefore, at sites where significant levels of other oxidised nitrogen compounds might occur, an alternative technique for measuring NO_2 might be preferential. One option would be to use an instrument that photolytically converts NO_2 to NO, followed by chemiluminescence detection of NO (e.g. Ref. [24]). Alternatively, a mid-infrared quantum cascade laser (QCL) spectrometer can also be used [25]. In this type of instrument the mid-infrared QCL passes multiple times through a cell to allow for a long optical path (typically > 50 m), with the exiting beam being focussed on an infrared detector and NO_2 concentration is obtained from the attenuation of IR radiation at a very narrow spectral band specific to NO_2 absorption.

At the Welgegund atmospheric research station, two of the above-mentioned types of NO_2 measurement instruments are used. Firstly, a QCL instrument is used for NO_2 flux measurements, which was chosen since ultrafast NO_2 analysis is required for eddy covariance flux measurements. However, this instrument is not simple to maintain and operate continuously. Secondly, the conventional instrument that catalytically converts NO_2 to NO on a heated molybdenum surface followed by chemiluminescence detection is used. This instrument was chosen since the technical expertise to service this instrument on-site was acquired by operational personnel at Welgegund, which significantly reduced the risks of large data gaps. To illustrate the importance of on-site servicing of this instrument, one of the weaknesses of the commercial instrument used at Welgegund is considered. It has been observed that the reaction cell window of the afore-mentioned instrument becomes coated over time, which significantly reduces the response time of the instrument. Figure 1 indicates a test conducted during calibration of the NO/NO_x instrument at Welgegund, after the reaction cell window was allowed to build up a significant coating. In addition to the zero calibration, the NO/NO_x instrument is usually calibrated at three different span (calibration) gas concentrations up to 50 ppb. In this example it is evident that after the 50 ppb calibration gas entered the instrument, the NO_x concentration increased to approximately

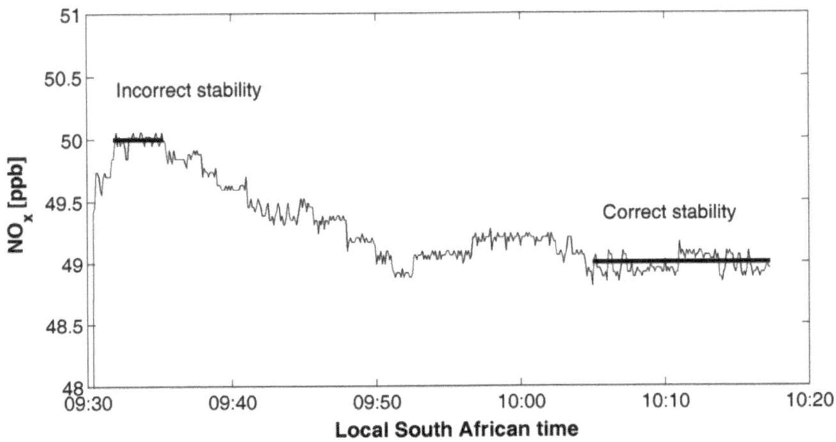

FIGURE 1 Example of 50 ppb NO_x span (or calibration) gas being recorded at Welgegund to indicate slow response due to a dirty reaction cell window. Usually NO_x measurements are conducted at 1 min resolution, but during calibration this is increased to a 6 s resolution.

50 ppb almost instantaneously. However, the concentration then slowly drifted down and only stabilised after about 35 min, at approximately 49 ppb. This slow response is due to two reasons, i.e. firstly, the afore-mentioned reaction cell window coating and, secondly, stabilisation of the gas concentration delivered by the calibration mixer that dilutes the calibration gas to the appropriate level. However, a significant portion of this slow response can be attributed to the reaction cell window being coated over time. The small concentration difference between approximately 49 and 50 ppb does not seem significant. However, the impact of this difference will depend on whether it affected the calibration slope or the offset setting of the instrument. If this difference is due to an incorrect slope setting, a relatively small percentage error will occur during normal ambient measurements, which will be within the error range allowed by most ambient air quality monitoring standards (and also within the accuracy of most calibration procedures). However, from an atmospheric science perspective the difference is not acceptable, especially if the 1 ppb underestimation is due to a constant offset of the analyser. Welgegund is a regional background site with a median NO_2 concentration below 5 ppb. Air masses arriving at the site from the west, which reflect the relatively clean regional background, have a median of approximately 1.5 ppb. Therefore, a 1 ppb offset error would cause an inaccuracy of 20–70% in the NO_x readings for most of the time. Hence, the afore-mentioned slow response and the possible effect on concentrations measured at 1 min resolution are not acceptable, especially if it is considered that the response will deteriorate even further if the instrument is not regularly serviced.

The above-mentioned reaction cell window is cleaned by operational personnel at Welgegund every 3 months to ensure acceptable response time.

Figure 2(a) indicates the coating typically observed on a dirty reaction cell window, while Figure 2(b) presents a cleaned reaction cell window. From this example, it is evident that on-site technical expertise and available service backup should receive due attention when selecting an instrument to conduct automated continuous monitoring.

2.3 Measurement Accreditation

If accreditation is required for a specific measurement method or for all the measurements conducted at a specific atmospheric measurement site, the site must be managed according to standard operating procedures (SOPs) for the relevant measurements and/or other general management aspects (e.g. change management, data storage, etc.). Accreditation for atmospheric measurements focussing specifically on relatively standard measurement methods, e.g. as applicable to ambient air quality, is recommended to ensure data credibility. However, for sites that focus on scientific research for which the instrumentation is usually not standard and/or common, accreditation is much less viable. The peer review process during scientific publication of studies based on continuous monitoring data plays an important role in ensuring data quality. Additionally, international best practices should be considered, e.g. the Global Atmosphere Watch (GAW) program under the auspices of the World Meteorological Organisation (WMO) has well-defined criteria for many measurements. Irrespective of accreditation status, following well-defined SOPs is recommended.

A common misconception is that it is sometimes assumed that accreditation alone is enough to guarantee data quality. The measurement of NO/NO_x is again considered as a practical example. At a certain measurement station, which was accredited for NO/NO_x measurement (using conventional NO_2 catalytic conversion to NO on a heated molybdenum catalyst and chemiluminescence detection), it was observed that the NO/NO_x instrument was

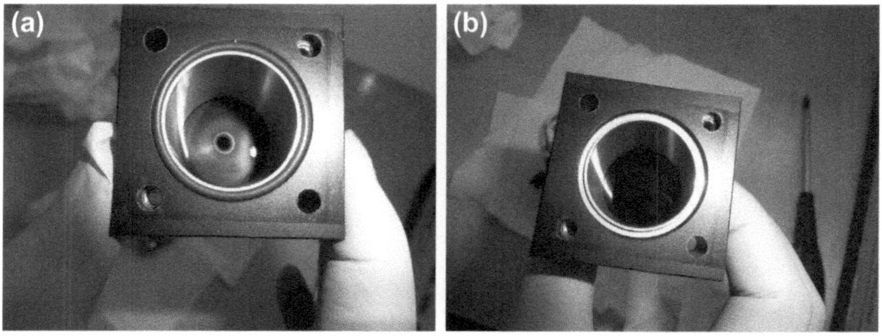

FIGURE 2 A dirty reaction cell window (note the bulls-eye effect caused by the coating) of a conventional NO/NO_x instrument (a) and a cleaned reaction cell window (b).

calibrated between 0 and 500 ppb. The annual median NO_2 concentration for this site was approximately 7 ppb. Figure 3(a) presents observations made during ambient measurements and during a zero check, while Figure 3(b) indicates a calibration span check with 500 ppb calibration gas. The first error that was made at this site is that the ambient concentrations and calibration span range do not correspond. Therefore, most of the measurements conducted will be within the noise level of the instrument. Secondly, the ambient measurements (10:30–11:00 in Figure 3(a)) indicate an offset error of at least 17 ppb, which is 2.4 times larger than the annual median NO_2 concentration. Thirdly, after introduction of the zero air (approximately 11:15–11:45 in

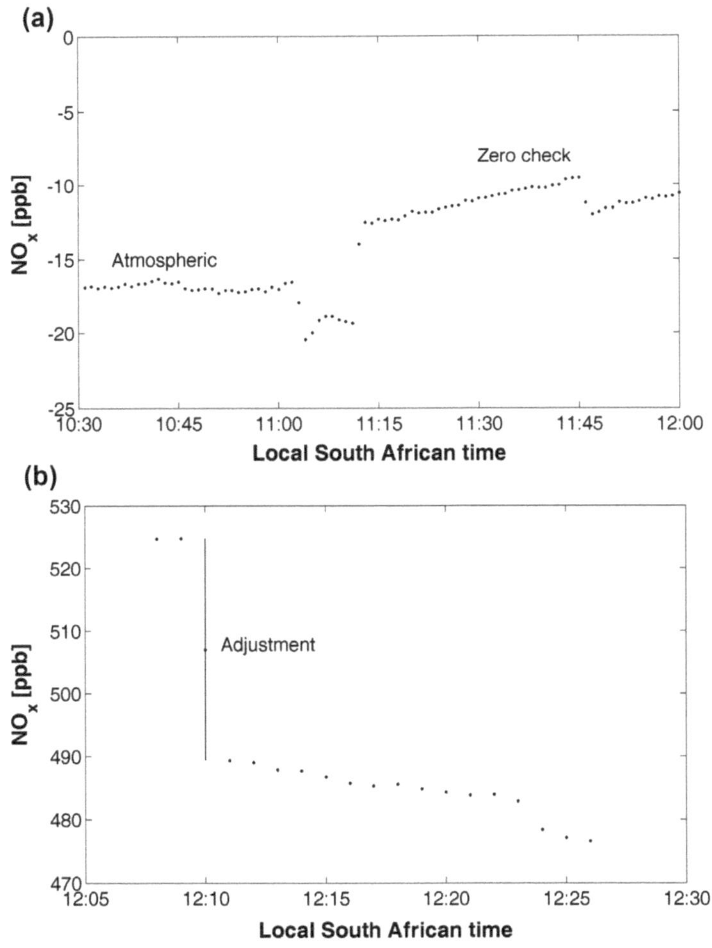

FIGURE 3 Problems associated with NO_x ambient and zero checks (a), as well as calibration at 500 ppb (b) in automated NO_x measurements.

Figure 3(a)) the NO_X concentration increased, which indicates that the zero air was not clean and contained NO_X. And lastly, 2 min after the 500 ppb span gas was introduced the span coefficient was adjusted from approximately 5% overestimation to approximately 5% underestimation, before the instrument reading had stabilised (as indicated in Figure 3(b)). After this adjustment the NO_X concentration dropped from 490 to approximately 475 ppb, indicating that the instrument suffered severely from delayed response time, as was discussed in Section 2.2 of this chapter. Notwithstanding these errors, the site operators followed the SOPs specified in the accreditation precisely but lacked insight into the fundamental analytical principles. During the above-mentioned calibration, the measurement frequency of 1 min was maintained, in contrast to the 6 s frequency that is applied at Welgegund during calibration (see Figure 1). Although not an error, the coarser measurement frequency during calibration makes it difficult to establish whether the detection limit of the instrument is correct [26], since the noise is reduced by the coarser measurement frequency.

2.4 Site Selection

The reason for conducting automated continuous monitoring will also influence site selection. For instance, if human health is the main motive (as dictated by for instance ambient air quality legislation) it stands to reason that the monitoring station should be located in an area with a high population density to ensure that measurements are representative of the pollution level exposure of the residents. However, if determining long-term regional and/or hemispherical trends is one of the main scientific objectives, as is the case with the WMO GAW program, the monitoring stations should be positioned far from local sources. In Figure 4 the locations of the WMO GAW stations are indicated on a global map, which illustrate the deliberate positioning of these sites in somewhat remote areas and/or areas that are not significantly impacted by local pollution.

Also indicated on Figure 4 are the Cape Point GAW, Welgegund and Elandsfontein (used during the EUCAARI project) stations, as well as the Johannesburg-Pretoria megacity within a regional southern African context. A number of large point sources (including coal-fired power station, petrochemical operations and pyrometallurgical smelters) in the northern interior of South Africa are also indicated on this map. Since the dominant wind directions at the Cape Point GAW station are between south-west to south-east, it is evident from Figure 4 that this station was positioned specifically to be representative of the Southern Hemisphere oceanic background [27–29], which is useful for tracking long-term trends of atmospheric species in the Southern Hemisphere. Venter et al. for instance reported a slow downward trend in observed gaseous elemental mercury at Cape Point for the measurement period 2007–2011 [27]. In contrast, the temporary EUCAARI

FIGURE 4 The locations of the WMO GAW stations indicated on a global map, as well as the Cape Point GAW, Welgegund and Elandsfontein atmospheric measurement stations in South Africa. Very large point sources (dots – pyrometallurgical smelters, triangles – coal-fired power stations and diamonds – petrochemical operations) in the northern interior of South Africa and the Johannesburg-Pretoria megacity (greyscale area) are also indicated within a regional southern African context.

monitoring station (continuous monitoring conducted for 2 years) that was positioned at Elandsfontein was in an area where nine large coal-fired power stations, as well a large petrochemical operations and several pyrometallurgical smelters occur within an approximately 70 km radius [6,7]. As previously mentioned, this area has been identified as a NO_2 hotspot [23]. Measurements conducted close to large point sources (e.g. at Elandsfontein) can be used to characterise the respective sources for model parameterisations. However, changes in operation of large point sources emitting NO_x, e.g. shutdown or commissioning of additional large point sources within the area where Elandsfontein is situated, would lead to almost instantaneous decrease or increase in measured ambient NO_x. The permanent Welgegund measurement station was positioned to be regionally representative, i.e. removed from large local sources, while still being impacted by the important source regions in South Africa [2–5]. In the section between north and south towards the west of Welgegund, there are almost no large point sources and the population density decreases towards the west. Therefore, air masses arriving at Welgegund that had passed over this sector are representative of the regional background interior of South Africa. In the sector between north and south towards the east from Welgegund, there are several source regions that are significantly impacted by anthropogenic activities, e.g. the Western Bushveld Complex with 11 pyrometallurgical smelters within a 60 km radius, the Johannesburg-Pretoria megacity with more that 10 million inhabitants, the Vaal Triangle with numerous petrochemical operations, pyrometallurgical smelters and a coal-fired power station, as well as the Mpumalanga Highveld where all the large point sources mentioned for the EUCAARI station at Elandsfontein occur [7]. In contrast to the positioning of the afore-mentioned stations, automated continuous monitoring within the Johannesburg-Pretoria megacity (indicated with the grey-scaled area in Figure 4) would mainly yield locally relevant data. In order to be representative of the larger area, a monitoring network with an adequate number of stations to make provision for substantial differences in sources over relatively small regional scales would be required for the megacity. For instance, atmospheric emissions from suburban (well established formal settlements), urban (city centre), vehicle dominated (next to a highway), urban industrial (industrial parks), as well as semi- and informal settlements, or any combination of the afore-mentioned, will differ substantially.

2.5 Data Logging Requirements

The motive for conducting automated continuous monitoring also significantly affects the type and extent of information that is logged. Measurement of aerosol black carbon (BC) is considered as an example here. According to the latest report of the Intergovernmental Panel on Climate Change the climatic effect of BC is regarded as the second most important after CO_2 [1]. However, in contrast to CO_2, the atmospheric lifetimes of aerosol particles are relatively short. Therefore, automated continuous monitoring of BC over long temporal and widespread spatial scales is required to accurately determine its potential climatic effect. In South Africa, large scale savannah and grassland fires, as well as other sources such as household combustion and industries, contribute to relatively high atmospheric BC levels [30].

Currently the most commonly used instruments for automated continuous monitoring of BC are the filter-tape-based Aethalometer [31], multi-angle absorption photometer (MAAP) [32,33] and the single-filter-based particle soot absorption photometer (e.g. [34]). However, BC cannot by definition be unequivocally measured with these instruments. Therefore, it is customary to refer to the measured carbonaceous light absorbing aerosol component as equivalent BC. For direct BC detection, more sophisticated methods are required, e.g. for instance a single particle soot photometer (SP2) based on intra-cavity laser-induced particle incandescence [35]. For the purpose of the discussions in this chapter, the term BC will be used and measurement of BC by MAAP will be discussed further.

The fundamental functioning of a MAAP [33] was recently summarised ([36] and references therein). This entails that in addition to the typical transmission measurement, signals scattered to 130° and 165° are also measured. The measured values are then used as input in a radiative model that considers the particle-loaded filter, the aerosol filter layer alone and the blank filter alone [37]. The final output of the radiative model is the single-scattering albedo and the optical depth of the aerosol loaded filter layer that matches the measured transmission and scattered signals. From these values, the absorption coefficient is calculated, by also considering the filter spot area and the air volume sampled. In addition to the afore-mentioned MAAP method that utilises the instrument's internal algorithm, the absorption coefficient can also be determined from the photodetector raw signals in a post processing procedure. However, the typical output formats of the commercial MAAP instruments do not give the raw signals as output and these have to be logged separately, if required. These raw signals include the photodetector response signal at 0° (transmittance), and signals at 130° and 165° (reflectance).

Since the impact of BC is mostly related to climatic effects, there are not many countries that include BC limits in the ambient AQ standards. Obviously BC will form part of aerosol mass, for which most countries do have ambient air quality limits in their legislation (e.g. PM_{10} and $PM_{2.5}$ limits as presented in Table 1). However, due to the importance of BC it has become quite common for ambient air quality monitoring stations to also measure it. Due to the relatively complex nature of the above-mentioned fundamental process involved in the measurement of BC with a MAAP, most ambient air quality monitoring stations only log the final calculated BC values. In contrast, measurement stations focussing on scientific research might also log some of the above-mentioned additional raw signals, as well as other parameters such as air flow and error messages. Logging of this additional, more fundamental information, recently enabled Hyvärinen et al. [36] to indicate that the commercial MAAP measurements had an artefact at high atmospheric BC loading that was related to the filter spot change, i.e. when a new blank filter spot is moved into the sampling head the observed BC concentration decreases. Since the raw signals were logged, these authors could quantify the artefact and propose a method to rectify it, thereby improving the quality of future BC measurements with the MAAP instruments. Figure 5 indicates how MAAP BC measurement at stations with high BC loading, i.e. (a) Gual Pahari (India) and (b) Welgegund (South Africa), could be corrected, as suggested by the afore-mentioned authors [36].

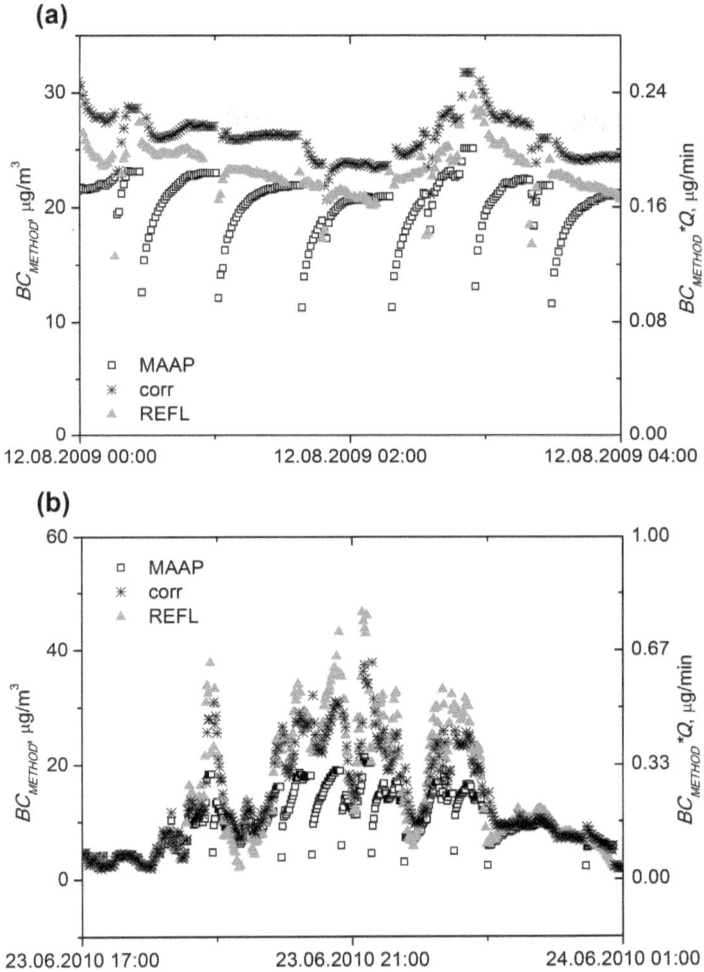

FIGURE 5 Examples of how logging of additional information, i.e. raw signals, could be used to calculate and correct a BC artefact observed at sites with high atmospheric BC loading. The normal MAAP BC (MAAP), the corrected (corr) and BC calculated from the reflectance signal are indicated for Gual Pahari, India (a) and Welgegund, South Africa (b) (from Ref. [36]). (The primary y-axis indicates BC per volume air, while the secondary y-axis indicates it per time unit). *Figures courtesy of Hyvärinen and co-authors [36].*

Logging additional raw signals, error messages and operational parameters such as cell temperatures and flows of automated continuous monitoring devices does significantly increase the load of data that has to be transferred, maintained and processed. These aspects will be discussed in the next section.

3. DATA HANDLING, QUALITY AND PROCESSING

3.1 Working with Large Data Sets

If the suite of measurements conducted at the Welgegund atmospheric research station with the associated monitoring frequencies (as presented in Table 2) is considered as an example, and it is further taken into account that additional raw signals, error messages and operational parameters (e.g. measurements cell temperatures, pressures and flows) are also logged, it becomes evident that data sets associated with such comprehensively equipped long-term research stations are large. In fact, it is challenging to process even a single day's data with a normal spread sheet program, since it becomes too laborious. Additionally, some of the spread sheet-based programs, which were not specifically designed for this type of data processing, have limitations with regard to the size of the data set that can be handled. As an example, the PM_1 single-scatter albedo (SSA), which is an indication of the direct climatic effect of aerosols, determined at a specific site is considered.

PM_1 SSA was determined from aerosol scattering and absorption measurements for approximately 6 years, together with numerous other parameters. These measurements were converted into 30 min averages, which then formed a data set with >105,000 data lines and tens of columns for the different species considered. However, the analysts handled the data set in a normal spread sheet program and therefore decided to further reduce the length of the data set by applying larger averaging periods. From this compacted data set, it was concluded that the SSA had a downward trend over the 6 year period. For SSA, a value of one implies that all particle extinction is due to scattering, while an SSA of zero implies that all extinction is due to absorption. Therefore, the afore-mentioned downward trend implied an increased contribution from absorbing species, such as BC that implies climate warming, which could be a very important deduction. However, replotting the original 30 min average data set with an appropriate mathematical/statistical programming language program, without additional averaging, resulted in a different observation. This simple scatter plot, with the associated trend line, is presented in Figure 6. As is evident, the data set indicated three distinct periods. During approximately the first 2 years, the SSA was very scattered with numerous, probably erroneously (the values should between 0 and 1), too high values. During the next approximately 1.5 years the situation improved somewhat with maximum SSA values being roughly 1.2. For the last 2.5 years of measurements, the SSA values were in the correct range. This example indicates the importance of considering all the data, since processing that make it easier to handle in normal spread sheet programs may possibly hide incorrect data that could significantly influence deductions. It is therefore critical to handle these large data sets associated with comprehensively equipped long-term research stations with the appropriate data processing

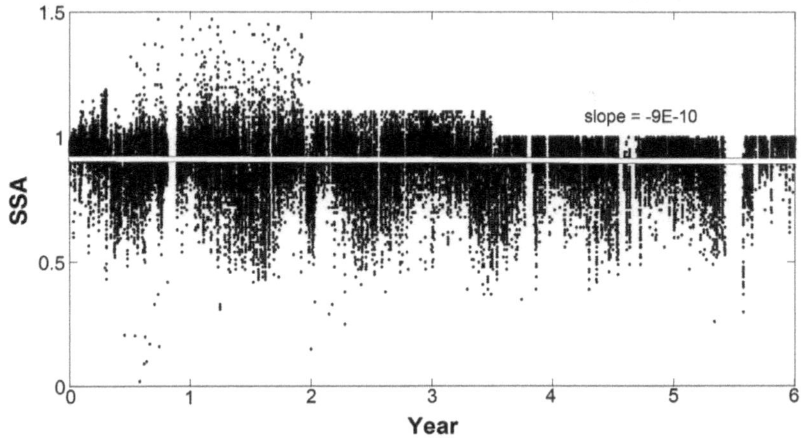

FIGURE 6 Six year PM_1 SSA indicating an apparent downward trend (white line, with slope indicated), which was an artefact, caused by the inability to work with large data sets.

tools. Such programs will also enable the identification of trends and structures within the data with the application of statistical methods such as principal component analysis, clustering and multiple-linear regression analysis (e.g. [27]).

3.2 Data Quality

3.2.1 Importance of Regular Checks and Data Transfer

Many aspects can be considered as critical to produce quality data during automated continuous monitoring of atmospheric species. Two relatively simple aspects, i.e. regular inspection of the instruments and data, as well as the ability to immediately act on errors/concerns, have proven to be vital in the operation of the Welgegund atmospheric research station. Currently, routine maintenance is performed weekly, according to set SOPs, as well as to address *ad hoc* identified issues. These scheduled visits take place, irrelevant if there were any problems or not. Additionally, every night the entire data set from Welgegund is transferred with third generation mobile telecommunications technology (3G) to a server. Data transfer once a day with 3G is used, instead of more regular transfer since the site is somewhat remote and better connectivity is not yet available. This daily data are then visualised with an appropriate processing program and reviewed the next morning to identify problems/concerns. Diagnostic data such as instrument temperatures, pressures, flows, relative humidity and error messages are also checked, since these are very important in ensuring proper instrument function and therefore data quality.

Another tool used to ensure data quality is an appropriate live-feed viewer that allows viewing diagnostic and signal data as they are being generated by instruments. An example screenshot demonstrating how actual live flux data, radiation, as well as soil temperature and moisture can be viewed remotely is presented in Figure 7.

If any data and/or instrumental operational errors/concerns are identified with either the daily data overview or live-feed viewing, the Welgegund research station is visited as soon as possible. The contributions that effective data transfer, daily data evaluation, live-feed viewing, the ability to immediately visit the site if issues are identified and the existence of an instrument workshop to assist with more technical issues cannot be underestimated.

3.2.2 Maintaining an Electronic Diary

Keeping of an electronic diary wherein notes and observations are made is also important. The Welgegund electronic diary is transferred together with all the other data daily to a server to prevent the loss of information logged in it. Every site visit is logged in the diary with the arrival and departure times, as well as the names of personnel visiting the site being logged (Figure 8). The

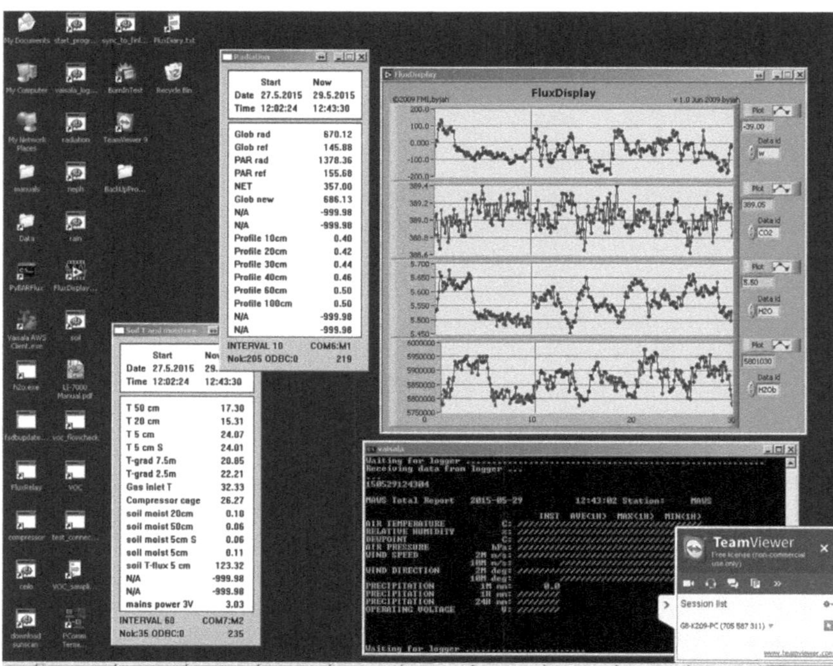

FIGURE 7 Example of real-time view of data being recorded by automated continuous monitoring devices at Welgegund.

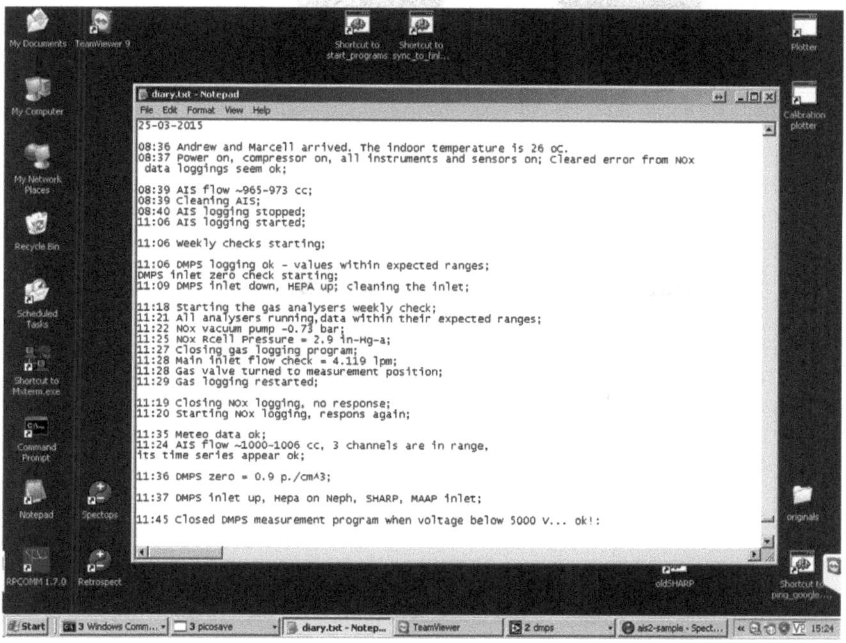

FIGURE 8 Example of a diary file recorded at Welgegund.

personnel names are just as important as the times, especially if the site is maintained by several personnel. If data/instrument abnormalities are identified later, the appropriate person can be consulted to resolve the matter. Secondly, all activities with associated time stamps are recorded, even routine checks. Thirdly, *ad hoc* procedures are described in detail, e.g. the installation of a new systems or casting of a cement floor. Abnormal observations or events of interest, e.g. nearby savannah or grassland fires, are also logged since these are vital in understanding abnormally high/low concentrations of species monitored. Even simple diagnostics such as indoor air temperature and operation of the air conditioning units are logged. Additionally, the electronic diary also serves as an institutional memory of how to solve problems that had occurred in the past.

3.3 Data Cleaning

Data cleaning, or at least verification of data quality, of a large data set acquired during the operation of a comprehensively equipped long-term atmospheric station conducting automated continuous monitoring is a topic that is too broad to attempt to cover within the scope of this chapter. However, it will briefly be illustrated how the electronic diary is used to enhance data quality and how power outages are handled.

As indicated in the Section 3.2.2, all site visits to the Welgegund atmospheric research station are logged in an electronic diary. Once a year when the data is cleaned, the arrival and departure times of site visits during which work was performed that could have affected the measurements are transferred from the electronic diary into a simple data file. This data file is then read into an appropriate data processing program and all the data collected during these site visits are cut out of the data set to ensure that the interferences caused by the site visits are not included in the data set. Additionally, the artefacts caused by power failures are also removed from the data set. Obviously all instruments are protected by operating through uninterrupted power supplies (UPSs) but these are mainly to ensure safe shutdowns and cannot be used to operate the instruments for significant periods of time. As an example, the NO_X concentration and the voltage of the station's power indicator are presented in Figure 9(a). In this figure, every spike in the NO_X concentration indicates a power failure. In Figure 9(b) one such power outage, indicated with the small box in Figure 9(a), is enlarged. As is evident from this example, just after a power outage the NO_X concentration decreases. This occurs since the instrument is running on the UPS at that stage, but without the pump functioning, therefore at no flow rate. This is then followed by a gap in the data, where after a significant spike in the NO_X data is observed after the power is restored. Therefore, it is clear that the instrument needs to stabilise after a power failure before accurate readings are achieved. During the annual data clean-up, these power-related data artefacts are also removed in a semi-automated manner from the data set to ensure that the data are not biased by such events.

4. CONCLUSIONS

In order to understand the sources, transformations, transport, fates and impacts of atmospheric species, automated continuous monitoring is vital, both on extended spatial and temporal scales. The examples presented in this chapter clearly illustrate how the motive for conducting monitoring influences the operational philosophies applied. Probably the two most common motives are legislative requirements and scientific research interest. If ambient air quality monitoring (a legislative requirement) is the main motive, the species included in monitoring and the techniques applied are usually set in the appropriate legislation. In contrast, if scientific research interest is the main motive, the species included will be determined by the scientific question(s) considered. During the selection of species to be included in monitoring and the instruments that will be used, due consideration should be given to on-site technical expertise, ease of use, reliability and availability of backup and/or spares in addition to analytical selectivity and sensitivity thereof. These often forgotten issues are especially important if the measurements will be conducted in a country or region that is far from the instrument manufacturers and/or technical experts. Following SOPs and obtaining accreditation add

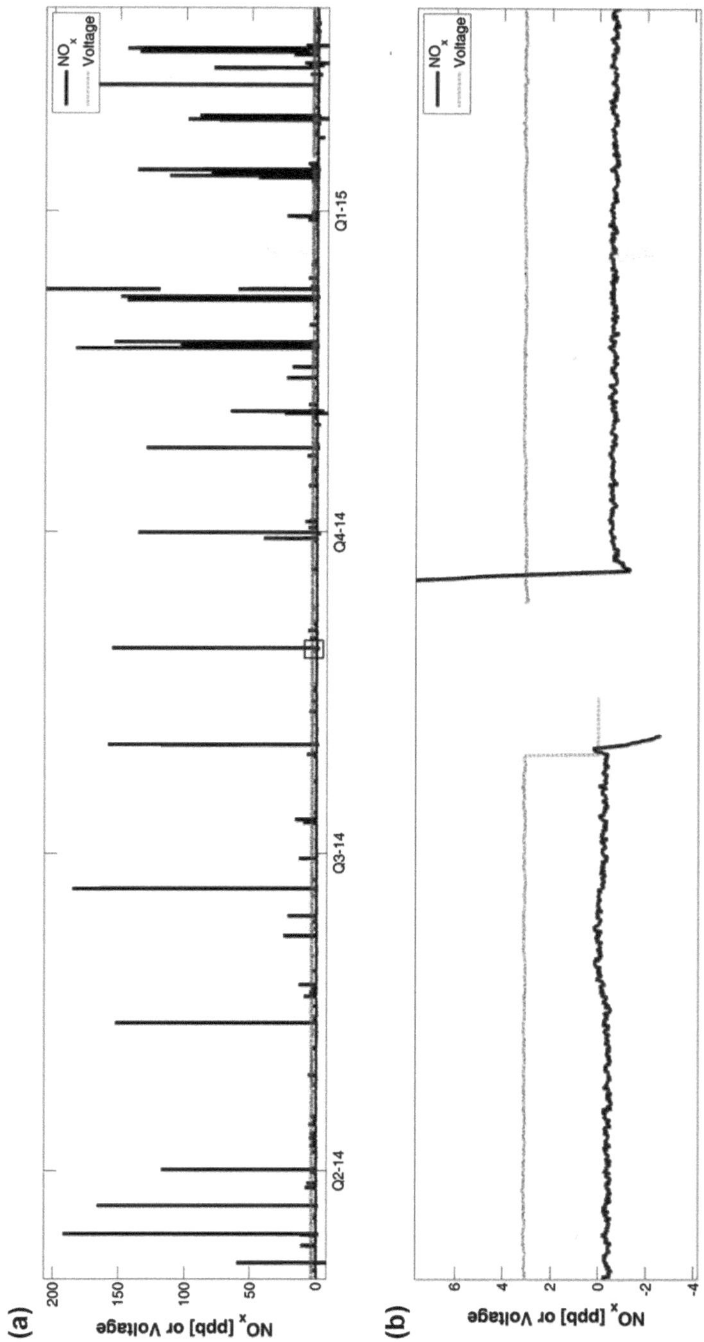

FIGURE 9 NO_x concentration and voltage of the power indicator at Welgegund, from the middle of the first quarter in 2014 to middle of the first quarter in 2015. The area indicated with the small box in (a) is enlarged in (b).

significant value, but insight into the fundamental principles of the analytical techniques applied remains critical and SOPs cannot be followed blindly. The monitoring motive also influences site selection, as well as the type and volume of data logged. Especially, continuously operating long-term research orientated measurement stations should consider logging additional raw signals, error messages and operational parameters (e.g. measurement cell temperatures, pressures and flows) in order to resolve fundamental research questions. Considering the afore-mentioned, it is evident that data sets associated with such comprehensively equipped long-term research stations are large which require effective data processing tools. Regular online and on-site checks of the instruments/data and the ability to immediately act on concerns, as well as keeping of an electronic diary and data cleaning to remove artefacts, are also essential to ensure data quality.

REFERENCES

[1] Technical summary, in: T.F. Stocker, D. Qin, G.-K. Plattner, M. Tignor, S.K. Allen, J. Boschung, A. Nauels, Y. Xia, V. Bex, P.M. Midgley (Eds.), Climate Change 2013: The Physical Science Basis. Contribution of Working Group I to the Fifth Assessment Report of the Intergovernmental Panel on Climate Change, Cambridge University Press, Cambridge, UK and New York, NY, USA, 2013.

[2] P. Tiitta, V. Vakkari, P. Croteau, J.P. Beukes, P.G. Van Zyl, M. Josipovic, A.D. Venter, K. Jaars, J.J. Pienaar, N.L. Ng, M.R. Canagaratna, J.T. Jayne, V.-M. Kerminen, H. Kokkola, M. Kulmala, A. Laaksonen, D.R. Worsnop, L. Laakso, Atmos. Chem. Phys. 14 (2014) 1909–1927, http://dx.doi.org/10.5194/acp-14-1909-2014.

[3] K. Jaars, J.P. Beukes, P.G. Van Zyl, A.D. Venter, M. Josipovic, J.J. Pienaar, V. Vakkari, H. Aaltonen, H. Laakso, M. Kulmala, P. Tiitta, A. Guenther, H. Hellén, L. Laakso, H. Hakola, Atmos. Chem. Phys. 14 (2014) 7075–7089, http://dx.doi.org/10.5194/acp-14-7075-2014.

[4] V. Vakkari, V.-M. Kerminen, J.P. Beukes, P. Tiitta, P.G. Van Zyl, M. Josipovic, A.D. Venter, K. Jaars, D.R. Worsnop, M. Kulmala, L. Laakso, Geophys. Res. Lett. 41 (2014) 2644–2651, http://dx.doi.org/10.1002/2014GL059396.

[5] W. Booyens, P.G. Van Zyl, J.P. Beukes, J. Ruiz-Jimenez, M. Kopperi, M.-L. Riekkola, M. Josipovic, A.D. Venter, K. Jaars, L. Laakso, V. Vakkari, M. Kulmala, J.J. Pienaar, J. Atmos. Chem. 72 (2015) 43–64, http://dx.doi.org/10.1007/s10874-015-9304-6.

[6] M. Kulmala, A. Asmi, H.K. Lappalainen, U. Baltensperger, J.-L. Brenguier, M.C. Facchini, H.-C. Hansson, Ø. Hov, et al., Atmos. Chem. Phys. 11 (2011) 13061–13143, http://dx.doi.org/10.5194/acp-11-13061-2011.

[7] L. Laakso, V. Vakkari, A. Virkkula, H. Laakso, J. Backman, M. Kulmala, J.P. Beukes, P.G. Van Zyl, P. Tiitta, M. Josipovic, J.J. Pienaar, K. Chiloane, S. Gilardoni, E. Vignati, A. Wiedensohler, T. Tuch, W. Birmili, S. Piketh, K. Collett, G.D. Fourie, M. Komppula, H. Lihavainen, G. de Leeuw, V.-M. Kerminen, Atmos. Chem. Phys. 12 (2012) 1847–1864, http://dx.doi.org/10.5194/acp-12-1847-2012.

[8] K. Korhonen, E. Giannakaki, T. Mielonen, A. Pfüller, L. Laakso, V. Vakkari, H. Baars, R. Engelmann, J.P. Beukes, P.G. Van Zyl, A. Ramandh, L. Ntsangwane, M. Josipovic, P. Tiitta, G. Fourie, I. Ngwana, K. Chiloane, M. Komppula, Atmos. Chem. Phys. 14 (2014) 4263–4278, http://dx.doi.org/10.5194/acp-14-4263-2014.

[9] J. Backman, A. Virkkula, V. Vakkari, J.P. Beukes, P.G. Van Zyl, M. Josipovic, S. Piketh, P. Tiitta, K. Chiloane, T. Petäjä, M. Kulmala, L. Laakso, Atmos. Meas. Tech. 7 (2014) 4285–4298, http://dx.doi.org/10.5194/amt-7-4285-2014.

[10] E. Giannakaki, A. Pfüller, K. Korhonen, T. Mielonen, L. Laakso, V. Vakkari, H. Baars, R. Engelmann, J.P. Beukes, P.G. Van Zyl, M. Josipovic, P. Tiitta, K. Chiloane, S. Piketh, H. Lihavainen, K.E.J. Lehtinen, M. Komppula, Atmos. Chem. Phys. 15 (2015) 5429–5442. http://dx.doi.org/10.5194/acp-15-5429-2015.

[11] L. Laakso, H. Laakso, P.P. Aalto, P. Keronen, T. Petäjä, T. Nieminen, T. Pohja, E. Siivola, M. Kulmala, N. Kgabi, M. Molefe, D. Mabaso, D. Phalatse, K. Pienaar, V.-M. Kerminen, Atmos. Chem. Phys. 8 (2008) 4823–4839. http://dx.doi.org/10.5194/acp-8-4823-2008.

[12] T. Petäjä, V. Vakkari, T. Pohja, T. Nieminen, H. Laakso, P.P. Aalto, P. Keronen, E. Siivola, V.-M. Kerminen, M. Kulmala, L. Laakso, Aerosol Air Qual. Res. 13 (2) (2013) 421–435. http://dx.doi.org/10.4209/aaqr.2012.08.0207.

[13] L. Laakso, J. Merikanto, V. Vakkari, H. Laakso, M. Kulmala, M. Molefe, N. Kgabi, D. Mabaso, K.S. Carslaw, D.V. Spracklen, L.A. Lee, C.L. Reddington, V.-M. Kerminen, Atmos. Chem. Phys. 13 (2013) 1957–1972. http://dx.doi.org/10.5194/acp-13-1957-2013.

[14] V. Vakkari, H. Laakso, M. Kulmala, A. Laaksonen, D. Mabaso, M. Molefe, N. Kgabi, L. Laakso, Atmos. Chem. Phys. 11 (2011) 3333–3346. http://dx.doi.org/10.5194/acp-11-3333-2011.

[15] A.D. Venter, V. Vakkari, J.P. Beukes, P.G. van Zyl, H. Laakso, D. Mabaso, P. Tiitta, M. Josipovic, M. Kulmala, J.J. Pienaar, L. Laakso, S. Afr. J. Sci. 108 (9/10) (2012). http://dx.doi.org/10.4102/sajs.v108i9/10.1059. Art. #1059, 10 pages.

[16] A. Hirsikko, V. Vakkari, P. Tiitta, H.E. Manninen, S. Gagné, H. Laakso, M. Kulmala, A. Mirme, S. Mirme, D. Mabaso, J.P. Beukes, L. Laakso, Atmos. Chem. Phys. 12 (2012) 3951–3967. http://dx.doi.org/10.5194/acp-12-3951-2012.

[17] V. Vakkari, J.P. Beukes, H. Laakso, D. Mabaso, J.J. Pienaar, M. Kulmala, L. Laakso, Atmos. Chem. Phys. 13 (2013) 1751–1770. http://dx.doi.org/10.5194/acp-13-1751-2013.

[18] A. Hirsikko, V. Vakkari, P. Tiitta, J. Hatakka, V.-M. Kerminen, A.-M. Sundström, J.P. Beukes, H.E. Manninen, M. Kulmala, L. Laakso, Atmos. Chem. Phys. 13 (2013) 5523–5532. http://dx.doi.org/10.5194/acp-13-5523-2013.

[19] R.T. Gierens, L. Laakso, D. Mogensen, V. Vakkari, J.P. Beukes, P.G. van Zyl, H. Hakola, A. Guenther, J.J. Pienaar, M. Boy, S. Afr. J. Sci. 110 (5/6) (2014). http://dx.doi.org/10.1590/sajs.2014/20130108. Art. #2013-0108, 12 pages.

[20] http://www.epa.gov/air/criteria.html, (accessed 17.08.15).

[21] A.S. Lourens, J.P. Beukes, P.G. van Zyl, G.D. Fourie, J.W. Burger, J.J. Pienaar, C.E. Read, J.H. Jordaan, S. Afr. J. Sci. 107 (1/2) (2011). http://dx.doi.org/10.4102/sajs.v107i1/2.269. Art. #269, 8 pages.

[22] K.S. Collett, S.J. Piketh, K.E. Ross, S. Afr. J. Sci. 106 (5/6) (2010). http://dx.doi.org/10.4102/sajs.v106i5/6.220. Art. #220, 9 pages.

[23] A.S.M. Lourens, T.M. Butler, J.P. Beukes, P.G. van Zyl, S. Beirle, T.K. Wagner, K.-P. Heue, J.J. Pienaar, G.D. Fourie, M.G. Lawrence, S. Afr. J. Sci. 108 (9/10) (2012). http://dx.doi.org/10.4102/sajs.v108i11/12.1146. Art. #1146, 6 pages.

[24] M. Steinbacher, C. Zellweger, B. Schwarzenbach, S. Bugmann, B. Buchmann, C. Ordóñez, A.S.H. Prevot, C. Hueglin, J. Geophys. Res. 112 (D11307) (2007) 1–13. http://dx.doi.org/10.1029/2006JD007971.

[25] D.D. Nelson, J.H. Shorter, J.B. McManus, M.S. Zahniser, Appl. Phys. B Lasers Opt. 75 (2002) 343–350. http://dx.doi.org/10.1007/s00340-002-0979-4.

[26] D.A. Skoog, D.M. West, F.J. Holler, S.R. Crouch, Fundamentals of Analytical Chemistry, ninth ed., Brooks/Cole, Cengage Learning, USA, 2014.

[27] A.D. Venter, J.P. Beukes, P.G. van Zyl, E.-G. Brunke, C. Labuschagne, F. Slemr, R. Ebinghaus, H. Kock, Atmos. Chem. Phys. 15 (2015) 10271–10280. http://dx.doi.org/10.5194/acp-15-10271-2015.

[28] E.-G. Brunke, C. Labuschagne, B. Parker, H.E. Scheel, S. Whittlestone, Atmos. Environ. 38 (2004) 5693–5702. http://dx.doi.org/10.1016/j.atmosenv.2004.04.024.

[29] F. Slemr, E.-G. Brunke, S. Whittlestone, W. Zahorowski, R. Ebinghaus, H.H. Kock, C. Labuschagne, Atmos. Chem. Phys. 13 (2013) 6421–6428. http://dx.doi.org/10.5194/acp-13-6421-2013.

[30] T.C. Bond, S.J. Doherty, D.W. Fahey, P.M. Forster, T. Berntsen, B.J. DeAngelo, M.G. Flanner, S. Ghan, B. Kärcher, D. Koch, S. Kinne, Y. Kondo, P. Quinn, M.C. Sarofim, M.G. Schultz, M. Schulz, C. Venkataraman, H. Zhang, S. Zhang, N. Bellouin, S.K. Guttikunda, P.K. Hopke, M.Z. Jacobson, J.W. Kaiser, Z. Klimont, U. Lohmann, J.P. Schwarz, D. Shindell, T. Storelvmo, S.G. Warren, C.S. Zender, J. Geophys. Res. Atmos. 118 (2013) 5380–5552. http://dx.doi.org/10.1002/jgrd.50171.

[31] A.D.A. Hansen, H. Rosen, T. Novakov, Sci. Total Environ. 36 (1984) 191–196. http://dx.doi.org/10.1016/0048-9697(84)90265-1.

[32] A. Petzold, H. Kramer, M. Schönlinner, Environ. Sci. Pollut. Res. 4 (2002) 78–82.

[33] A. Petzold, M. Schönlinner, J. Aerosol Sci. 35 (2004) 421–441. http://dx.doi.org/10.1016/j.jaerosci.2003.09.005.

[34] T.C. Bond, T.L. Anderson, D. Campbell, Aerosol Sci. Technol. 30 (1999) 582–600.

[35] M. Stephens, N. Turner, J. Sandberg, Particle identification by laser-induced incandescence in a solid-state laser cavity, Appl. Opt. 42 (2003) 3726–3736. http://dx.doi.org/10.1364/AO.42.003726.

[36] A.-P. Hyvärinen, V. Vakkari, L. Laakso, R.K. Hooda, V.P. Sharma, T.S. Panwar, J.P. Beukes, P.G. van Zyl, M. Josipovic, R.M. Garland, M.O. Andreae, U. Pöschl, A. Petzold, Atmos. Meas. Tech. 6 (2013) 81–90. http://dx.doi.org/10.5194/amt-6-81-2013.

[37] G. Hänel, Beitr. Phys. Atmos. 60 (1987) 241–247.

Chapter 7

Novel and Emerging Air-Sampling Devices

Jonathan Grandy,* Saba Asl-Hariri and Janusz Pawliszyn
Department of Chemistry, University of Waterloo, Waterloo, ON, Canada
**Corresponding author: E-mail: jgrandy22@gmail.com*

Chapter Outline

1. Introduction to Emerging Microextraction Technologies 209
2. SPME Devices and Techniques 210
2.1 Basic Fundamentals of Equilibrium-Based Microextraction Methods 211
2.2 Selection and Characteristics of SPME Sorbent Coatings for Air Sampling 212
2.3 Description of SPME Fibre Geometry and Basic Operation 215
3. Calibration of SPME for On-site Sampling 216
3.1 External Calibration of Gaseous Samples Using Portable Standard Gas Generating Vials 216
3.2 Interface Model Based on Diffusive Laws 218
3.3 Kinetic Calibration Techniques 219
4. Needle Trap Devices 221
4.1 Description of NTD Geometry and Extraction Mechanism 222
4.2 Active Sampling 227
4.3 Passive TWA-Based Sampling 228
4.4 Characterisation of Particulate Matter Using Needle Trap Sampling 229
4.5 Application of NT Sampling 231
4.5.1 Combination of SPME and NTD for Characterisation of Free and Total Analyte Distribution 233
5. Concluding Remarks 233
References 234

1. INTRODUCTION TO EMERGING MICROEXTRACTION TECHNOLOGIES

Growing global concerns with regard to the detrimental interactions between humanity and the environment have spawned a particular need to quickly and

Comprehensive Analytical Chemistry, Vol. 70. http://dx.doi.org/10.1016/bs.coac.2015.10.003

easily characterise chemical pollutants entering our atmosphere at the low parts per billion to parts per trillion (ppt) levels. Fortunately, the advent of modern analytical instrumentation, which continues to couple high resolution mass spectrometry to powerful chromatographic techniques, is making it increasingly easy to reach these goals. In particular, gas chromatography (GC) in combination with mass spectrometry detection has proven to be a staple in the modern determination of various volatile and semi-volatile organic compounds (VOCs, SVOCs) such as organic solvents, polycyclic aromatic hydrocarbons (PAHs), petroleum by-products, polychlorinated dibenzo-*p*-dioxins (PCDDs) and various other atmospheric contaminants sourced from industrial processes. However, even with the most efficient separation technologies and most powerful mass analysers, one simply cannot get any meaningful scientific data if the preceding sampling and sample preparation steps were haphazardly performed. Furthermore, it would be preferential if the chosen sampling technique avoids the use of additional organic contaminants, such as solvents, as the waste thus generated defeats the underlying purpose of environmental analysis. These are the goals of solid phase microextraction (SPME) techniques which, in contrast to some traditional sorbent trap methods, require no desorption solvent, combine sampling with sample pretreatment and can be directly coupled to analytical instrumentation with full automation [1–3].

2. SPME DEVICES AND TECHNIQUES

Indeed, since first being introduced in 1989, SPME techniques have continued to grow in popularity due to a solvent-free, easy-to-handle and miniaturised design [4,5]. The original SPME fibre employed a variety of polymeric, organic sorbents, such as polyethylene glycol (PEG), coated upon a fused silica core [5]. In the first ever reported SPME extractions, these fibres were shown to effectively extract 2-naphthol from aqueous solution in a form that could be easily injected onto GC-FID (flame ionisation detector) instrumentation for detection [5]. Although not much has fundamentally changed with regard to this initial SPME design, the modern SPME fibre is now accessible with a wide variety of sorbent materials depending on the target analyte matrix [1,4,6]. Furthermore, many distinct SPME architectures including the needle trap device (NTD), membrane extraction with a sorbent interface and the thin film microextraction membrane have since been developed with air sampling in mind [1,4,6–9]. However, regardless of the chosen architecture and sorbent phase, many of the physio-chemical properties governing said samplers remain the same. Hence, many of the fundamentals described for the classical fibre design continue to hold true for these additional configurations; therefore, the basic principles required for the application of SPME techniques towards environmental air sampling are discussed in the following section.

2.1 Basic Fundamentals of Equilibrium-Based Microextraction Methods

Much like what is observed with any other phase partitioning interface, microextraction methods are governed by a standard thermodynamic equilibrium between the analyte concentration in the sorbent phase, C_f, and that in the sample, C_s, as dictated by a distribution constant K_{fs} giving rise to Eqn (1) [1]:

$$K_{fs} = \frac{C_f}{C_s} \tag{1}$$

Furthermore, by performing simple algebraic rearrangement and knowing that the sum of the analyte extracted and that which remains in the sample at equilibrium has to equal the amount initially in the system, we can generate Eqn (2) where n is the amount of analyte extracted, C_0 is the initial concentration of the sample, K_{fs} is the fibre–sample distribution coefficient, V_s is the volume of the sample and V_f is the fibre volume [1,3,4]:

$$n = C_0 * \frac{K_{fs} V_s V_f}{K_{fs} V_f + V_s} \tag{2}$$

Favourably, when sampling of ambient air is performed on-site, it can generally be assumed that the sample volume is infinite ($V_s \ggg V_f$) such that the sample volume can be factored out and eliminated from Eqn (2) giving rise to the much simpler Eqn (3) [1,3,4]:

$$n = C_0 K_{fs} V_f \quad \rightarrow \quad C_0 = \frac{n}{K_{fs} V_f} \tag{3}$$

This simplified expression gives rise to one of the major advantages of using SPME for the atmospheric analysis of VOCs. In this way, the concentration of a given analyte may be calculated using a known K_{fs} at a given temperature as long as equilibrium has been achieved. Hence equilibrium calibration acts as a calibration curve-free method for the determination of VOCs in gaseous samples. It is worth noting, however, that one would still need a spiked analyte amount versus response curve for the analytical instrument employed (for GC analysis this can be generated by performing liquid injection of a known amount of standard).

Typically, SPME extractions are described by being in either the equilibrium or pre-equilibrium regime. Furthermore, when less than 50% of the equilibrium amount has been extracted, the extraction can still be considered as a zero sink where the kinetics of analyte uptake remain linear [1,3,4]. When in this linear regime, the rate of analyte uptake $\left(\frac{dn}{dt}\right)$ and therefore the sensitivity is dictated purely by kinetics which are in turn controlled by the surface area A, of the coating, the thickness of the stagnant, 'boundary' layer (δ) of air surrounding the coating, the static diffusion constant D_s of the analyte in this air and finally the analyte concentration in the sample. These values give rise to Eqn (4) [1,3,4]:

$$\frac{\mathrm{d}n}{\mathrm{d}t} = C_s\left(\frac{D_s A}{\delta}\right) \tag{4}$$

A wide variety of factors can greatly affect the values of both the boundary layer thickness and the distribution coefficient regardless of the SPME mode being performed. With air sampling, the two most important factors would have to be the air temperature, which alters both values, and the linear velocity of the sample (air flow rate) which shrinks the boundary layer thus resulting in faster extraction kinetics [1]. When temperature is increased, the analyte-dependant diffusion coefficient also increases thus enhancing the extraction kinetics; however, the opposite may be said about the distribution constant meaning less analyte will be sorbed at equilibrium [1,3,4]. Furthermore, a similar phenomenon is seen with regard to the molecular weight of the analyte. Generally speaking as molecular weight increases, the partial pressure of the analyte in air drops. That combined with an increasing similarity in structure to that of the polymeric coating results in an increase in the fibre constant, K_{fs}. However, larger molecules also have a much lower static diffusion coefficient meaning that the kinetics will also decrease.

These relationships can give rise to a number of difficulties when trying to optimise a microextraction method as compromises must be made between sampling kinetics and equilibrium sensitivity. Depending on the experimental goal and target analyte(s), the analyst will need to properly select an appropriate sorbent coating and, if possible, control the sample temperature and flow rate. Although these relationships may not necessarily limit SPME application, they do most certainly require a reasonable understanding of the fundamentals that govern them in order to use microextraction to its fullest.

2.2 Selection and Characteristics of SPME Sorbent Coatings for Air Sampling

As previously eluded to there are currently a wide variety of SPME sorbents available for gaseous sampling (Table 1). Generally speaking, polymeric sorbents have seen increased use in the extraction of organic analytes from otherwise complex sample matrices. Much like liquid–liquid extraction, a polymeric material can be chosen to more closely match the structure and polarity of the analyte of interest so as to increase the affinity of said analyte for the coating. The use of 100 μm thick polydimethylsiloxane (PDMS) as a sorbent has been, by far, the most widely used SPME extraction phase worldwide [1]. The nonpolar structure of PDMS makes it highly appropriate for the extraction of similarly nonpolar organic analytes [10]. Furthermore as

TABLE 1 Analyte-Dependant Selection of Solid Phase Microextraction Polymeric Sorbents for Gaseous Sampling

Sorbent Material (Coating Thickness)	Analyte Properties	Example of Analyte(s)	Extraction Mode	Comments	References
PDMS (100 μm)	Nonpolar, highly volatile	BTEX at mid to high concentration	Absorption	Most commonly used, poor sensitivity	[1,10]
PDMS (7 μm)	Nonpolar, semi-volatile	PCBs, PAHs (high concentration)	Absorption	Faster equilibrium, lowest sensitivity	[1,11]
PEG (60 μm)	Polar, broad volatility	Alcohols, carbonyls	Absorption	More common for aqueous sampling	[1,12]
DVB/PDMS (65 μm)	Bipolar semi-volatile	PCBs, PCDDs	Adsorption	Best for SVOCs	[1,13]
Car/PDMS (75 μm)	Bipolar highly volatile	THMs, BTEX	Adsorption	Strongest sorbent, difficult to desorb SVOCs	[1,14,15]
DVB/Car/PDMS (50 μm/30 μm)	Bipolar, broad volatility	Untargeted, broad-range screening	Adsorption	Multipurpose	[1,16,17]

PDMS, polydimethylsiloxane; PEG, polyethylene glycol; DVB, divinylbenzene; Car, Carboxen; BTEX, benzene, toluene, ethylbenzene, xylene; PCBs, polychlorinated biphenyls; PAHs, polycyclic aromatic hydrocarbons; PCDDs, polychlorinated dibenzo-*p*-dioxins; THMs, trihalomethanes; SVOCs, semi-volatile organic compounds.

compound molecular weight, and hence boiling point, increases, the kinetics of extraction decrease concurrently [1,4]. Hence the use of a thinner, 7 μm, PDMS coating is regarded preferential for SVOCs as it is more likely for equilibrium to be achieved with shorter sampling times, allowing for the ability to perform a simpler equilibrium-based calibration technique [1,4,11]. However pure PDMS coatings are not without their downsides. For starters their nonpolar nature makes them inappropriate for the sampling of more polar analytes such as short chain alcohols and carbonyl compounds [1,12]. For these analytes the choice of a more polar, PEG coating would be ideal [1,12]. Much like PDMS, PEG coatings are prepared by cross-linking of a relatively low molecular weight pre-polymer that gives a liquid-like sorbent where absorption is the predominant extraction mechanism. As such, these sorbents exhibit linear analyte uptake with increasing analyte concentration [1]. However, exposing these sorbents to exceedingly high levels of high-affinity VOCs (mid to high ppm level) can eventually cause the sorbents to swell and break off of the solid support. Despite their ease of use, these absorptive liquid coatings just do not offer the sensitivity and sorbent strength demonstrated by solid sorbent particles [13–17].

The use of solid sorbent particles suspended in PDMS has been shown to vastly increase the sensitivity of SPME techniques, while their bipolar nature reduces analyte selectivity resulting in much more multipurpose coatings [13–17]. This strength stems from the general porosity of the sorbent. Divinylbenzene (DVB) particles, for instance, are normally considered to be mesoporous (20–500 Å pore diameter) with a porosity of 1.54 mL g^{-1} [1]. As such, these mid-strength particles have found a particular niche for the analysis of atmospheric SVOC compounds such as PCDDs and PAHs, commonly found in fly ash from industrial smokestacks [13]. It is worth noting that DVB/PDMS fibres will still extract any analyte that has been classically analysed using PDMS alone [15]. Furthermore because of their bipolar nature, these fibres can also be applied for the sampling of carbonyls and other polar compounds. Although DVB/PDMS will also extract highly volatile compounds, those fibres prepared with the microporous (2–20 Å pore diameter), Carboxen (Car), activated carbon particles have been proven to be much more sensitive for this application [1,14,15]. Indeed, the Car/PDMS fibre is currently the strongest SPME sorbent commercially available for environmental air analysis. As such, these fibres are the best choice when a given analyst only wishes to target the most volatile pollutants such as benzene, toluene, ethylbenzene and xylenes (BTEX) from car exhaust or trihalomethanes emitted from chlorinated water [14,15]. However, the strength of the Carboxen sorbent has also been shown to have a considerable downside as well. When less volatile components are extracted it can prove very difficult, if not impossible, to thermally desorb them from the carbon particle. Hence, to account for this limitation the multifunctional DVB/Car/PDMS fibre was developed.

Commonly referred to as the 'sandwich' fibre, DVB/Car/PDMS devices are prepared such that the stronger Carboxen sorbent is at the centre of the fibre, protected within the DVB/PDMS layer [1]. This arrangement allows the higher-affinity SVOC compounds to be trapped by the weaker DVB sorbent well before it ever reaches the activated carbon at the centre of the fibre. Furthermore, any volatile component which would typically be weakly extracted by DVB can permeate through the overcoating and instead be adsorbed by Carboxen. Hence, the sandwich fibre is particularly useful when broad spectrum untargeted analysis is performed [16,17]. Such a coating would be very useful in applications where an analyst wishes to screen a range of compounds in an ambient air sample in a single analytical run.

2.3 Description of SPME Fibre Geometry and Basic Operation

As previously mentioned, of all the currently available SPME architectures, the fibre-based geometry is by far the most popular configuration worldwide. Before use, these fibres are typically placed within a plunger-operated fibre holder as shown in Figure 1. This assembly is used to move the delicate fibre coating (A) in and out of the protective stainless steel needle (B). As such, it is always important to keep the sorbent coating withdrawn into the needle unless sampling or desorption in a GC injector is being performed. Furthermore, the fibre should always be kept deep inside the protective needle anytime the assembly is pushed passed a barrier such as the septum of a GC inlet, only exposing the fibre for thermal desorption after the septum has been pierced. Although this clarification may seem obvious, it represents one of the most common errors performed by first time SPME users.

Commercial SPME fibres for GC applications are generally 2 cm long and possess a coating thickness of anywhere between 7 and 100 μm (more

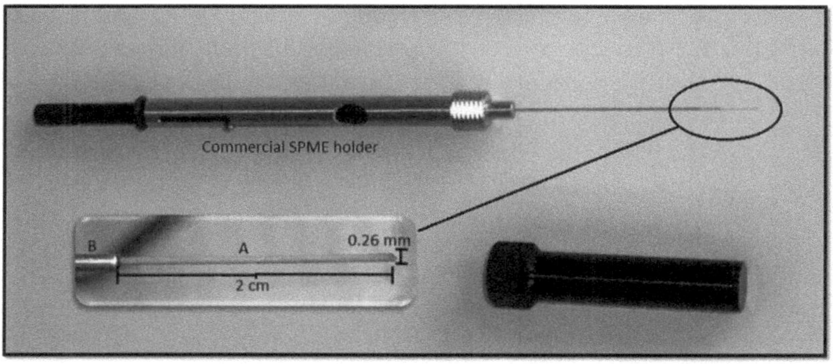

FIGURE 1 Configuration of commercially available solid phase microextraction (SPME) fibres and a conventional fibre holder, highlighting A: the exposed sorbent coating (65 μm thick DVB/PDMS) B: the protective needle sheath.

commonly between 60 and 100 μm) [1]. Using a 65 μm thick DVB/PDMS fibre on a 0.120 mm diameter fused silica core as an example, we obtain a sorbent volume of 0.880 mm^3 and a cylindrical surface area of 15.6 mm^2 [1]. These factors become exceedingly important in determining the sensitivity of a given SPME technique depending on whether the extraction is in the equilibrium regime or kinetic regime, respectively.

3. CALIBRATION OF SPME FOR ON-SITE SAMPLING

Although calibration is not required for performing untargeted, qualitative, screening analyses of atmospheric air, choosing the proper calibration method becomes imperative if one wishes to perform quantitation. Furthermore, for on-site sampling, it becomes increasingly difficult, or impossible, to control the environmental conditions such as air, temperature and velocity. In these cases, it is always necessary to measure these values such that they can be replicated back in the lab if classical calibration approaches are to be used. Furthermore, many important constants such as the fibre–sample distribution coefficient and the analyte static air diffusion coefficient are dependent on these values [1,3,4]. Moreover, extraction time and fibre desorption time must be optimised prior to going on-site so as to ensure the best analytical signal possible [1,18].

SPME techniques can be calibrated using many classical approaches including external calibration, internal calibration and internal standardisation [1,19]. However, with the exception of external calibration, none of these methods can be used for on-site air monitoring. Interestingly enough, recent advancements in standard gas/headspace generation have made it increasingly easy to perform targeted quantitation of atmospheric pollutants using external calibrants [20,21]. As previously discussed, equilibrium-based calibration can be very easily performed for any target analyte that achieves equilibrium in a reasonable time [1,19]. However, this may prove impractical for large PAHs and other heavy analytes that may require in excess of an hour to reach equilibrium. For these compounds it would indeed be preferential to use one of the available diffusion or kinetic-based calibration methods [1,19].

3.1 External Calibration of Gaseous Samples Using Portable Standard Gas Generating Vials

Of all the calibration methods available for SPME, external calibration requires the least amount of understanding of extraction fundamentals to perform properly. As long as the same extraction conditions used during calibration are replicated during the actual on-site sampling, reliable quantitation should be achievable [19]. To further ensure analytical accuracy in this case, it is recommended that the actual calibration be performed after analytical sampling. This way it can be ensured the sampling conditions such as temperature are properly matched in the laboratory. Mind you the greatest difficulty imposed when performing this type of calibration would be in standard preparation. A large number of standards would have to be prepared in aqueous solution, relying on Henry's law to determine standard gaseous concentration. Fortunately, the recent development of a highly reusable standard gas/headspace generating vial, shown in Figure 2, could facilitate streamlined external gaseous calibration.

These vials, which consist of a multiple analyte spiked, high capacity sorbent, have been shown to load a reproducible amount of VOCs and SVOCs onto DVB/PDMS/Car, SPME fibres, even after hundreds of extractions had been performed [20,21]. Moreover, where these vials can be prepared with intra and inter-batch reproducibility, it would be possible to prepare entire calibration sets for common target analytes [20,21]. Such sets would be incredibly useful for environmental laboratories in which high-throughput analyses are performed. However, the equilibrium between the sorbent phase and vial headspace is still heavily controlled by temperature [21]. This would not be a problem, however, as an entire calibration could be performed from an autosampler heater block using an SPME-enabled autosampler on GC-MS.

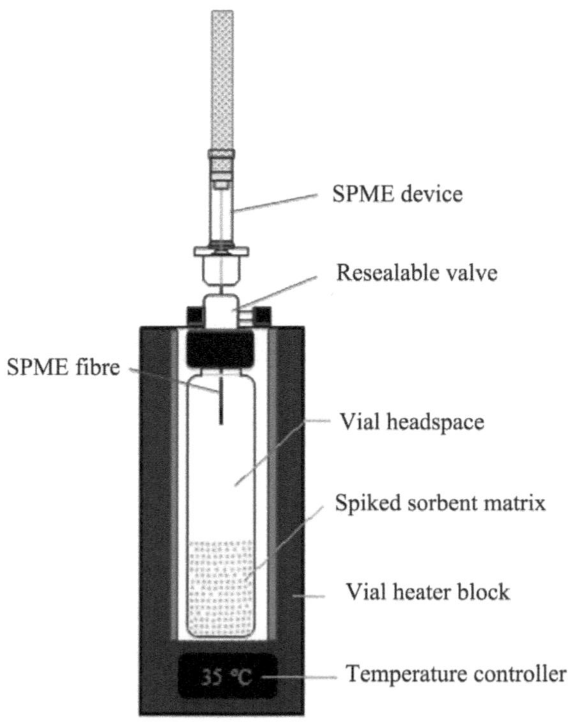

FIGURE 2 Solid phase microextraction (SPME) extraction from an in-vial standard gas generating system, thermally controlled with a portable block heater assembly.

It is important to note that as the matrix cannot be matched, external calibration will only yield the free (non-particulate bound) analyte concentrations. Furthermore, this method may still prove impractical for the quantitation of uncommon contaminants as it takes a great deal of time to prepare a batch of these gas-generating vials. For such niche analysis it may therefore be far more practical to employ a diffusive technique instead.

3.2 Interface Model Based on Diffusive Laws

The interface model is an example of a diffusion-based calibration technique (Figure 3). This method relies on careful regulation of the boundary layer thickness so as to control the extraction kinetics. As such, it is important that the extraction mode remains in the zero sink, linear regime of the equilibrium profile and that both the air temperature and velocity are constant and known [1,3,19]. Furthermore, the diffusion coefficient of the target analyte at the given temperature must be known, or derived empirically. If these three values are known, the interface model can be employed

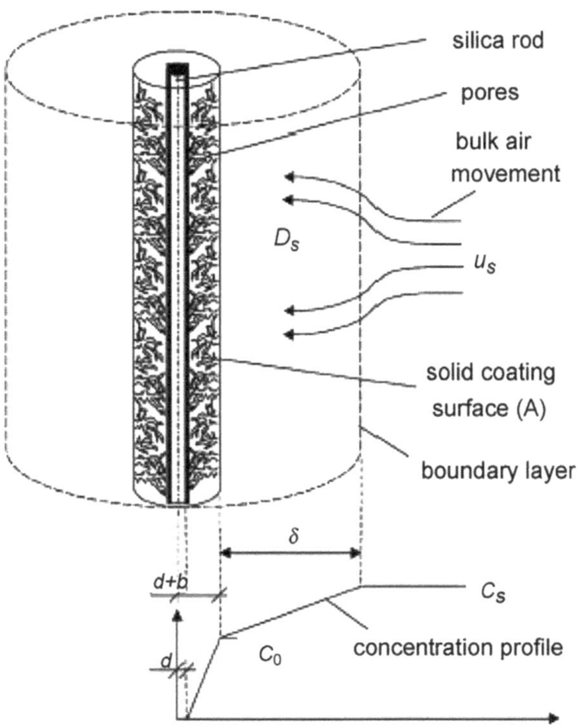

FIGURE 3 Schematic representation of the interface model of solid phase microextraction extraction with a porous fibre coating highlighting the presence and importance of the gas flow rate-dependant boundary layer thickness δ. *As reproduced from Ref. [19] Figure 5, p. 190.*

using Eqn (5), where C_s is the sample concentration, n is the absolute amount of analyte extracted, B_3 is the geometric factor of a cylinder which has a value of 3, D_s is the sample diffusion coefficient, A is the surface area of the coating, t is the sampling time and δ is the boundary layer thickness which is itself described by Eqn (6). Equation (6) further shows that the boundary layer thickness is determined by dividing the fibre diameter, d, by the Reynolds number, Re, and the Schmidt number, Sc. These constants are further described by Eqns (7) and (8) respectively with u_s representing the linear velocity of the sampled air and v signifying kinematic viscosity of the matrix (air) [1,19]:

$$C_s = \frac{n\delta}{B_3 D_s A t} \tag{5}$$

$$\delta = 9.52 \frac{d}{\mathrm{Re}^{0.62}\mathrm{Sc}^{0.38}} \tag{6}$$

$$\mathrm{Re} = 2\frac{u_s d}{v} \tag{7}$$

$$\mathrm{Sc} = \frac{v}{D_s} \tag{8}$$

Fortunately, what the interface modal lacks in mathematical simplicity is mitigated by a simplified analytical process that even allows the user to perform quantitation of an untargeted compound after it has been identified. This is because, much like equilibrium calibration, diffusive techniques remain calibration curve-free such that if a nanogram versus response relationship is known for the compound, and the prerequisite factors have been measured, quantitation may still be performed.

Another popular diffusive calibration technique is time-weighted average (TWA) sampling. However this topic is further covered in Section 4.3 and is therefore not discussed here.

3.3 Kinetic Calibration Techniques

Kinetic calibration can be performed in one of two ways. The first of these involves pre-loading a known amount of a deuterated analogue of the target compound onto the fibre prior to performing an analytical extraction. Then, when the fibre is exposed to the sample matrix, this deuterated analogue will desorb from the fibre in an inverse pattern to that of the corresponding uptake of the target analyte as described by Eqn (9) where q_0 is the initial amount of standard loaded, q is the amount of standard remaining after the extraction, n_e is the amount of analyte that would be extracted at equilibrium and n is the actual amount of analyte extracted [1,19]. This equation can be further rearranged to generate Eqn (10) thus calculating the concentration of the target analyte [1,19]:

$$\frac{n}{n_e} + \frac{q}{q_0} = 1 \tag{9}$$

$$C_0 = \frac{q_0 n}{K_{fs} V_f (q_0 - q)} \tag{10}$$

Despite the ease of use of Eqn (10), the method is rather impractical overall when actually employed. Firstly, it requires a standard be preloaded onto the fibre before the extraction; hence, it is inappropriate for untargeted analysis. Secondly, deuterated standards are exceedingly expensive and their availability is limited to a select list of compounds.

To avoid this limitation, standard-free kinetic calibration may be used instead [1,19,22]. In this method two separate extractions with times t_1 and t_2 in the linear regime are used to estimate the amount of analyte that would be extracted at equilibrium, n_e, by comparing the actual extracted amounts, n_1 and n_2,giving rise to Eqn (11) [1,19,22]:

$$\frac{t_2}{t_1} \ln\left(1 - \frac{n_1}{n_e}\right) = \ln\left(1 - \frac{n_2}{n_e}\right) \tag{11}$$

Solving for n_e in this expression is exceedingly complex; however, Ouyang et al. who originally conceptualised Eqn (11) have suggested a simpler way to solve it. By arranging it into Eqn (12) with the addition of the surrogate constant Y, it is possible to then use an Excel table to determine n_e. With all other variables known from the experiment, one can prepare a table of estimated values for n_e substituting them into the Eqn (12) to solve for Y. At whatever point Y is found to be zero will indicate the true value for n_e [22]. Alternatively it could be possible to just use Excel's goal seek function to determine n_e when Y is equal to 0. Furthermore, once n_e has been calculated it just needs to be substituted back into Eqn (3) to calculate the sample concentration [1,19,22].

$$Y = \frac{t_2}{t_1} \ln\left(1 - \frac{n_1}{n_e}\right) - \ln\left(1 - \frac{n_2}{n_e}\right) \tag{12}$$

Indeed, the use of standard-free kinetic calibration may prove to be the future of calibration for SVOCs in air. Not only is no deuterated standard required, there is an additional benefit that the only required constant is the fibre sample partitioning coefficient which can always be calculated experimentally if it is not already available in the literature. However, like all standard equilibrium-driven SPME techniques, it is only sensitive to the free analyte concentration in the sample. Fortunately by employing NTDs as well, the total analyte concentration can be determined [1,7].

4. NEEDLE TRAP DEVICES

A number of authors have reported and summarised the development and progress of needle-based extraction methods [3,6]. The relatively new five types of needle-based extraction techniques are shown in Figure 4. To date, one of the most commercially successful of these methods is SPME. In this technique, the microfibre coated with extraction phase is exposed to the sample matrix [1,23]. SPME is a solvent-free technique that combines sampling, isolation and concentration. Due to its simplicity of use, short sampling times, elimination of matrix interferences, reusability of fibres, convenient instrument introduction, easy automation and availability of a wide variety of commercial fibres [3], SPME has been widely employed in a variety of applications. Following the success of SPME, other needle extraction techniques were proposed and evaluated. For example, pieces of GC capillary columns housed inside the needle which is named in-tube or in-needle capillary adsorption trap [24,25]; the internally coated needle called solid phase dynamic extraction [26,27]; single drop microextraction [28] which describes a microlitre volume liquid phase extraction within a needle; and a packed needle with sorbent particles named the needle trap device [29,30]. As with other needle-based methods, the NTD is a solventless, one-step sample preparation technique that can be easily coupled to analytical instrumentation and be automated. In addition, this approach eliminates errors associated with sample transportation and storage, which can consequently result in more accurate and precise analytical data. This technique is also useful for a wide variety of on-site applications, including sample preparation of compounds with different chemical and physical properties, as well as varying volatilities. The major

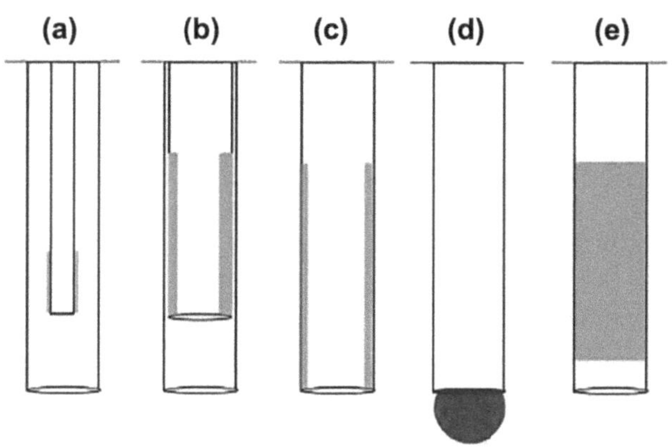

FIGURE 4 Schematics of: (a) solid phase microextraction (SPME) fibre located in a needle; (b) in-tube SPME or in-needle capillary adsorption trap (INCAT); (c) solid phase dynamic extraction (SPDE); (d) single drop microextraction (SDME); and (e) a sorbent-packed needle trap device (NTD). *As reproduced from Ref. [7] Figure 1, p. 678.*

difference between NT and other mentioned methods is its exhaustive extraction nature that can be easily calibrated. Furthermore, NT is a valuable sample preparation technique that, based on its geometry, allows it to act as a filter for collection of particles. In this chapter, the evolution and progress of the NTD as an active and passive sampler is summarised. In addition, the various applications of NTs investigated to date, as well as their filtration property, are reported and summarised.

4.1 Description of NTD Geometry and Extraction Mechanism

Generally, air samples consist of complex matrices with low levels of target analytes and the presence of several interferences [31]. As such, a sample preparation step is usually necessary to extract target analytes from complex matrices. As sample preparation is one of the main sources of uncertainty in air sample analysis [32], it is important that a given sample preparation procedure can adequately eliminate interfering compounds from the matrix. It should be noted that the common accurate sample preparation methods are time-consuming. Therefore there is increasing scientific interest in the development of accurate, cost-effective, rapid and environmentally friendly sample preparation techniques [33,34]. Commercially available procedures, which employ stainless steel containers or nylon bags, have been used for sampling of trace gaseous contaminants and aerosols in air [35] (refer to Chapter 4). However, these approaches require long sample preparation times, multiple steps, different adsorbents with high breakthrough volumes and a large amount of solvents. To address these concerns, sample preparation techniques such as SPME [1] and NTDs [36] have been developed and optimised for analysis of pollutants in air samples in order to provide solventless techniques, to automate systems, for convenient introduction to analytical instruments, and to integrate extraction steps [37]. Combining these sample preparation technologies allows for quantitative analysis of gaseous and particulate-bound compounds in atmospheric samples [38].

The use of a syringe needle packed with Tenax was first documented in the 1970s to sample airborne VOCs [36]. In 1996, a similar approach was developed for the determination of gaseous trace organic compounds in urban air and industrial atmospheres using needles packed with charcoal and silica gel [29]. However, the proposed design suffered from a major drawback due to the need for a modified inlet system for the GC instrument, which was required due to the large size of the needles used and the need for a dedicated carrier gas purge line.

The first practical and successful application of an NT suitable for automation and on-site analysis was introduced by Pawliszyn's research group in 2001 [30]. An NT device containing a 23-gauge stainless steel needle measuring 40 mm in length and containing 5 mm of quartz wool packing was used to trap particulate matter. Air samples were then drawn through the NT.

Next, the NT was placed directly into the injector of the GC, followed by injection of clean air into the injector, which aided introduction of thermally desorbed analytes to the column. In 2005, Pawliszyn and co-workers proposed two new designs of NT that afford more efficient extraction and superior packing sorbents [39]. One type of NT called Shinwa, shown in Figure 5, was designed with small side hole in its sealed conical tip in which sorbents are filled segmentally from the tip with PDMS, DVB and Car. Quartz wool is packed between the tip and side hole. Sampling is performed by drawing air through the NT via attachment of a syringe pump or gas-tight syringe. In this configuration, desorption is accomplished by applying inert gas, as thermal expansion can facilitate the desorption process. The second type of NT uses a configuration where the NT is packed with Car near the blunt tip of the NT and with the side hole positioned above the sorbent, as illustrated in Figure 6. In this design, desorption is performed by directly injecting the blunt tip of the NT into the GC injector equipped with a narrow neck liner. As can be seen in Figure 7, the restriction of the liner seals the blunt-tip NT. Subsequently, the carrier gas enters from the side hole and passes through the sorbent to assist the transfer of analytes into the column. The side hole NT desorption approach has been shown to boast a higher efficiency compared to previous NT designs and desorption methods [41]. Recently, a new prototype of NT with an extended tip (as shown in Figure 8) was introduced by Pawliszyn's research group to improve the desorption efficiency of the blunt-tip NT. In this new prototype, the shape and diameter of the tip was modified to increase the efficiency of the sealing with restriction of the narrow neck liner [40].

Various studies have been conducted to date with aims to increase the desorption efficiency of the NT devices; in these, several desorption methods

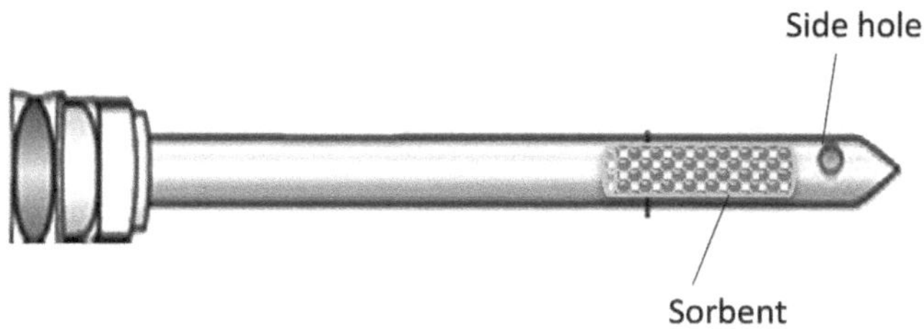

FIGURE 5 Schematic illustration of a Shinwa needle trap.

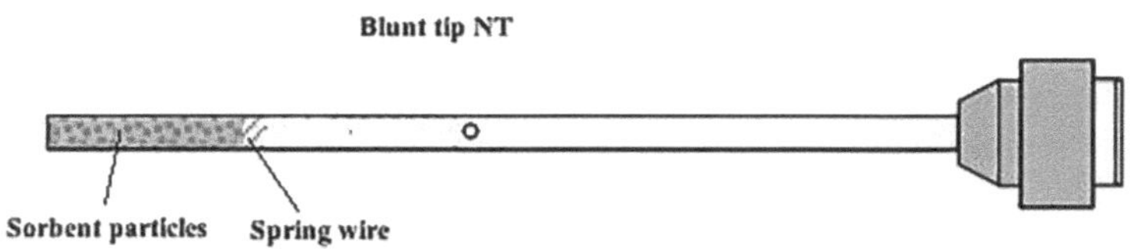

FIGURE 6 Schematic design of a blunt-tip needle trap (NT) with side hole.

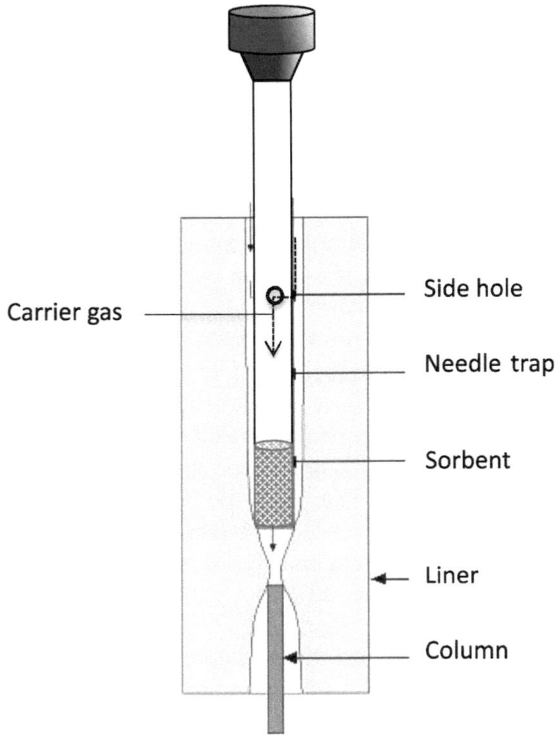

FIGURE 7 Schematic illustration of the narrow neck liner design and blunt-tip needle trap during the desorption process.

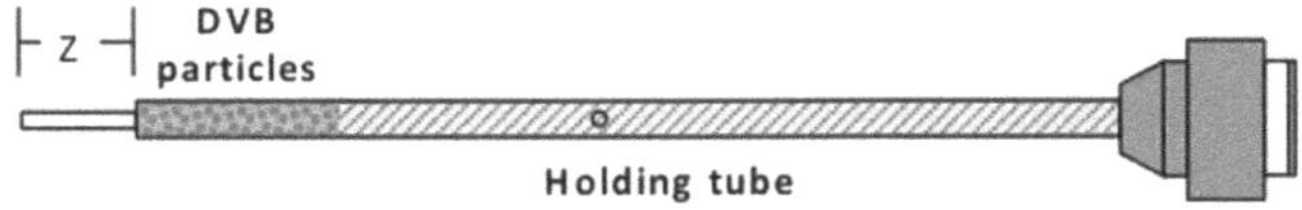

FIGURE 8 Schematic design of extended tip needle trap with side hole and holding tube for immobilisation of sorbent particles.

have been examined, including air-assisted desorption [37], thermal expansion [24], inert gas-assisted desorption [36], water vapour-assisted desorption [25] and diversion of carrier gas flow through the side hole of the NTs [39]. In air-assisted desorption, clean air is drawn through the syringe and expelled when the NT is injected into the GC. Conversely, in the thermal expansion technique, the NT is inserted into the injector, and desorption occurs based on the injector temperature profile. In related work, thermal expansion studies conducted by Wang et al. demonstrated carryover effects dependent on exposure of the sorbent to the heated zone [39]. Moreover, the design of the injector set-up and the geometry of the NT were concluded to be critical parameters in determining optimal desorption conditions. In work related to inert gas-assisted desorption, several authors have demonstrated that thermal desorption may be successfully obtained with the use of an inert gas injection through the NT device. The limitations of this technique include the introduction of oxygen through the sorbent bed due to penetration of air into the inert gas syringe and insufficient gas volumes for desorption of analytes. Studies performed by Warren et al. and Zhan et al. [40,42] demonstrated that in order to achieve complete desorption (no carryover), an aid-gas should be directed through the needle trap packing, either through carrier gas or through gas-tight assisted desorption [42]. Thus, if a good seal is created between the outer surface of the needle and the inner surface of the liner, the carrier gas is exclusively driven through the side hole of the needle, passing through the sorbent, then finally migrating alongside the extracted analytes through the needle tip. The sealing system on the first side-hole NTs relied entirely on the tapered shape of the needle's tip. However, inefficient desorption of analytes and carryover issues revealed the weakness of this design.

The efficient needle/liner prototype described a metal/metal seal between the tip of the needle and the bore of the metal liner. In this design, as shown in Figures 9 and 10, the outer diameter of the needle tip (OD 0.495 mm) fits precisely on the bottom section of the GC-liner, which has a smaller diameter (ID 0.500 mm) than the upper part of the liner. A conical guiding system allows for smooth insertion of the needle tip into the smaller section of the liner [40,42]. Since this design guarantees a better seal with the narrow neck liner, the carrier gas is forced to only pass through the sorbent bed, as seen in Figure 10. In addition, metal liners were proven to be more efficient in transferring heat evenly throughout the full length of the packing.

Several groups have worked on the development of sorbent-packed needles or similar devices [24]. Some of the sorbents that have been used for the analysis of VOCs include Carboxen (Car), DVB, Porapak Q™ and Carbopack X [19,43]. Several factors, such as pore size and shape, surface area, and particle size, can affect the ability of the analyte to access and interact with the surface of the adsorbent [44]. Moreover, because of the special shape of the needle, sorbents used for NTs must have suitable physical characteristics with regard to size, hardness and shape (spherical), as well as adequate mechanical and thermal stability. Generally, the design of the NT geometry must guarantee several factors: exhaustive extraction (active sampling), negligible breakthrough during sampling and efficient desorption [37].

The breakthrough volume of NTs, which describes the capacity of the NT, is related to the length and density of packing, the concentration of the analyte of interest and the affinity of the analyte to the sorbent [44]. In addition, the breakthrough volume is inversely proportional to the sampling rate. However, while the sampling rate and the concentration of target analyte are constant, the breakthrough volume of the NT is proportional to the sampling volume. As such, the volume of air that passes through the NT device, namely the

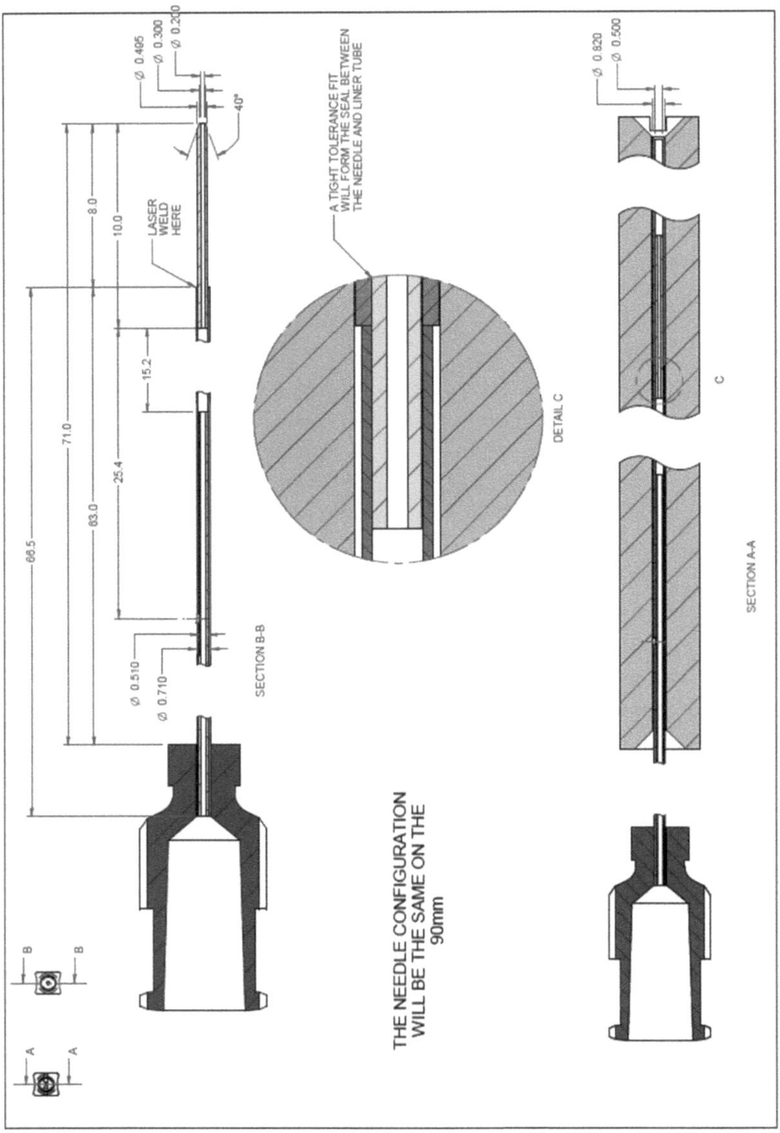

FIGURE 9 Schematic illustration of the needle trap and matching narrow neck liner.

FIGURE 10 Schematic illustration of the extended tip needle trap with narrow neck liner.

sampling volume, should be accurately measured; in other words, it is important to select an appropriate sampling volume and sampling flow rate due to the breakthrough volume. Different types of sorbent and packing lengths have specific breakthrough volumes depending on the properties of the analytes of interest and the surface area of the sorbent particles. Due to the limited surface area, the sorbent can be easily saturated at high concentrations and/or large sample volumes. However, several factors, such its simple operation, rapid analysis, and elimination of dilution or sample preparation steps, in addition to its relatively low cost, make the NT method a viable alternative for environmental sampling [45]. Moreover, the NT can be easily calibrated by controlling the sampled volume (*V*) and determining the amount extracted (*n*) via instrumental analysis, as shown in Eqn (13).

$$n = C_0 V \tag{13}$$

4.2 Active Sampling

Active sampling is performed by drawing the air sample through the sorbent bed of an NT using a pump or gas-tight syringe (Figure 11), although manual sampling with a gas-tight syringe is not reproducible enough especially for higher sampling volumes. The volume of air passed through the NT is called the sampling volume and should be accurately measured. It is important to select an appropriate sampling volume and sampling flow rate due to the breakthrough volume, as previously discussed. Different types of sorbent and packing lengths have specific breakthrough volumes depending on the

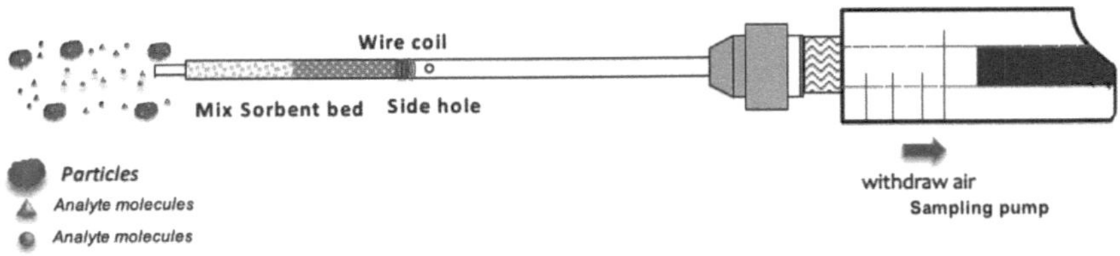

FIGURE 11 Schematic illustration of active sampling by drawing the air through the extended tip needle trap with side hole. *As adapted from Ref. [37] Figure 1, p. 5890.*

properties of analytes and surface area of the sorbent particles. It is important to emphasise that at high sampling flow rates like 20 mL min^{-1} or more, the amount of adsorbed analytes decreases because of the channelling effect and most of the analytes pass through the sorbent bed without being adsorbed.

4.3 Passive TWA-Based Sampling

Collecting of air samples can be carried out via either active or passive sampling. The second approach is achieved by the passive diffusion of the analytes over an extended period of time (Figure 12 and refer to Chapter 2 on

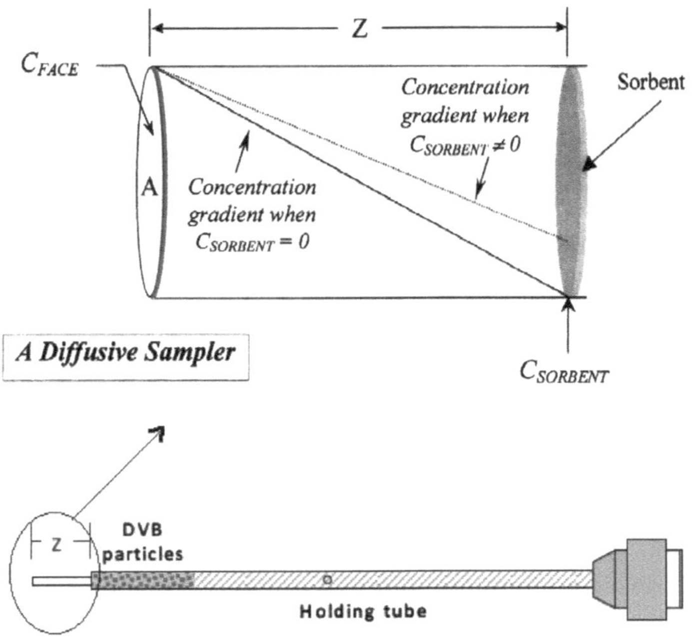

FIGURE 12 The concentration gradient of an analyte produced between the opening of the needle and of the position of the sorbent coating. *Z* is the path length and *A* is the surface area of the opening. *As reproduced from Ref. [46] Figure 2, p. 457.*

passive sampling). The basic principle of TWA sampling relies on the diffusive movement of the analyte molecules (defined by the diffusion coefficient D_s) to the tip of needle, and further to the extraction phase [1,19]. During this process, a linear concentration profile exists in a defined distance (Z) between the small opening of the needle, characterised by surface area (A), and the position of the extraction phase. The average concentration over the sampling period, C_{avg}, can then be determined by analysing the amount of loaded analyte (n), during the sampling time (t), by use of Eqn (14) which is based on Fick's first law of diffusion [1,19].

$$C_{\text{avg}} = \frac{nZ}{AD_s t} \tag{14}$$

In passive sampling, the extraction phase should remain a zero sink for a target analyte. Moreover, the concentration of the analyte in the bulk of the sample matrix and at the face of the needle opening should always be equal [1,46].

4.4 Characterisation of Particulate Matter Using Needle Trap Sampling

The collection of aerosol particles by filtration is the most common sampling method. However, the NT is able to act as a filter and trap particulate matter. To determine the particle collection efficiency of a needle trap, we have to describe the mechanical collection mechanisms. There are four mechanical mechanisms, which control aerosol trapping by NT, namely (1) interception, (2) inertial impaction, (3) diffusion and (4) gravitational settling. These mechanisms are illustrated in Figure 13. Inertial impaction occurs when a given particle, by its inertia, is unable to adjust its direction and deviates from the original gas streamline. Capture by interception occurs when a particle that follows the gas stream has insufficient inertia to depart from the original gas stream. Such particles will deposit as they come within a one-particle radius of the fibre surface. Collection of particles by interception is related to the ratio of the particle diameter to the sorbent diameter as well as the packing density. Diffusion is caused by a Brownian motion that is sufficiently strong to move the particles from the gas stream to the fibre. Typically, this mechanism is only significant for particles smaller than a few tenths of a micrometre. Contribution of gravitational settling is negligible compared to these three mechanisms. The efficiency of NT particle sampling is related to the packing density, diameter ratio of sample particles to sorbent particles, diffusion coefficient of particles and sorbent particle size. The collection efficiency of needle traps for sampling particles ranging in size from 0.01 to 50 μm with NTs packed with different size of sorbent particles has been studied. It was found that packed NT with 150 μm sorbent particles was able to trap

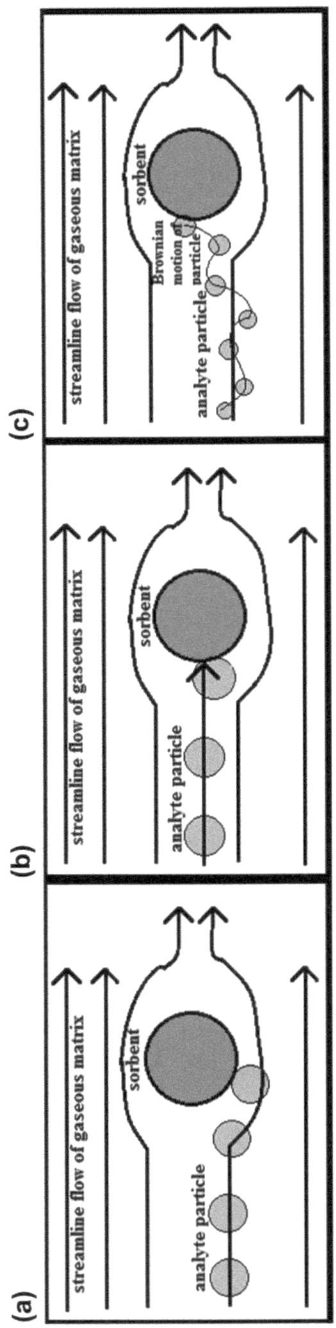

FIGURE 13 Schematic explanations of interception, impaction and diffusion onto solid sorbent particles. (a) Interception of particulate by sorbent particle; (b) Extraction of particulate by sorbent impaction; (c) Diffusion of analyte particle to sorbent.

particles larger than 0.5 μm, while the packed needle with smaller particle size could collect the smaller-sized particles more efficiently. It should be noted that the particle collection efficiency of needles packed with porous or fibrous material is significantly higher than that of granular sorbent particles, and the porosity of the frit also influences the trapping of particles.

4.5 Application of NT Sampling

Numerous applications of NTs have been introduced and are summarised in Table 2. With the development of needle trap theory, several sorbent materials and desorption methods have been proposed to extend the applications of NTs. Needle trap techniques have been widely used for the analysis of VOCs [47], biogenic volatile organic compounds [37], and breath [48], as well as for determinations of free and total concentrations of target analytes [49]. Early applications of NT studied the amount of VOCs in gaseous and water matrices. Berezkin et al., investigated the amount of benzene and toluene in tobacco smoke using the NT packed with Tenax. It was found that the results were in good agreement with other methods. Subsequently, Kubince and co-workers developed the NT packed with Porapack Q to determine the concentration of BTEX in aqueous samples. Based on the experimental results, the detection limit of NT was comparable to other methods. The work performed by Berezkin and co-workers showed that the needle trap has a high sorbent capacity and is comparable to other extraction methods like SPME and NIOSH. In 2005, Pawliszyn and co-workers proposed and developed the multi-bed sorbent for the extraction of a wide range of volatile compounds using a method based on thermal and carrier gas-assisted desorption. In addition, experimental results proved that the detection limit of NT was comparable to techniques such as purge and trap. Concurrently, Jinno's group packed NT devices with co-polymer particles to determine specific organic compounds in the gas phase [50]. Moreover, several applications have been introduced for NTs, such as analysis of formic and acetic acid [51], determination of PAHs [52] and analysis of emitted biogenic compounds from trees [53].

In addition to the mentioned applications, the use of NTs for trace gas analysis in breath has been widely documented in the literature. In 2005, Jinno and co-workers investigated acetone in breath [54]. An alveolar sampling method using NTs was later developed by Miekisch and co-workers [55]. The same group introduced the application of NTs in GC × GC characterisation of breath samples. Currently, different research groups have reported the use of NT technology towards the determination of various compounds in breath, such as acetone, isoprene and cancer biomarkers [56–58].

TABLE 2 Summary of Air Sampling Applications of Needle Trap Devices

Sorbent Type	Instrument	Application	Sampling Mode	Desorption Mode	References
Quartz wool and DVB	GC-MS	Allethrin in mosquito coil	Active	Air assisted/TD	[38]
DVB/Car	GC-MS	Tobacco smoke	Active	Carrier gas assisted/TD	[36]
DVB/Car/PDMS	GC-MS	BVOCs	Active	Carrier gas assisted/TD	[37]
PDMS/Car	GC-MS	Toluene in polymer lab	Passive	Carrier gas assisted/TD	[37]
Car 1000	GC-MS	VOC in air	Passive	Carrier gas assisted/TD	[32]
Charcoal and silica gel	GC-FID	Trace organic compounds in breath	Active	Carrier gas assisted/TD	[29]
DVB/Car	GC-MS	Breath analysis	Active	Carrier gas assisted/TD	[39]
DVB	GC-FID	BTEX in aqueous samples	Active	Thermal expansion	[31]
Carbopack X	GC-MS	Volatile breath biomarker	Active	Thermal expansion	[59]
Car	GC-FID	Volatile organic compounds	Passive	Inert gas assisted	[29]

PDMS, polydimethylsiloxane; PEG, polyethylene glycol; DVB, divinylbenzene; Car, Carboxen; BVOCs, Biogenic volatile organic compounds; BTEX, benzene, toluene, ethylbenzene, xylene; VOC, volatile organic compound.

4.5.1 Combination of SPME and NTD for Characterisation of Free and Total Analyte Distribution

One of the most interesting applications of the NT method is its combination with SPME to determine free and particulate-bound analyte concentrations. Koziel et al. [30] and Niri et al. [38] have demonstrated that NT can act both as a filter and as an extraction phase. Therefore, it is possible for NT to extract free molecules and particulate-bound analytes simultaneously. Koziel et al. packed NT with 5 mm quartz wool to investigate the amounts of triamcinolone acetonide in breath samples, as well as quantities of PAHs in diesel exhaust [30]. Niri et al. reported successful qualitative analysis of free and total concentrations of allethrin in mosquito coil smoke [38]. Both groups reported high extraction efficiencies for NT for semi-volatile compounds. This can be related to the higher distribution coefficient of semi-volatile compounds between gaseous and particulate phases compared with the low distribution coefficient of volatile compounds. While volatile compounds tend to be released in the gas phase, which can then easily diffuse to the boundary layer of the SPME fibre, the diffusion of particles through the boundary layer is much lower. A comprehensive study was performed by Li et al. on the validation of NT and SPME devices for sampling aerosols: using theoretical calculations and experimental data, the authors demonstrated that NT is capable of extracting both gaseous and particulate analytes, while SPME can only extract the gas phase molecules. Theoretical calculations were validated experimentally through determination of PAHs found in smoke emitted from barbecues and cigarettes [49]. By choosing appropriate sorbent particle size and sampling flow rates, particles with different size ranges were collected.

5. CONCLUDING REMARKS

The use of SPME techniques stands as a great choice when performing on-site sampling of environmental air as it combines sampling and sample preparation into one easy-to-perform step that requires no organic solvent. However, despite this ease of handling, SPME still requires that the user understands the fundamental principles governing equilibrium-based extractions in order to fully optimise the experimental parameters to attain optimum results. Choice of an appropriate sorbent phase for a given set of analytes cannot be understated, especially when one wishes to attain ppt level detection. Furthermore, with the exception of NTD techniques, calibration of SPME for quantitative analysis is not always as cut and dry as more classical exhaustive sampling procedures. However, this does not mean that calibration cannot be performed if the user understands the choices available to them. Indeed calibration methods such as equilibrium calibration, the interface model of diffusion-based calibration as well as more empirical approaches such as standard free kinetic calibration have all been shown to be more than appropriate for air sampling. Furthermore, if needle trap techniques are used instead, calibration becomes exceptionally easy when active extractions are performed. This simplicity for NTD air sampling stems from the concurrent trapping of airborne particles in combination with the free small organic compounds, which gives an exhaustive examination of the total amount of analyte in a gaseous matrix. Additionally, when combined with classical fibre-based SPME, it is possible to fully characterise the distribution of organic analytes between any airborne particulate matter and the gas phase of an atmospheric sample. Surely, with these factors in mind, many SPME methods will continue to emerge to assist in the analysis of ambient air.

REFERENCES

[1] J. Pawliszyn, Theory of solid-phase microextraction, in: Handbook of Solid Phase Microextraction, 2012.

[2] G. Ouyang, J. Pawliszyn, J. Chromatogr. A 1168 (2007) 226–235.

[3] J. Pawliszyn, Anal. Chem. 75 (2003) 2543–2558.

[4] B. Bojko, E. Cudjoe, G.A. Gómez-Ríos, K. Gorynski, R. Jiang, N. Reyes-Garcés, et al., Anal. Chim. Acta 750 (2012) 132–151.

[5] R. Belardi, J. Pawliszyn, Water Pollut. Res. J. Can. 24 (1) (1989) 179–191.

[6] S. Risticevic, V.H. Niri, D. Vuckovic, J. Pawliszyn, Anal. Bioanal. Chem. 393 (2009) 781–795.

[7] H.L. Lord, W. Zhan, J. Pawliszyn, Compr. Sampl. Sample Prep. 2 (2012) 677–697.

[8] A. Segal, T. Górecki, P. Mussche, J. Lips, J. Pawliszyn, J. Chromatogr. A 873 (2000).

[9] R. Jiang, J. Pawliszyn, Anal. Chem. 86 (2014) 403–410.

[10] W.-H. Cheng, W. Zhan, J. Pawliszyn, J. Chin. Chem. Soc. 60 (2013) 1027–1032.

[11] D.W. Potter, J. Pawliszyn, Environ. Sci. Technol. 28 (1994) 298–305.

[12] A. Spietelun, Talanta 87 (2011) 1–7.

[13] M.R. Criado, I.R. Pereiro, R.C. Torrijos, Talanta 63 (2004) 533–540.

[14] S. Valencia, J. Marín, G. Restrepo, Water Sci. Technol. Water Supply 13 (2013) 499.

[15] P.Y. Schüpfer, C.K. Huynh, J. Occup. Environ. Hyg. 5 (2008) 490–500.

[16] J. Zhou, Y. Han, H. Zhuang, T. Feng, B. Xu, Food Anal. Methods 8 (2014) 1661–1672.

[17] I. Fettig, S. Krüger, J.H. Deubel, M. Werrel, T. Raspe, C. Piechotta, J. Forensic Sci. 59 (3) (2014) 743–749.

[18] M. Jia, J. Koziel, J. Pawliszyn, Field Anal. Chem. Technol. 4 (2000) 73–84.

[19] G. Ouyang, J. Pawliszyn, Anal. Chim. Acta 627 (2008) 184–197.

[20] G.A. Gómez-Ríos, N. Reyes-Garcés, J. Pawliszyn, J. Sep. Sci. 36 (2013) 2939–2945.

[21] J.J. Grandy, G.A. Gómez-Ríos, J. Pawliszyn, J. Chromatogr. A 1410 (2015) 1–8.

[22] G. Ouyang, J. Sep. Sci. 31 (2008) 1167–1172.

[23] I.-Y. Eom, A.-M. Tugulea, J. Pawliszyn, J. Chromatogr. A 1196–1197 (2008) 3–9.

[24] M. McComb, Talanta 44 (1997) 2137–2143.

[25] S. Shojania, R. Oleschuk, M. McComb, H. Gesser, A. Chow, Talanta 50 (1999) 193.

[26] J. Lipinski, Fresenius. J. Anal. Chem. 369 (2001) 57–62.

[27] M.A. Jochmann, X. Yuan, T.C. Schmidt, Anal. Bioanal. Chem. 387 (2007) 2163–2174.

[28] L. Xu, C. Basheer, H.K. Lee, J. Chromatogr. A 1152 (2007) 184–192.

[29] T. Qin, Talanta 44 (1997) 1683–1690.

[30] J.A. Koziel, M. Odziemkowski, J. Pawliszyn, Anal. Chem. 73 (2001) 47–54.

[31] D. Djozan, M.H. Sorouraddin, J. Norouzi, M.A. Farajzadeh, Microchim. Acta 177 (2012) 81–88.

[32] S.S. Albaseer, K. Mukkanti, R.N. Rao, Y.V. Swamy, TrAC Trends Anal. Chem. 30 (2011).

[33] C. Bendicho, I. De La Calle, F. Pena, M. Costas, N. Cabaleiro, I. Lavilla, TrAC Trends Anal. Chem. 31 (2012) 50–60.

[34] M. Farré, S. Pérez, C. Gonçalves, M.F. Alpendurada, D. Barceló, TrAC Trends Anal. Chem. 29 (2010) 1347–1362.

[35] D.K.W. Wang, C.C. Austin, Anal. Bioanal. Chem. 386 (2006) 1089–1098.

[36] H.L. Lord, W. Zhan, J. Pawliszyn, Anal. Chim. Acta 677 (2010) 3–18.

[37] S. Asl-Hariri, G.A. Gómez-Ríos, E. Gionfriddo, P. Dawes, J. Pawliszyn, Anal. Chem. 86 (2014) 5889–5897.

[38] V.H. Niri, I.-Y. Eom, F.R. Kermani, J. Pawliszyn, J. Sep. Sci. 32 (2009) 1075–1080.

[39] A. Wang, F. Fang, J. Pawliszyn, J. Chromatogr. A 1072 (2005) 127–135.

[40] J.M. Warren, J. Pawliszyn, J. Chromatogr. A 1218 (2011) 8982–8988.

[41] I. Eom, A. Tugulea, J. Pawliszyn, J. Chromatogr. A 1197 (2008) 3–9.

[42] W. Zhan, J. Pawliszyn, J. Chromatogr. A. 1260 (2012) 54–60.

[43] J.M. Warren, D.-R. Parkinson, J. Pawliszyn, J. Agric. Food Chem. 61 (2013) 492–500.

[44] Y. Gong, I.-Y. Eom, D.-W. Lou, D. Hein, J. Pawliszyn, Anal. Chem. 80 (2008) 7275–7282.

[45] I. Eom, V.H. Niri, J. Pawliszyn, J. Chromatogr. A 1197 (2008) 10–14.

[46] A. Khaled, J. Pawliszyn, J. Chromatogr. A 892 (2000) 455–467.

[47] A. Wang, F. Fang, J. Pawliszyn, J. Chromatogr. A 1072 (2005) 127–135.

[48] I. Ueta, A. Mizuguchi, M. Okamoto, H. Sakamaki, M. Hosoe, M. Ishiguro, et al., Clin. Chim. Acta 430 (2014) 156–159.

[49] X. Li, G. Ouyang, H. Lord, J. Pawliszyn, Anal. Chem. 82 (2010) 9521–9527.

[50] Y. Saito, I. Ueta, K. Kotera, M. Ogawa, H. Wada, K. Jinno, J. Chromatogr. A 1106 (2006) 190–195.

[51] B.J. Johnson, S.C. Huang, A. Wong, L. Yao, Microchem. J. 49 (1) (1994) 78–84.

[52] M. Ogawa, Y. Saito, I. Ueta, K. Jinno, Anal. Bioanal. Chem. 388 (2007) 619–625.

[53] N. Reyes-Garcés, G.A. Gómez-Ríos, É.A. Souza Silva, J. Pawliszyn, J. Chromatogr. A. 1300 (2013) 193–198.

[54] I. Ueta, Y. Saito, M. Hosoe, M. Okamoto, H. Ohkita, S. Shirai, et al., J. Chromatogr. B Anal. Technol. Biomed. Life Sci. 877 (2009) 2551–2556.

[55] M. Mieth, J.K. Schubert, T. Gro, B. Sabel, S. Kischkel, P. Fuchs, et al., Anal. Chem. 82 (2010) 2541–2551.

[56] M. Mieth, S. Kischkel, J.K. Schubert, D. Hein, W. Miekisch, Anal. Chem. 81 (2009) 5851–5857.

[57] P. Trefz, S. Kischkel, D. Hein, E.S. James, J.K. Schubert, W. Miekisch, J. Chromatogr. A 1219 (2012) 29–38.

[58] I. Ueta, Y. Saito, M. Hosoe, M. Okamoto, H. Ohkita, S. Shirai, et al., J. Chromatogr. B 877 (2009) 2551–2556.

[59] M. Mieth, J.K. Schubert, T. Gröger, B. Sabel, S. Kischkel, P. Fuchs, et al., Anal. Chem. 82 (2010) 2541–2551.

Section III

Analysis of Air Pollutants

Chapter 8

Spectroscopic and Chromatographic Techniques and Methodologies for the Determination of Metals, Metalloids and Ions in Atmospheric Aerosols

Patricia Smichowski[1,2,*] and Darío Gómez[1]
[1]*Comisión Nacional de Energía Atómica, Gerencia Química, Buenos Aires, Argentina;* [2]*Consejo Nacional de Investigaciones Científicas y Técnicas (CONICET), Buenos Aires, Argentina*
**Corresponding author: E-mail: patricia.smichowski@gmail.com*

Chapter Outline

1. Introduction 240
2. Physical Aspects of Atmospheric Particles 241
2.1 Size Distribution 242
2.2 Morphology 243
3. Chemical Profile of Aerosols 244
3.1 Chemical Composition 244
3.2 Chemical Markers of Emission Sources 245
3.2.1 Geological Elements 245
3.2.2 Traffic-Related Elements 246
3.2.3 Marine Aerosols 253
3.2.4 Secondary Aerosols 253
4. Analytical Techniques 254
4.1 Spectroscopic Techniques for Elemental Analysis 255
4.1.1 General Considerations 255
4.1.2 Atomic Absorption Spectrometry 256
4.1.3 Atomic Fluorescence Spectroscopy 258
4.1.4 Plasma-Based Techniques 259

Comprehensive Analytical Chemistry, Vol. 70. http://dx.doi.org/10.1016/bs.coac.2015.09.007

4.1.5 X-Ray Fluorescence, Particle-Induced X-Ray Emission and Neutron-Activation Analysis 260
4.2 Ion Chromatography for Anions and Cations 261
5. Concluding Remarks 263
References 264

1. INTRODUCTION

Atmospheric chemistry, as other environmental disciplines, has benefited from the development of innovative analytical techniques and methodologies in the last decades. Nowadays there is concern, not only in the scientific community, but also in regulatory agencies and the general public about the importance of the presence of a variety of toxic and potentially toxic pollutants in atmospheric aerosols.

Atmospheric aerosol is a dispersion of solid and liquid particles suspended in air. In spite of this definition, it is a common practice to refer the term aerosol only to the particles composing the dispersion. These particles occur in various shape and size distributions depending on their emission sources and fate in the atmosphere.

Today, it is well known that problems related to human exposure, health risks, effects on biological systems, materials and building damage can be ascribed to air pollution. In connection with their optical properties, aerosols play an important role in the global climate, the radiative forcing of climate and the Earth's radiative balance.

Primary particles are emitted directly by the source while secondary ones are formed in the atmosphere. The chemical composition of aerosols usually differs for different sizes and even between particles within the same size range. For gaseous pollutants, only levels of concentration are sufficient to characterise their abundance and to specify their properties and reactions. This situation is different when airborne particles are considered because of their complexity as mixtures. In this context, the characterisation of airborne particulate matter (APM) implies a broader concept. Atmospheric particles can be characterised according to mass concentration, chemical composition (metals, metalloids, carbonaceous components, organic and ionic species), chemical form (valence or the type of molecular association), size distribution, morphology and their source types.

The potential toxicity or hazards produced in a given situation depends on a combination of a variety of factors. Some of them are related to the physical and chemical characteristics. Other factors are associated with meteorological conditions (e.g. wind-speed range) or total human exposure and health status of the population. Epidemiological studies have revealed consistent association between ambient concentrations of inhalable and respirable particles, and their elemental composition with adverse respiratory health effects. In this regard, the size and shape of particles are critical when the extent to which airborne particles penetrate the respiratory tract is considered. While coarse inhalable particles deposit in the upper respiratory tract, the fine particles can be deposited in deeper areas of the lung and the finest particles can potentially reach the alveolar region.

Considering their behaviour in the atmosphere, particles can travel long distances (e.g. transboundary transport of pollutants between regions, episodic transport of desert sand to another continent). As a result, populations are potentially exposed to metals, metalloids, ions, organic and carbonaceous compounds in airborne particles that often are well above natural background levels. Trace and ultra-trace elements are atmospheric pollutants and their determination in air particulates, and especially in the corresponding inhalable fractions, represents an important parameter in evaluating their possible implications for human health and presents a challenge for most analytical techniques for their reliable determination. Much research has focused on the determination of traffic-related elements such as Cu, Mo, Pd, Pt, Rh, Sb and Zn in atmospheric aerosols. Their determination in particulate-based environmental matrices (e.g. APM, road dust, topsoil, fly and volcanic ashes) has also become an analytical challenge, since many of these elements are present at pg m^{-3} levels.

The scenario depicted previously is indicative of the importance raised in the scientific community to gain knowledge with respect to atmospheric aerosol composition. It is not by chance that many laboratories and research groups are currently devoted to study the composition of aerosols using different analytical techniques.

Due to the diversity of the chemical composition of atmospheric aerosols, it is impractical to cover all of these in this chapter. For this reason, the discussion is restricted to key components such as ions, metals and metalloids collected using filter-based methods. The most used instrumental analytical techniques that can be employed, alone or hyphenated, that have played a crucial role in the analysis of atmospheric aerosols, are briefly reviewed. The research done in the last decades in the field of analysis of atmospheric aerosols using ion chromatography (IC), flame atomic absorption spectrometry (FAAS), electrothermal atomic absorption spectrometry (ETAAS), atomic fluorescence spectrometry (AFS), X-ray fluorescence (XRF), proton-induced X-ray emission (PIXE), neutron activation analysis (NAA) and plasma-based techniques such as inductively coupled plasma optical emission spectrometry (ICP-OES) and inductively coupled plasma−mass spectrometry (ICP-MS), is briefly described.

2. PHYSICAL ASPECTS OF ATMOSPHERIC PARTICLES

Primary particles are directly released into the atmosphere by natural sources such as sea-spray, some wildfires and wind erosion, while anthropogenic sources include stationary and mobile combustion, dust entrainment and mechanical wear of diverse materials. Secondary particles are formed by gas–solid reactions in the atmosphere such as the formation of new particles by nucleation and condensation of gaseous precursors.

The presence of both primary and secondary aerosols in different spatial and temporal scales is associated with local, regional and global environmental problems including impacts on human health associated with air quality deterioration, climate change because of direct and indirect effects on solar radiation and acidification through dry and wet deposition of sulphates and nitrates [1,2].

2.1 Size Distribution

Since APM exhibits a wide variety of shapes, a standardised length is customarily used for size characterisation. This metric, known as aerodynamic diameter (d_a), is defined as the diameter of a sphere with a density of 1 g cm^{-3} with the same settling velocity in calm air to that of the particle in question, under the prevailing conditions of temperature, pressure and relative humidity. Atmospheric particles range in size over four orders of magnitude from molecular clusters to about tens of microns in d_a.

Size distribution plays a relevant role regarding the effects of atmospheric particles on climate, health and visibility. Size distribution functions are expressed in terms of number, surface area, volume or mass of particles. Typical mass and volume distributions present a multimodal nature, which can exhibit significant variability according to the characteristics of the sampling site (e.g. remote, rural or urban) and period of interest (e.g. diurnal and seasonal trends). Early studies of size distribution functions identified two distinct modes delimited at about 2 μm [3], which serve to distinguish between fine particles or $PM_{2.5}$ ($d_a < 2.5$ μm) and coarse particles or PM_{coarse} or $PM_{2.5-10}$ (2.5 μm $\leq d_a <$ 10 μm). Origin and chemical composition of the particles composing the fine and coarse modes are normally quite different. The fine particle mode has been further subdivided into ranges containing smaller particles, namely the accumulation range (0.1 μm $\leq d_a <$ 1–2 μm) and the nuclei range ($0.01 < d_a < 0.1$ μm), this last one corresponding to the denominated ultrafine particles. More recently, particles with $d_a < 0.01$ μm have received increasing attention [4].

Coarse particulate matter is, in general, emitted as fully formed particles from natural phenomena (e.g. pollen becoming airborne, sea-spray, volcanic eruptions) and human activities (e.g. construction/demolition, dust from tyres and brake linings, fly ash and soot agglomerates from combustion, soil dust resuspension). They can also be formed by reaction of gases on pre-existing particles and by expansion with humidity of hygroscopic fine particles. Fine particles derive greatly from physical and chemical transformations in the atmosphere; in addition, they may contain primary particles, particularly those from incomplete combustion products and metal oxides from volatilised metals. Ultrafine particles are produced by condensation of hot vapours emitted from combustion and other high-temperature processes, or by spontaneous nucleation or condensation of low vapour pressure compounds on other small particles [5]. The lifetime of fine particles can be from days to weeks while the coarse fraction has a lifetime of hours to days.

2.2 Morphology

It is beyond the scope of this chapter to provide a thorough coverage of the morphological diversity of APM. This section briefly discusses the main morphological characteristics of atmospheric particles, which appear in a variety of shapes including spheres, cubes, flakes, fibres and irregular geometries. Aggregate particles are a special case of irregular particles typically composed of similarly sized particles attached together. Two main types of aggregate particles are commonly distinguished by the shape of their external envelope, being either not far from spherical (compact aggregates) or highly nonspherical aggregates (often termed fractal) [6]. Ultrafine particles are often emitted from combustion and other high-temperature processes in the form of fractal-like aggregates composed of solid nanoparticles.

Much of the elemental carbon in the atmosphere is emitted in the form of aggregates [7] being soot or black carbon, arising from incomplete combustion as an agglomerate of individual spherules. Tulmova et al. [8] examined morphological and elemental properties of fresh soot particles produced from a variety of fuels and processes. The authors observed distinct differences in morphological properties with diesel soot particles appearing as the most irregular among the particles analysed. They observed that wood combustion in the flaming phase produced fractal-like particles while the smouldering phase (i.e. the slow, low-temperature, flameless form of combustion) produced a significant number of spherical particles. Energy-dispersive spectroscopy (EDS) analysis of soot particles produced from burning wood and rice straw showed that they contained potassium, identified as a potential marker for biomass burning emissions. Freshly generated soot particles are initially hydrophobic and are unlikely to facilitate water vapour nucleation. With age, soot particles may acquire coatings such as acids, salts and organics, and the soot surface may become partially oxidised. The sulphate coatings and partial oxidation products may make the soot hydrophilic, providing soot a potential to act as cloud condensation nuclei [9].

Spherical particles arise from both natural sources (e.g. pollen) and human activities, particularly as fly ash from combustion processes. Fly ash is the mineral by-product of the combustion of fossil fuels, particularly coal and residual fuel oil. When these fuels are burned, the residual materials and various trace metals that they contain are released unchanged or thermally transformed, mostly as spherical particles in the coarse mode.

Dust particles are uniquely and irregularly shaped; they can be inhomogeneous, form agglomerates, be composed of anisotropic materials and have a preferred orientation [10]. They range in size from a few nm to ~100 μm and the size distribution is typically divided into clays ($d_a < 2$ μm), silt ($2 < da < 50$ μm) and sand ($d_a > 50$ μm). Airborne sand particles are expected to have a short residence time in the atmosphere; however, long-range transport of sand has been registered from Australian, East Asian and the Sahara deserts.

3. CHEMICAL PROFILE OF AEROSOLS

Unlike gaseous pollutants, for which only concentration is sufficient to characterise their abundance and to specify their properties and reactions, atmospheric aerosols require a broader approach to identify and determine the concentrations of its various components.

3.1 Chemical Composition

The chemical composition of primary aerosols is mainly determined by the corresponding sources while that of secondary particles depends on the formation processes. Primary anthropogenic particles are derived from stationary and mobile sources. Stationary sources include activities such as fuel combustion, industrial processes, construction and demolition, waste disposal and recycling and biomass burning. Mobile sources belong to the different transportation modes that emit particles of both exhaust and non-exhaust origin. Non-exhaust emissions from road transportation include tyre, brake and road surface wear, corrosion products and resuspension of dust from paved and unpaved roads [11].

As mentioned before, the chemical composition of aerosol varies with particle size. Fine particles, typically rich in C and sulphates, may also contain NH_4^+ and compounds of metals and metalloids such as Cd, Cu, Fe, Mn, Ni, Pb, V and Zn. The relative concentration of rare-earth elements (REEs) in fine particles is indicative of emissions from oil-fired power plants and refineries. Organic compounds with very low saturation vapour pressure at ambient temperature usually constitute a significant fraction of ultrafine particles. On the other hand, nitric acid is usually inhibited from entering fine particles already containing sulphuric acid because this acid has a lower saturation vapour pressure and a greater solubility [3]. Because of their large surface area, fine particles can act as carriers of adsorbed materials, particularly organic compounds.

Coarse particles, formed by mechanical processes, may contain crustal components such as Al, Ca, Fe, Mg, Si and Ti; base cations (Ca^{2+}, K^+, Mg^{2+}, Na^+) and anions (Cl^-, CO_3^{2-}, SO_4^{2-}). Sea salt aerosols are primarily in the coarse mode and contain predominantly Cl^- and Na^+ plus other ions such as Ca^{2+}, K^+, Mg^{2+}, Br^- and SO_4^{2-}. Gaseous HNO_3 and other gases may condense directly onto pre-existing coarse particles. Evaporated metals, especially Fe, emitted from high-temperature processes typically condense and integrate into fly ash particles. Coarse mode particles can be altered by physical and chemical interactions with gaseous or liquid contaminants [11].

3.2 Chemical Markers of Emission Sources

Chemical tracers/markers are single species or a group thereof and/or concentration ratios that are able to distinguish between APM sources. Specific trace elements, elemental/organic carbon, and distinct organic molecules have been widely used in receptor model (RM) analysis to identify sources of APM and quantify their apportionment. Contributing major, minor and trace elements present in the Earth's crust, exhaust and wear emissions from road vehicles and sea-spray that serve as markers of the most common APM sources, as well as the analytical techniques used for their determination, are discussed in the following sections. The accuracy of analytical determinations is crucial for a reliable estimation of source contributions to APM mass [12].

A range of chemical markers has been identified and employed in receptor modelling studies. Elemental tracers such as Al, Cr, Fe, Ni, Pb and Zn have been used since the inception of these models. A drawback is that a single trace element may be emitted from several sources, making the accurate assessment of the contribution of each source to APM difficult. An example is the presence of Fe in particles from natural sources (wind-driven topsoils) and from wear of steel-based materials. In the last two decades, research has also focused on the identification and development of organic molecular markers such as levoglucosan for biomass burning [12].

The source signature refers to the mix of tracer elements and/or molecular markers used for identification of sources. One of the most important steps in RM analysis is the selection of signatures and profiles that best represent the sources considered.

3.2.1 Geological Elements

Fine particles detached from the land surface by wind erosion can enter the atmosphere as resuspended dust. Wind erosion is a natural process that commonly occurs in deserts and on coastal sand dunes and beaches. Under dry weather conditions, it can also occur in agricultural regions where vegetation cover is reduced. Wind-driven dust from topsoil, paved and unpaved roads, construction/demolition activities, and long-range transport of desert sand are typically responsible for the presence of Earth's crustal materials in atmospheric aerosols, which contain a group of chemical elements, most noticeably Al, Ca, Fe, Mg, Si and Ti.

The elemental profiles of natural and anthropogenic crustal matter-based particles exhibit significant overlap, which often causes difficulties in differentiating them. However, it is in principle possible to distinguish these various sources based on the differences of their trace element profiles relative to natural soils and dusts, for example, traffic-related particles present in road dust are enriched in trace elements such as Cu, Mo, Sb (brake dust) and Zn (tyre wear) [13].

In the case of resuspended crustal contributions, it is often necessary to differentiate the natural mineral dust arising from the local topsoil, from long-range transported dust and from crustal materials derived by urban activities [14]. For this purpose, source profiles and elemental signatures of local (topsoil and road dust) and regional contributors (e.g. desert airborne dust) are extremely useful. Two examples are briefly described here.

In their characterisation of the local topsoil contribution to APM in the area of Rome (Italy), Pietrodangelo et al. [14] observed little resuspension of large mineral crystals such as feldspar, pyroxene and quartz, which preferentially include major elements like Al, Ca, Fe and Si and have a relatively large density. However, the small size mineral phases, which preferentially include transition metals, were wind-driven and contributed to the local aerosol. The authors identified that the ratios of Mg/Ca and Ti/Ca, allowed a large number of dust profiles from the desert in Africa to be differentiated from both the nearby topsoil contribution to the local PM_{10} and from the urban road dust.

In their study of aerosol concentration levels over the Tibetan Plateau in the province of Qinghai (China), Zhang et al. [15] were able to associate the set (Ca, K, Si) with local sources and the set (Al, Fe, Ti) with wind-driven material transported from remote sources. The authors based their conclusions on the relatively high proportion of coarse particles enriched in Ca, K and Si and that fine particles contained A1, Fe and Ti in near-crustal proportions.

3.2.2 Traffic-Related Elements

In cities and mega-cities (>10 million inhabitants), vehicular traffic is the main source of several polluting elements (called traffic-related elements, TREs) that can be considered as emerging contaminants. In this context, research groups all over the world have carried out studies to assess the influence of traffic on atmospheric pollution in urban areas.

During many years it was assumed that fuel combustion was the main and primary mechanism responsible for particle formation and also metal emission in this regard. Studies acknowledged that other important processes are responsible for metal emission ascribed, among others, to the wear of different parts of a vehicle. We can include as sources or potential sources: wear products from tyres, brake linings, bearings, road surface, corrosion of different components of the vehicle, and resuspension of soil and road dust. The widespread range of problems associated with the study of emissions of TREs includes the identification of the sources of such contaminants as well as of the deposition processes, traffic density, seasonal conditions, environmental contamination, transport, transformation in the atmosphere, biological activity, bioaccumulation, possible health risk and, last but not least, the challenges inherent to analytical determinations [16].

Many of the elements that are mainly released as particulate matter such as Ca, Ce, Cu, Fe, Hg, Mn, Mo, Ni, Pb, Pd, Pt, Rh, S, Sb, V and Zn have been ascribed to vehicular traffic and were detected not only in APM, but also in related matrices such as road dust, soil, grass plants growing near roads, river sediments and other matrices [17]. As an example, platinum group elements (PGEs) are an ongoing theme of study and their importance is reflected by the fact that they have been detected in alpine and polar snow, providing clear evidence of their global environmental impact [18]. Road surfaces are not free of compounds containing metals. They are generally composed of either concrete or aggregates with a bituminous binder. Abrasion of road surfaces generates particulate matter of mineral origin. The health risk associated with TREs deserves special attention for their bioaccumulation in aquatic microorganisms as they play an important role in the food chain. The following sections briefly consider some of the most studied elements such as antimony, lead, PGEs and others.

3.2.2.1 Antimony

Antimony and its compounds are natural components of the Earth's crust. In spite of this, it is well known that Sb is a potentially toxic trace element that may produce adverse effects to humans and the environment and has no known physiological functions. As other trace elements, Sb travels through the atmosphere as part of its global biogeochemistry cycling processes. After being emitted to the atmosphere, Sb, as with other trace elements, is transported long distances at both local and regional scales. The presence of atmospheric Sb is mainly due to anthropogenic inputs, the contribution of which is 100- to 200-fold higher than its emission from natural sources [16]. It is believed that when Sb is released into the atmosphere as an aerosol, it is oxidised to Sb_2O_3 by reaction with airborne oxidants [16]. More recently, Sb was related to traffic and identified as a TRE. Different parts of vehicles contain Sb alloys and other Sb compounds. Antimony as Sb_2S_3 (up to 7% of Sb) is a component of brake pads, and serves as a lubricant to reduce vibrations and to improve friction stability. A brake pad is composed of a polymer matrix of more than 10 different compounds consisting of fibres, fillers and modifiers held together by a binder. Antimony (as Sb_2O_3) is also a flame retardant in the vulcanisation of rubber. Due to its stabilising properties, Sb is used in a number of alloys for motor bearings. Many other Sb compounds are responsible for its presence in APM. As an example, road traffic emission of Sb has also been ascribed to the use of certain organic Sb-containing compounds such as Sb-dialkyldithiocarbamates used in greases and motor oil.

Other elements such as Al, As, Cu, Cr, Hg, Mn, Ni, Sn and Sr were also detected not only in brake pads, but also in dust. Emissions of Sb and other associated elements are a function of traffic density while its concentration in APM also depends on the distance between the sampling site and the emission zone. In general terms, concentrations of Sb in urban aerosols from different cities varied from 0.23 to 107 pg m^{-3} [16].

Information on the distribution of Sb species in APM is scarce. The development of Sb speciation techniques in APM presents further complex analytical challenges, and attention has only been paid to this area of analytical research recently. The combination of specific extraction procedures, anion exchange chromatography, and/or size-exclusion chromatography in combination with plasma-based techniques or atomic fluorescence spectrometry (AFS) and hydride generation (HG) allowed the separation and quantification of inorganic and organic Sb species in APM samples [19,20]. Thirteen fractions of size-classified APM were collected in Japan for subsequent determination of Sb species [19]. The study demonstrated that Sb(III) dominated the coarse fraction whereas Sb(V) was distributed in fine and coarse fractions. The speciation analysis of Sb in urban particulate matter and road dust collected in Valparaiso (Chile) was reported by Quiroz and coworkers [21]. Inorganic species of Sb were measured and the authors noted that up to 70% of the total Sb extracted from samples was Sb(III) giving clear evidence of the anthropogenic origin of this element. Combining several characterisation techniques such as SEM (scanning electron microscopy)-EDS, ICP-MS and synchrotron radiation X-ray absorption (SR-XAS), Varrica et al. [22] reported that the samples of wheel rims dust, road dust and APM were composed of a mixture of Sb(III) and Sb(V) in different relative abundances in Italy. Antimony was also identified as a TRE in size-fractionated road dust samples collected in Buenos Aires (Argentina) exhibiting a significant enrichment factor (272) in the finest fraction, <37 μm [23].

3.2.2.2 Lead

Lead of mainly anthropogenic origin is transported by the atmosphere and the fluxes of this metal largely reflect the relevant anthropogenic source regions, transport and removal mechanisms. For many years, automobile emissions were the major source of urban APM exposure to this metal. Then alkyl-Pb compounds were banned as gasoline antiknock additives in more than 100 countries across the world. A major factor in banning Pb was the introduction of catalytic converters that are used to reduce vehicle exhaust emissions of gaseous air pollutants such as NO_x, CO and aromatic hydrocarbons. Automotive catalysts become rapidly inactive when Pb from the additives is deposited on the active catalyst sites.

The effect of strategies for the reduction of Pb emissions from the beginning of the 1980s to the mid-1990s is reflected by the available estimation of worldwide Pb emissions from mobile sources, i.e. 248,000 tons in 1983, more than 116,700 tons in 1989 and more than 88,700 tons in 1995 [11].

Lead is no longer added to gasoline and diesel oil in most countries. Nevertheless, Pb is still detected in APM at low levels because it is a natural component of fuels used in cars and trucks. When studies related to the determination of levels of Pb in fuels and vehicular emissions are compared, the decrease in emissions after the phase out of leaded gasoline and the level of vehicular emissions arising from the natural Pb content in gasoline and diesel oil is noticeable.

Different atomic and plasma-based techniques can be applied to the determination of Pb in gasoline and APM. The determination of trace metals in PM_{10} urban aerosols collected in Buenos Aires (Argentina) by plasma-based techniques corroborated a significant decrease of Pb content since the elimination of Pb-additives in 1995 [24]. Other studies focused on the determination of Pb in diesel and gasoline by ICP-MS after sample clean up using hollow fibre solid phase micro-extraction [25]. Reported values of Pb (in $\mu g\ L^{-1}$) detected in diesel samples were 153 ± 4 and 60 ± 25, whilst for unleaded gasoline concentrations up to 61.9 ± 1.7 were found. More recently, research has also focused on employing Pb isotopes (Pb^{206}, Pb^{207} and Pb^{208}) to identify sources of metal and metalloid contamination in APM [26]. ICP-MS analysis of atmospheric aerosol allowed for the identification of two distinct isotopic signatures: one prevalent in fine particles ($<1\ \mu m\ d_a$) while the other corresponded to coarse particles as well as particles in all size ranges from a nearby urban environment.

Nowadays, in countries where Pb is no longer added to gasoline, its emissions and, as a consequence, its presence in APM is mainly ascribed to wear rather than fuel combustion.

3.2.2.3 Platinum Group Elements

The term PGEs is ascribed to six elements that are all transition metals lying in the d-block of the periodic table namely, Ir, Os, Pd, Pt, Rh and Ru. Palladium, Pt and Rh belong to the six elements with the lowest abundance in the Earth's crust. Due to their unique physical and chemical properties, this group of elements is critical to many emerging applications, mainly in the field of catalysis. The catalytic activity of Pt, often in combination with Pd and Rh, is exploited in the oil industry and in the pharmaceutical industry for the synthesis of ammonia and other chemical products.

More recently, one of the major uses of three of the PGEs (Pd, Pt and Rh) is in car catalytic converters to abate the emission of C_xH_y (hydrocarbons), CO and NO_x, into less toxic substances; carbon dioxide and water, carbon dioxide and nitrogen and oxygen, respectively. Catalytic converters can be considered as the prime and mobile source of PGEs detected in APM and other environmental matrices such as road dust. These converters consist basically of an alumina or cordierite monolith treated with an Al_2O_3 wash coat which in turn contains rare Earth oxides and 0.10–0.15% (w/w) Pt, Pd and/or Rh to catalyse the removal of the pollutants previously mentioned [27]. To a large extent, atmospheric pollution by PGEs can be traced back to the release of micro particles from the converters as a result of surface ablation during car operation.

Vehicular catalytic converters were adopted in the United States in the early 1970s and in the 1980s in Europe. Since then, a number of studies have been published in scientific journals dealing with the increasing concentration of PGEs in different countries ([16] and references therein). Catalytic converters have been used for more than 40 years and, for this reason, there exists sufficient experimental evidence to support the view that PGEs are emitted from car catalytic converters in particulate form, primarily in the 0 oxidation state and also as oxides.

Many research groups all around the world have carried out studies on PGEs in atmospheric aerosols and environmental matrices affected by their presence. In this context, in 1997 the European Union launched an international joint research project named CEPLACA (Environment and Climate Programme of the European Union), to carry out a series of studies aimed at assessing concentrations of PGEs in environmental matrices, including APM. The study also investigated the bioaccumulation and toxicity of environmental Pt and other PGEs in humans and exposed organisms. Another objective was to assess the bioaccumulation and toxicity of PGEs at the new environmental concentrations in exposed organisms. These studies showed that brand-new catalytic converters can release highly variable concentrations of PGEs (with differences of up to two to three orders of magnitude). It was concluded that the soluble portion of PGEs was ~5% of the total amount released to the atmosphere [28].

The determination of PGEs in APM is a challenge for most instrumental analytical techniques due to the insufficient detection power which resulted in the need to employ very sensitive techniques such as SF-ICP-MS (sector field-inductively coupled plasma–mass spectrometry). This technique alone or in combination with ultrasonic nebulisation is a fully adequate technique for PGEs determination due to the very low concentrations of these elements in urban aerosols. In addition, and to minimise the risk of contamination, it is recommended that sample manipulation be carried out in an extremely clean environment such as a class-100 clean room when available. In general terms, reported Pt concentrations vary from 0.1 to 130 pg m^{-3} [16]. Other techniques have been used including a preconcentration or separation technique to reach the sensitivity required for the levels of PGEs in most environmental matrices.

The determination of PGEs in APM and road dust collected in different cities confirms their widespread distribution all around the world. In Germany, Zereini et al. determined Pt and Rh concentrations by adsorptive voltammetry in APM samples collected from 1988 to 1998 [29]. The study confirmed that PGEs are not only present in areas close to emissions (e.g. roads), but can be transported over longer distances. In a collaborative study, Bocca and coworkers reported the first values of Pt and Rh in PM_{10} for a city of Latin America [30]. Analyses were performed by high-resolution-ICP-MS (HR-ICP-MS) and concentrations (in pg m^{-3}) varied from 2.3 to 47.7 for Pt and 0.3 to 16.8 for Pd. Recently, Kowalska et al. applied a very sensitive adsorptive stripping voltammetry technique to the determination of Pt and Rh in environmental samples contaminated by automotive exhausts after an ion exchange matrix separation using a Cellex-T sorbent [31].

There is currently a wealth of information that confirms that PGEs can bioaccumulate. The presence of PGEs in a variety of environment compartments, their increasing accumulation and health risk was reviewed by Ravindra et al. [32]. More and more studies are available dealing with the levels of PGEs in foods of animal and vegetal origin. The concentrations of PGEs detected in food are probative of the fact that particulate matter can enter the food chain. More studies are necessary and consideration should be given, therefore, to PGEs monitoring in key food and environmental matrices and to investigate their potential chronic adverse effects on humans and biota. Recently, demonstrating that there is a permanent and continuous interest in this topic, Ruchter and Sures [33] reported a study on the distribution of Pt and other TREs in sediments and clams.

3.2.2.4 Additional Traffic-Related Elements

Different materials are components of brake pad composites. A series of major, minor and trace metals are also present in brake pads as follows: Ba (as $BaSO_4$) is a popular cheap filler in brake linings that can be found at levels up to 40%. In different proportions, these metals are also components of brake pads: Cu, Fe, Mo (MoS_2), Ni, Sn and Zn.

Fuel additives also contain TREs. They are used for different purposes such as: fuelborne catalyst, to increase engine efficiency reducing the amount of soot that leaves the exhaust from diesel trucks and buses, etc. Another important element is Ce as even when the oxide (CeO) is inert, this compound can oxidise carbon at high temperatures. Cerium is an extremely effective fuel oil additive for lighter boiling or distillate fuels. One of the compounds added to distillate fuels is a carboxylate complex of Ce. The use of a Pt–Ce bimetallic fuel additive has been proposed as a cost-effective way for reducing particulate matter and also NO_X emissions from diesel-powered engines. Krishna and Makkee [34] studied and characterised by X-ray diffraction (XRD) analysis soot containing Ce-, Pt-, Ce-, Fe-, and Cu-fuelborne catalyst generated in a diesel engine under various reaction conditions. The study showed that nitrate species were involved in the oxidation of Ce- and Pd-Ce soot. Recently, Potekim et al. [35] reported XRD and transmission electron microscopy results indicating the formation of Pt-rich Pt-Co alloy nanoparticles supported over an alumina surface.

Manganese is an effective and most commonly used metal additive for improving combustion and suppressing smoke in residual and distillate oils. The addition of Mn to fuels is in connection to its property of impeding the formation of SO_3. Manganese is a gasoline octane enhancer added to oil as $Mn_3(PO_4)_2$. Methylcyclopentadienyl Mn tricarbonyl is a motor vehicle fuel additive used as antiknock agent. In spite of the potentially increasing levels of

Mn in airborne particles resulting from automotive exhausts, only a few studies have been conducted to specifically study this metal. In spite of its toxicological implications, few research papers have been devoted to specifically study Mn in APM. A large-scale population-based exposure study (~1000 participant periods) was conducted in Toronto (Canada) to estimate the distribution of 3-day average personal exposures to particulate matter ($PM_{2.5}$ and PM_{10}) and Mn [36]. In this case, NAA was the technique chosen for sample analysis.

Zinc is a TRE because tyres, belts, hoses, oil seals, trim and mountings contain Zn. The most important contribution is from tyres because ZnO has been used for over a century in the vulcanisation process. In addition, tyres contain small but significant amounts of Cd. Its function is the activation, with stearic acid, of sulphur cross-linking of rubber enhancing the ability to absorb frictional heat and improving tyre performance as well as the combination of beneficial properties such as strength and flexibility. During the last 15 years, the content of Zn in APM from samples collected in different countries using different analytical methodologies and techniques was reported.

Nomizu and coworkers [37] developed a high-speed digital signal processing system as an individual particle analyser combined with ICP-MS to achieve signal profiling with a good time resolution and a lower detection limit for Zn determination in APM. Zinc concentrations in PM_{10} samples collected in Buenos Aires (Argentina) and determined by ICP-OES varied from 19.8 to 166 ng m^{-3} [24]. Although the content of Fe, Mn, Pb and Zn measured was considerably lower than the corresponding concentrations reported for Mexico DF (México), Santiago (Chile) and São Paulo (Brazil), the PM_{10} mass concentration was equivalent. FAAS in conjunction with FI was employed for monitoring Zn in the soluble fraction using a dynamic extraction procedure [38]. The procedure was applied to the determination of water-soluble Zn in aerosol samples from two urban sites in Austria.

Gasoline is an important source of TREs. Some metals present in gasoline catalyse oxidative reactions in hydrocarbon mixtures, deteriorating their thermal stability and impairing their use as fuel. Metals in oil products may also cause corrosion and catalyst poisoning. Sulphur in the gasoline is an important contributor to SO_x emissions and it also poisons the low-temperature activity of automotive catalytic converters. Levels of V in the fluid catalytic cracking (FCC) catalyst affects the hydrogen transfer activity. Vanadium poisoning destroys the zeolite component of the FCC catalyst and for this reason the content of both metals in fuels should be controlled. Their presence in APM is indicative of fossil fuel combustion. The content of V in APM has been studied more. The ICP-OES determination of V in size-fractionated APM samples collected in an industrial city in Argentina showed a mean concentration of V of 13 ± 4 ng m^{-3} [39]. Shafer et al. [40] reported the chemical speciation of V in particulate matter emitted from diesel vehicles and urban atmospheric aerosols. The authors developed and applied different strategies for the determination of total, extractable as well as V(IV) and V(V) using miniaturised extraction/separation methods and magnetic sector-ICP-MS for detection. The study found that V(IV) was the dominant species in diesel PM emissions.

Several kinds of gasoline contain ~0.2–3.3 ng g^{-1} of Hg. Londonio et al. [41] linked the presence of Hg in Buenos Aires street dust with exhaust emissions from vehicular liquid fuel combustion. Mercury was determined by flow injection-cold vapour-AAS (FI-CV-AAS) and concentrations ranged between 0.1 μg g^{-1} and 4.6 μg g^{-1} and the mean concentration value was found to be 1.09 μg g^{-1}.

3.2.3 Marine Aerosols

Sea-spray aerosol is mechanically produced from the interaction of wind stress at the ocean surface. Sea salt consists predominantly of Cl and Na and carries also chemical species containing elements such as Br, Ca, Fe, I, K, Mg and S [42]. Sea salt aerosol (SSA) intervenes in various physical and chemical atmospheric processes at urban, regional and global scales. It influences radiative transfer directly by scattering solar radiation and indirectly by forming cloud condensation nuclei and increasing cloud-top reflectivity [43]. Halogens, particularly Cl and Br, originally bound in the SSA, play an important role in atmospheric chemistry once they are released to the atmosphere. Chlorine may be mobilised from sea salt particles by: (1) the release of gaseous HCl through the reaction of NaCl with gaseous acids, particularly HNO_3 or H_2SO_4; (2) the formation of gaseous Cl_2 through reactions of OH^- with Cl^- at the air–water interface of deliquesced NaCl; and (3) heterogeneous reactions at the surface of sea salt particles, especially with nitrogen oxides that can promote the release of compounds such as ClNO [44].

For air quality assessment in coastal areas, it is necessary to take into account the large gradients in SSA levels along sea-inland directions. This feature is considered in a number of regional models, which include a description of sea salt. However, the verification of these models is severely hampered by the limited number of available ground-based measurements worldwide [45]. This scarcity of data may be associated with the relatively low attention that air pollution researchers, traditionally more concerned with health impacts, have paid to the relatively benign SSA [46]. Despite this situation, recent studies in Europe and the United States compiled data regarding concentrations of sea salt markers measured in aerosols collected at a significant number of ground-based monitoring sites [46].

3.2.4 Secondary Aerosols

Gaseous precursors of secondary aerosols such as NH_3, SO_2, NO_x and volatile and semivolatile organic compounds (VOCs and SVOCs) arise from multiple natural and anthropogenic sources. Nitrate- and sulphate-rich secondary aerosols are typically associated with NH_4^+. Fossil fuel combustion is a major contributor of SO_2 and NO_x, precursors of particulate sulphates and nitrates, respectively. Natural precursors of sulphates include H_2S and SO_2 emitted by volcanoes and forest fires, and reduced sulphur gases emitted by oceans and wetlands. Agricultural activities such as animal husbandry and application of nitrogen fertilisers are significant emitters of NO_x and NH_3. Ammonia reacts rapidly with nitric and sulphuric acids to generate light-scattering secondary inorganic aerosols.

Regarding nitrates, it is well known that sampling artefacts may appear under filter-only sampling. Negative sampling artefacts are typically associated with evaporative losses of aerosol nitrate while positive sampling artefacts occur when gaseous HNO_3 is retained in the filter through adsorption or it reacts with alkaline aerosol components. Nitrate losses depend on the meteorological conditions; they increase under high temperature and/or low humidity. To avoid negative artefacts, denuders are used to remove HNO_3 and NH_3 from the sampled air. These devices, which are discussed in Chapter 5, can also be used to measure the concentration of the removed gases via analysis of the contents of the water draining from them. Filter-packs are used to remove HNO_3 and NH_3 evaporated from the collected particles, for example, a filter impregnated with H_3PO_4 is used to remove the evaporated NH_3 while another filter impregnated with NaF is used to remove evaporated HNO_3. More recently, a steam jet aerosol collector combined with an NH_4^+ conductivity detector and an IC system has been developed and satisfactorily applied to determine NO_3^- and NH_4^+ in aerosols [47].

Biological processes are the most important in producing organic precursors to secondary organic aerosol (SOA), which accounts for a large fraction of total organic particulate mass. Although anthropogenic hydrocarbons seem to make only a minor contribution to SOA formation, the fraction of SOA that is anthropogenic in origin is thus far uncertain [48].

4. ANALYTICAL TECHNIQUES

Analytical chemistry as well as environmental science is benefiting more and more from the development of sensitive, selective and innovative methodologies for the determination of metals and metalloids in environmental matrices as well as the identification, separation and quantification of the various forms under which an element may occur. The continuous growth of sophisticated instrumentation constitutes a key tool for analytical chemists.

In spite of the intrinsic advantages of each technique, the determination of metals and metalloids in atmospheric aerosols and other matrices such as road dust, soils or biomonitors can be problematic due to some points that should be considered.

In general terms, the concentration of chemical elements and ions are very low (pg m^{-3} or ng m^{-3}). In this context, before making a decision on selecting a technique, its analytical characteristics (sensitivity, accuracy, precision and robustness), potential interferences and especially the limits of detection that can be reached should be carefully assessed.

Another topic to consider is that sample preparation is a critical step and the greatest care should be devoted to the careful preparation of blanks and the minimisation of exogenous contamination. A detailed review of the different approaches and methodologies adopted for sample preparation of atmospheric aerosols is available [49].

The quality of analytical data is a point of prime importance in the analysis of atmospheric aerosols and has to be ensured by implementing strict quality control. In this regard, the use of specific certified reference materials (CRMs) is mandatory; in case they are not available, it is advisable to employ an alternative technique or to implement an inter-laboratory test.

In the following sections, we briefly describe classical, robust and sophisticated instruments used for metal determination in APM as well as the progress experienced in the last years in the development and improvement of analytical instrumentation.

4.1 Spectroscopic Techniques for Elemental Analysis

4.1.1 General Considerations

The selection of an adequate technique for APM analysis depends on the element/elements, concentrations, air sample volume collected, detection capability, potential interferences (spectral, matrix), accuracy and precision required, linear dynamic range, requirement for isotopic analysis, skill level of the operator, instrument cost and operating and maintenance costs of the instrument.

The important advances that most analytical techniques including atomic spectrometric and plasma-based techniques have experienced in the last decades are without any doubt attributed to improvements and innovations in instrumentation. The widespread use of preconcentration methods, flow injection (FI) facilities, slurry sampling, miniaturisation and automation of procedures has contributed to increased sensitivity, selectivity, improved performance, facilitating contamination control and reduced time of analysis.

Monoelemental techniques namely, FAAS, ETAAS and AFS, are useful and still used in APM characterisation for studying a particular element or a small group of elements. The outstanding developments undergone by all these techniques over the last years have marked a parallel and irreversible progress in the use of such techniques for environmental purposes from both the viewpoint of research activities, routine analysis and regulatory tasks.

The analysis of atmospheric aerosols is multielemental in nature and, for this reason, multielemental techniques are the most used for major, minor and trace element determination. Numerous analytical techniques are used to identify and measure metals and metalloids aimed at obtaining reliable results and correct evaluations in the field of atmospheric chemistry. Environmental assessment is benefiting from development of innovative analytical methodologies where multielemental techniques such as plasma-based techniques namely, ICP-OES, Q-ICP-MS (quadrupole-ICP-MS), HR-ICP-MS, DCP-OES (direct current plasma–optical emission spectrometry), MIP (microwave-induced plasma), XRF and PIXE have played a relevant role.

The coupling of HG methods to atomic and plasma-based techniques introduced advantages that deserve to be mentioned. HG has been employed for over a century for determining As by the Marsh reaction or the Gutzeit test. The cold vapour atomic absorption (CV-AAS) technique for Hg determination was introduced in 1968 [50]. Since then, a massive application of vapour generation (VG) to AAS and then to other analytical techniques, due to the unquestionable advantages offered is clearly observed in the literature. Nowadays, VG methods based on derivatisation of acidified samples with sodium tetrahydroborate(III) as reductant have become, without any doubt, a useful method to determine typical hydride forming elements (As, Bi, Cd, Ge, Sb, Se) and elements such as Fe, Ni, Zn, In, Pb, Ni, Sb, Se, Sn, Pd, Pt, Te, Tl that can be considered as 'nonconventional' hydride forming elements. Mercury as cold vapour can also be determined by this technique and it is today the most popular method for Hg determination at trace levels. The main advantages of VG over the widely used pneumatic nebulisation of liquid samples are the enhancement of transport efficiency, approaching 100% for many elements, which implies an improvement in the detection power and the separation of the analyte from the matrix leading to an improvement of selectivity in many instances. The possibility of analyte preconcentration and/or selective determination of species using specific methodologies is an additional advantage [51].

In spite of the wide use of all these techniques for many years, it goes without saying that the bottleneck of all the techniques (FAAS, ICP-OES, ICP-MS) that use a nebulisation system for sample introduction, is precisely the nebulisation system. Samples, in a liquid form, are introduced into the nebuliser where they are converted into an aerosol which only represents 1–2% of the total sample that then emerges from the spray chamber and is transported into the instrument. This has resulted in the design of more efficient nebulisers (e.g. direct injection nebuliser, high-efficiency nebuliser (HEN), direct injection high-efficiency nebuliser).

In the following sections, the main characteristics and applications of the more commonly used spectroscopic techniques are summarised.

4.1.2 Atomic Absorption Spectrometry

Since the introduction of FAAS by Sir Alan Walsh in 1955 and the development of the first commercial instrument in the early 1960s, this technique has become an established technique and is used worldwide. AAS requires that the sample be introduced into the instrument in the liquid form and sample treatment is required.

Another less used alternative is to introduce the sample into the instrument as a slurry. The slurry sampling method has been extensively employed for the analysis of different kinds of samples, including APM, by atomic spectrometric techniques in order to simplify sample preparation procedures and to avoid inconvenience related to wet decomposition and dry-ashing procedures. Introduction of slurry samples combines the advantages of reducing sample preparation time, contamination and possible analyte losses due to volatilisation and/or retention by insoluble residues. In addition, slurry sampling avoids the use of hazardous and concentrated acids such as $HClO_4$ and HF. Recently, Silva et al. [52] reported the determination of Cu in APM using slurry sampling and chemical VG-AAS reaching a limit of detection (LOD) of 0.3 $\mu g\ L^{-1}$. Chemometric tools were useful to study the influence of several variables on the generation of the volatile species of Cu.

In general terms, FAAS is a robust, simple to operate and consolidated technique. It has relatively moderately priced instrumentation, good accuracy and adequate precision (0.1−1%) for minor and major elements determination. For this reason, FAAS instruments are available in many laboratories across the world. In spite of this, FAAS presents the disadvantage of monoelement operation capability. Typical detection limits are in the range $1-10^3\ \mu g\ L^{-1}$ that are not low enough for certain elements present in atmospheric aerosols. Further, FAAS has a limited working range and it is not free from interferences. In spite of this, FAAS has been and still is a useful tool for the determination of selected elements in APM. In 2004, a new approach was introduced into the market: high-resolution continuous source atomic absorption spectrometry (HR-CS AAS). The hollow cathode lamp (HCL) was replaced by a specially designed xenon short-arc lamp that is used as the radiation source. The instrumental design allows for the entire part of the spectrum transmitted by an intermediate slit to reach the detector; a linear CCD array that not only detects the analytical line, but also its spectral environment at high resolution. A significant advantage of HR-CS AAS is that only a single radiation source is required for all elements (instead of an HCL for each element) and all wavelengths over the entire spectral range from 190 to 900 nm can be used. It is also worth mentioning that the emission intensity of this lamp is much higher. In spite of this, the radiation intensity has no influence on sensitivity in AAS but it has a significant influence on the signal-to-noise ratio. This results in an improvement in detection limits by a factor of ~5 with respect to line source AAS. The same commercial company combined a flame and furnace instrument and this new instrument was introduced in 2007. Palladium was monitored in the $PM_{2.5}$ fraction of APM samples by HR-CS- electrothermal atomic absorption spectrometry (ETAAS). Samples were collected in Istanbul (Turkey) and Budapest (Hungary), and it was possible to measure Pd at pg m^{-3} levels [53]. This technique was successfully applied to the determination of Sb [54], Hg [55], TREs [56] in airborne particles.

ETAAS was developed by L'vov in 1950 and became commercially available in the early 1970s as a complementary analytical possibility to FAAS. It is also a monoelemental technique but much more sensitive in comparison to FAAS. ETAAS is an important tool for the accurate determination of trace metal content in APM, especially when a low mass or low volumes (5–40 μL) are available, owing to the low-detection limits reached (<5 $\mu g\ L^{-1}$), precision (0.5–5%), accuracy and selectivity. As it is a very sensitive technique, especially clean reagents are required to avoid blank problems. ETAAS is not free from spectral and non-spectral interferences and this is a point to be carefully studied. Spectrometers employ different background correction systems such as continuum source background correction. Interference effects from the associated major elements in complex matrices like particulate matter are reliably corrected with Zeeman-effect background correction. Savio et al. [57] used another approach such as self-reversal background correction (SR-ETAAS) for Pb determination in APM with a heavy matrix of silicon.

Flow injection facilities for online preconcentration, *in situ* collection of gaseous analytes (e.g. converted to hydrides) and solid/slurry sampling devices have given new potential to the technique and have also been used for APM analysis [49].

4.1.3 Atomic Fluorescence Spectroscopy

Atomic fluorescence spectrometry is a monoelemental technique that represents a suitable, although much less robust alternative to the other atomic and mass spectrometric techniques. Since AFS can be quenched by many molecules, an effective separation of the analyte from the matrix and/or a very efficient atomisation is required. In comparison to AAS, AFS is superior in terms of sensitivity based on the fact that the fluorescence signal has a very low background. Compared to ICP-MS, AFS for some specific elements has a very similar sensitivity. AFS has a relatively simple set-up and requires lower investment and maintenance costs. Manufacturers have included a chemical vapour generation (CVG) system in AFS instruments. As a drawback, it is limited to certain elements (e.g. As, Bi, Ge, Hg, Se, Sb, Sn, Te, etc.). Unlike ICP-MS, the AFS detector is more tolerant of external hydrogen formed during the CVG process. Hence, it is compatible with the online VG system, which makes it most suitable for measuring vapour-forming elements. The combination of ultrasound probe fast extraction and analysis by HPLC-HG-AFS enabled the selective determination of Sb(III) and Sb(V) in APM samples collected in heavy traffic streets from Buenos Aires (Argentina). The results showed the presence of both Sb species at similar concentrations in the ng m^{-3} level [20]. Using the same coupling, Sb speciation in road dust and urban particulate matter collected in Valparaiso (Chile) was also reported [21]. The study showed that up to 70% of total Sb extracted from road samples was Sb(III) indicating the anthropogenic origin of this element.

4.1.4 Plasma-Based Techniques

Inductively coupled plasmas are the most used plasma source in commercial instruments. In spite of this, other plasma sources have been developed such as direct current plasma and microwave-induced plasma, although these are much less used.

Inductively coupled plasma atomic emission spectrometry was introduced in 1975 and since then has gained worldwide acceptance due to its main characteristics namely, multielemental capacity, sensitivity, precision, wide dynamic range (up to five orders of magnitude) and possibility of introducing micro-volumes of sample. In ICP-OES, detection limits span from 10^{-2} to $10\ \mu g\ L^{-1}$ and precisions reached are between 0.3% and 2%. An additional advantage of this technique is the easy way that the ICP spectrometer can be coupled to other techniques such as HPLC, HG, FI or an electrothermal atomiser. Although ICP was developed as a multielemental source for optical emission, improvements and new developments in instrumentation allowed for the coupling of the ICP source to a mass spectrometer. Since it was first commercialised in 1983, ICP-MS has become a powerful and fast-growing technique for multielemental trace level and ultra-trace elemental analysis. This technique is fast, has a wide dynamic range (up to nine orders of magnitude) with the advantage of isotopic capabilities. The wide dynamic range of ICP-MS allows sample analysis from ppt ($ng\ L^{-1}$) up to levels of a few 100 ppm ($mg\ L^{-1}$).

Thanks to the low LOD reached for most elements ($10^{-3}\ \mu g\ L^{-1}$ in solution) it is possible to analyse small aerosol samples (100–1000 μg). Levels of metals, metalloids, and some nonmetals can be determined, in terms of atmospheric concentrations, at about $1\ pg\ m^{-3}$ for $100\ m^3$ of collected air. Precisions from 1 to 2% can be achieved. Because of the advantages described before, ICP-MS is the technique of choice for APM analysis.

Another significant advantage of ICP-MS instruments is the ability to provide isotopic information and the possibility to be interfaced with other techniques, namely HPLC, FI, HG, laser ablation, electrothermal vapourisation and capillary electrophoresis for different purposes. Although Q-ICP-MS instruments are extensively used in environmental analysis, the technique has evolved to more sophisticated instruments with additional advantages. Quadrupoles with collision cell, high-resolution magnetic sector field, multi-collector systems for high-precision isotope ratio determination and time-of-flight equipment are now available.

The main drawback of Q-ICP-MS is the formation of spectral interferences originating from atomic and molecular ions produced in the Ar plasma by matrix constituents of the sample and is especially serious at a mass-to-charge ratio $(m/z) \leq 80$. The use of high-resolution magnetic sector field ICP-MS (ICP-HRMS) allows the elimination of most polyatomic interferences. For this reason (in spite of its more elevated price) and for the very low levels ($\sim pg\ m^{-3}$) of PGEs such as Pd, Pt and Rh in APM samples, ICP-HRMS has been widely used for the determination of these traffic-related elements. Caroli and his group carried out a series of relevant studies aimed at assessing PGE levels in different environmental and biological matrices and exposure in urban populations [30,58,59].

4.1.5 X-Ray Fluorescence, Particle-Induced X-Ray Emission and Neutron-Activation Analysis

X-ray fluorescence also has multielemental capabilities. In comparison to atomic and plasma-based techniques is a non-destructive technique because dissolution of the sample is not required. XRF is fast and cost-effective, is applicable to a wide range of concentrations (percentages to few ppm) and is competitive in terms of sensitivity for many elements. A disadvantage is that the X-rays from elements of low atomic number are emitted at very low energy and are strongly adsorbed by the filter onto which APM samples are collected. The selection of the adequate filter material is a key point for this technique. In general terms, analysis by XRF is restricted to elements heavier than fluorine. Recently, energy dispersive XRF was applied to measuring the elemental composition of 660 samples of atmospheric aerosols collected at a pristine site in the Amazon Basin (Brazil) and 1416 samples collected at a site perturbed by land usage [60]. The study showed the presence of trace elements associated with biogenic aerosols, soil dust, biomass burning and sea salt, even for the very low concentrations. Another alternative to conventional XRF used for liquid or digested samples is total reflection X-ray fluorescence (TXRF). It is a surface elemental technique more adequate for ultra-trace analysis. Only 5−10 μL of sample are required for analysis. It is used much less than XRF for APM analysis. TXRF is basically an energy dispersive XRF technique and the instruments are designed in a special geometry. The unique instrumental configuration confers an advantage: the measurement background contributions are reduced by elimination of sample scattering and, as a consequence, result in increased elemental measurement sensitivity [61].

Another technique based on a similar principle and used for atmospheric aerosol analysis is PIXE but in this case, protons from a cyclotron or a Van de Graff accelerator are used. The limits of detection reached will depend on the irradiation time. The necessity of an accelerator has limited the use of PIXE in the characterisation of APM to a limited group of laboratories. Fittschen et al. [62] presented the elemental characterisation of atmospheric aerosols down to particle sizes of 0.015 μm. The procedure was thoroughly evaluated regarding bounce off effects and blank values. Additionally, the potential of total reflection X-ray fluorescence−X-ray absorption near-edge structure (SR-TXRF-XANES) for speciation of FeII/III down to amounts of 34 pg was studied in aerosols collected during 1 h in Hamburg (Germany). Fittschen has reviewed different strategies for ambient aerosol characterisation using synchrotron X-ray fluorescence [63].

Neutron activation analysis is a sensitive technique and, for this reason, is adequate for trace elements determination in APM. It offers two approaches namely, instrumental (INAA) and radiochemical (RNAA). It is a nuclear technique used for the qualitative and quantitative multielement determination of major, minor, and trace elements including rare elements. INAA is non-destructive while studies involving RNAA require the dissolution of the sample. NAA allows discrete sampling of elements as this technique disregards the chemical or physical form of the elements focussing solely on their nucleus. When sequential instrumentation is used, it is possible to determine quantitatively ~60 elements in a small amount of sample (5−100 mg). Neutron activation analysis sensitivity and accuracy depends on the concentration of a particular element and radionuclide parameters (i.e. parent isotope abundance, neutron cross-section, half-life and gamma ray abundance). Element sensitivities vary from 10^{-3} to 10^{-10} g per gram of sample and depend on a large number of factors, including neutron flux, sensitivity of the detector, irradiation time, and neutron cross-section of the elements in the sample. During the 1960s, NAA and XRF were the multielemental techniques of choice for atmospheric aerosols analysis. When plasma-based techniques and PIXE were launched onto the market, they were successfully adopted for this kind of studies. Accuracy of an NAA determination is usually between 2% and 10%, depending on the element analysed and its concentration in the sample. Main advantages include minimum sample preparation, no matrix effects, non-destructive characteristics (dissolution of the samples is not required for INAA), sensitivity (detection limits from 0.1 to 1×10^6 ng g^{-1} for more than 70 elements), very low sample masses are required (particularly useful for samples from rural or pristine areas), many samples can be irradiated simultaneously and counted later on a given decay schedule. The drawbacks and limitations are that a fission reactor is required as neutron source, some elements cannot be determined such as carbon, oxygen, nitrogen, fluorine and lead; with poor sensitivity for sulphur and high background for ^{24}Na and ^{82}Br possible, making the LODs worse. In addition, irradiations for several hours and long decay and counting times are required for this technique.

Levels of atmospheric heavy metal in aerosols collected in different cities of North Egypt using NAA and ICP-OES were reported [64]. The results revealed that the highest concentrations of PM_{10} and total suspended particulate matter were close to industrial areas and that the enrichment was more significant for elements of anthropogenic origin such as Ba, Ce, Sb and Zn.

4.2 Ion Chromatography for Anions and Cations

Assessment of the concentrations of water-soluble inorganic ions, in fine and coarse particles, is an important tool to gain information on source contribution, chemical transformation, and potential adverse effects for human health and the environment.

Inorganic ions such as fluorine, chlorine, nitrate, phosphate, bromide, sulphite, sulphate and weak organic acids namely, aldehydes, carbohydrates, carbonates, acetate, formate, pyruvate, malonate, oxalate and cations namely, ammonium, amines, aminoacids, Ca^{2+}, K^{+}, Mg^{2+} and Na^{+} have been identified and measured in APM samples from different origins [65]. Cations and anions can be used as markers of sources such as industrial activity, resuspended street dust, urban construction or natural sources such as sea salt sprays. It is important to remark that sulphate, nitrate, ammonium and chloride have been extensively studied for their capacity to affect climate change and control the acidification and conductivity of aerosols. Short-chain organic acids are formed through photochemical reactions in the atmosphere.

Without any doubt, the most used technique for the determination of ions is ion chromatography (IC). (The fundamentals of chromatography are covered in Chapter 9). It is a powerful technique that allows the separation and quantification of anions and cations using different separation and detection approaches.

The main advantages of IC are multi-species capabilities, small sample requirements, good selectivity and high linear response. The first step for subsequent IC analysis is the removal of deposited particles from filters using an adequate solvent (water being commonly used) following an established and optimised analytical working protocol. To reach high extraction efficiency, it is recommended to extract the samples in an ultrasonic bath for a fixed period of time (~20–60 min). Extraction with heating (40–60 °C) is a good alternative that has been used to further increase extraction efficiency. Extracts are then analysed by IC.

The mobile phase may be water, diluted organic acids or buffers either on their own or mixed with another in different proportions. When the composition of the mobile phase is the same during the chromatographic separation, the method is called *isocratic*. Alternatively, when the composition varies during the separation, this methodology is called *gradient*. In normal-phase liquid chromatography, the stationary phase is relatively polar and the mobile phase relatively non-polar, while reverse phase LC uses a non-polar stationary liquid and a polar mobile phase. In ion-exchange chromatography (used in atmospheric chemistry studies mainly for inorganic ions determination) the stationary phase is an ion-exchange resin and separation is governed by the strength of the interaction between ions and the exchange sites of the resin.

The selection of the column and the optimised chromatographic conditions are critical for the quality of the determinations. The appropriate column will depend on the characteristics of the sample and the ions to be separated. The most common separation mechanisms are ion exchange (organic and inorganic cations and anions), ion pair (surface-active anions and cations), ion exclusion (weak inorganic or organic acids, alcohols, aldehydes), and reverse phase (higher polar and ionic species). In ion-exchange columns, organic polymers are mostly employed as packing materials. Other alternatives are silica-based columns and resins. Polystyrene cross-linked resins, with a small amount of divinylbenzene to give mechanical stability, is a very commonly used material in the fabrication of ion-exchange columns. The surface of these columns contains ionisable sites (e.g. sulphonic acid or quaternary ammonium) that confer a positive or a negative charge to the column. Positive and negative sites require an oppositely charged ion for overall neutrality.

The selection of the appropriate detector will strongly depend on the ion. The most used detection systems in IC are conductivity, UV absorption, refraction index and electrochemical detection. Currently, plasma-based techniques are easily coupled and are widely used to detect certain types of ions.

The interest regarding cation and anion determinations in APM is reflected in some of the selected studies discussed here. Canepari et al. [66] optimised a microanalytical two-step sequential leaching procedure for the determination of non-volatile ions (NO_3^-, SO_4^{2-}, Cl^-, Na^+, Mg^{2+}, NH_4^+ and Ca^{2+}) by IC and elements in PM_{10} samples. The determination of water-soluble and insoluble compounds in size-classified APM collected in Buenos Aires (Argentina) was reported by Dos Santos et al. [67]. The investigation demonstrated that among the four ions determined by IC, only SO_4^{2-}, Cl^- and Na^+ made an important contribution to the water-soluble aerosol mass. The assessment of the seasonal distribution of airborne trace elements and water-soluble ions (Cl^-, PO_4^{3-}, NO_3^-, SO_4^{2-}, $HCOO^-$, CH_3COO^-, oxalate, succinate, Na^+, NH_4^+, K^+, Mg^{2+} and Ca^{2+}) in São Paulo (Brazil) made possible the identification of several sources though the determination of ions by IC. Fuel burning and/or biomass burning were identified as sources of (1) NO_3^-, $HCOO^-$, $C_2O_4^{2-}$, K^+, Mg^{2+}, Ca^{2+} and metals (Fe, Pb, Zn, Al, Ca, K and Mg), (2) gas-to-particle conversion (SO_4^{2-} and NH_4^+) and Cl^-, Na^+ originated from sea salt spray [68].

5. CONCLUDING REMARKS

The key role played by trace components of atmospheric aerosols involved in a variety of environmental processes has prompted a wealth of studies over the years aimed at identifying and quantifying these analytes in particulate-based environmental matrices.

During the last decades, more efficient and reliable analytical controls have been solicited by environmental agencies and governmental offices in different countries where programmes for monitoring particle emission, as well as the presence of key components in APM, have been established. In this regard, identifying the presence and determining concentration levels of chemical markers allow ascertaining whether chemical particle components are emitted by natural or anthropogenic sources, contributing to the formulation of cost-effective air pollution control policies.

Of no lesser importance is to assess, depending on the capabilities of each laboratory, the most adequate analytical techniques and methodology for the determination of trace components of sampled airborne particles deposited on filters. Hence, at present, many laboratories and research groups invest time and resources in the determination of these elements in APM using different analytical techniques and strive hard to develop specific fractionation procedures, such as separation of particles according to their d_a or sequential extraction procedures that allow distribution of a given element to be measured on the basis of its degree of leachability from complex matrices.

This chapter has summarised the progress achieved in the last decades regarding the quantification of major, minor and trace elements, particularly TREs, as well as ions, achieved by means of different spectroscopic and chromatographic methods using different approaches. It is expected that it will serve as a useful resource for those involved in these kinds of instrumental analytical studies.

REFERENCES

[1] F.J. Kelly, J.C. Fussell, Atmos. Environ. 60 (2012) 504–526.

[2] T. Takemura, Atmos. Chem. Phys. 12 (2012) 11555–11572.

[3] J.H. Seinfeld, S.N. Pandis, Atmospheric Chemistry and Physics: From Air Pollution to Climate Change, second ed., Wiley, New York, 2006.

[4] P. Kumar, L. Morawska, W. Birmili, P. Paasonen, M. Hu, M. Kulmala, R.M. Harrison, L. Norford, R. Britter, Environ. Int. 66 (2014) 1–10.

[5] M. Kulmala, H. Vehkamäki, T. Petäjä, M. Dal Maso, A. Lauri, V.-M. Kerminen, W. Birmili, P.H. McMurry, J. Aerosol Sci. 35 (2004) 143–176.

[6] P.F. DeCarlo, J.G. Slowik, D.R. Worsnop, P. Davidovits, J.L. Jimenez, Aerosol Sci. Tech. 38 (12) (2004) 1185–1205.

[7] C. Xiong, S.K. Friedlander, Proc. Natl. Acad. Sci. U.S.A. 21 (2001) 11851–11856.

[8] L. Tumolva, J.-Y. Park, J.-S. Kim, A.L. Miller, J.C. Chow, J.G. Watson, K. Park, Aerosol Sci. Tech. 44 (2010) 202–215.

[9] J.G. Slowik, K. Stainken, P. Davidovits, L.R. Williams, J.T. Jayne, C.E. Kolb, D.R. Worsnop, Y. Rudich, P.F. DeCarlo, J.L. Jimenez, Aerosol Sci. Tech. 38 (2004) 1206–1222.

[10] T. Nousiainen, J. Quant, Spectrosc. Radiat. 110 (2009) 1261–1279.

[11] P. Smichowski, D.R. Gómez, Analysis of Atmospheric Aerosols, Encyclopedia of Analytical Chemistry, Online, John Wiley & Sons, Ltd, New York, 2012.

[12] P. Pant, R.M. Harrison, Atmos. Environ. 49 (2012) 1–12.

[13] M. Viana, T.A.J. Kuhlbusch, X. Querol, A. Alastuey, R.M. Harrison, P.K. Hopke, W. Winiwarter, M. Vallius, M.,S. Szidat, A.S.H. Prévôt, C. Hueglin, H. Bloemen, P. Wåhlin, R. Vecchi, A.I. Miranda, A. Kasper-Giebl, W. Maenhaut, R. Hitzenberger, J. Aerosol Sci. 39 (2008) 827–849.

[14] A. Pietrodangelo, R. Salzano, E. Rantica, C. Perrino, Atmos. Environ. 69 (2013) 1–14.

[15] X.Y. Zhang, R. Arimoto, J.J. Cao, Z.S. An, D. Wang, J. Geophys. Res. 106 (2001) 18,471–18,476.

[16] P. Smichowski, D. Gómez, C. Frazzoli, S. Caroli, Appl. Spectrosc. Rev. 43 (2008) 22–48.

[17] G. Weckwerth, Atmos. Environ. 35 (2001) 5525–5536.

[18] C. Barbante, G. Cozzi, G. Capodaglio, K. Van de Velde, C. Ferrari, A. Veysseyre, C.F. Boutron, G. Scarponi, P. Cescon, Anal. Chem. 71 (1999) 4125–4133.

[19] A. Iijima, K. Sato, T. Ikeda, H. Sato, K. Kozawa, N. Furuta, J. Anal. At. Spectrom. 25 (2010) 356–363.

[20] A. Bellido-Martín, J.L. Gómez-Ariza, P. Smichowski, D. Sánchez-Rodas, Anal. Chim. Acta 649 (2009) 191–195.
[21] W. Quiroz, M. Cortés, F. Astudillo, M. Bravo, F. Cereceda, V. Vidal, M.G. Lobos, Microchem. J. 110 (2013) 266–272.
[22] D. Varrica, F. Bardelli, C. Dongarrà, E. Tamburro, Atmos. Environ. 64 (2013) 18–24.
[23] F. Fujiwara, R. Jiménez Rebagliati, J. Marrero, D. Gómez, P. Smichowski, Microchem. J. 97 (2011) 62–67.
[24] P. Smichowski, D.R. Gómez, L.E. Dawidowski, M.F. Giné, A.C. Sánchez Bellato, S.L. Reich, J. Environ. Monit. 6 (2004) 286–294.
[25] P.N. Nomngongo, J.C. Ngila, Spectrochim. Acta Part B 98 (2014) 54–59.
[26] O.I. Félix, J. Csavina, J. Field, K.P. Rine, A.E. Sáez, A. Betterton, Chemosphere 122 (2015) 219–226.
[27] J.C. Ely, R.N. Clive, C.F. Kulpa, M.A. Schneegurt, J.A. Seidler, J.C. Jain, Environ. Sci. Technol. 35 (2001) 3816–3822.
[28] S. Rauch, G.M. Morrison, An environmental case history of the platinum group metals, in: L. Ebdon, L. Pitts, R. Cornelis, H. Crews, O. Donard, P. Quevauviller (Eds.), Trace Element Speciation for Environment, Food and Health, The Royal Society of Chemistry, Cambridge, 2001.
[29] F. Zereini, C. Wiseman, F. Alt, J. Messerschmidt, Environ. Sci. Technol. 35 (2001) 1996–2000.
[30] B. Bocca, S. Caimi, P. Smichowksi, D. Gómez, S. Caroli, Sci. Total Environ. 358 (2006) 255–264.
[31] J. Kowalska, K. Kińska, J. Pałdyna, M. Czyżewska, K. Boder, B. Krasnodebska-Ostrega, Talanta 127 (2014) 250–254.
[32] K. Ravindra, L. Bencs, R. Van Grieken, Sci. Total Environ. 318 (2004) 1–43.
[33] N. Ruchter, B. Sures, Water Res. 70 (2015) 313–324.
[34] K. Krishna, M. Makkee, Top. Catal. 42-43 (2007) 229–236.
[35] D.I. Potemkin, E.Yu Filatov, A.V. Zadesenets, P.V. Snytnikov, YuV. Shubin, V.A. Sobyanin, Chem. Eng. J. 207-208 (2012) 683–698.
[36] E.D. Pellizzari, C.A. Clayton, C.E. Rodes, R.E. Mason, L.L. Piper, B. Fort, G. Pfeifer, D. Lynam, Atmos. Environ. 33 (1999) 721–734.
[37] T. Nomizu, H. Hayashi, N. Hoshino, N. Tanaka, H. Kawaguchi, K. Kitagawa, S. Kaneco, J. Anal. At. Spectrom. 17 (2002) 592–595.
[38] A. Mukhtar, A. Limbeck, J. Anal. At. Spectrom. 25 (2010) 1056–1062.
[39] F. Fujiwara, M. Dos Santos, J. Marrero, G. Polla, D. Gómez, L. Dawidowski, P. Smichowski, J. Environ. Monit. 8 (2006) 913–922.
[40] M.M. Shafer, B.M. Toner, J.T. Overdier, J.J. Schauer, S.C. Fraka, S. Hu, J.D. Herner, A. Ayala, Environ. Sci. Technol. 46 (2012) 189–195.
[41] J. Londonio, F. Fujiwara, R. Jiménez Rebagliati, D. Gómez, P. Smichowski, Microchem. J. 105 (2012) 77–82.
[42] F.J. Millero, Physicochemical controls on seawater, in: H.D. Holland, K.K. Turekian (Eds.), Treatise on Geochemistry, Elsevier, Amsterdam, 2004.
[43] Y.J. Yoon, P. Brimblecombe, Atmos. Chem. Phys. 2 (2002) 17–30.
[44] R. von Glasow, Nat. Geosci. 1 (2008) 292–293.
[45] A.M.M. Manders, M. Schaap, X. Querol, M.F.M.A. Albert, J. Vercauteren, T.A.J. Kuhlbusch, R. Hoogerbrugge, Atmos. Environ. 44 (2010) 2434–2442.
[46] W.H. White, Atmos. Environ. 42 (2008) 2319–2330.

[47] J. Slanina, H.M. ten Brink, R.P. Otjes, A. Even, P. Jongejan, A. Khlystov, A. Waijers-Ijpelaan, M. Hu, Y. Lu, Atmos. Environ. 35 (2001) 2319–2330.

[48] J.H. Kroll, J.H. Seinfeld, Atmos. Environ. 42 (2008) 3593–3624.

[49] P. Smichowski, D. Gómez, G. Polla, Sample preparation of atmospheric aerosols for elemental analysis and fractionation studies, in: M.A. Zezzi Arruda (Ed.), Trends in Sample Preparation, Nova Science Publishers, Inc., New York, 2007.

[50] W.R. Hatch, W.L. Ott, Anal. Chem. 40 (1968) 2085–2087.

[51] A.A. Menegario, P. Smichowski, P.S. Tonello, G. Polla, E.P. Oliveira, R.E. Santelli, Anal. Chim. Acta 625 (2008) 131–136.

[52] L.O.B. Silva, D.J. Leao, D.C. dos Santos, G.D. Matos, J.B. de Andrade, S.L.C. Ferreira, Talanta 127 (2014) 140–145.

[53] S. Atilgan, S. Akman, A. Baysal, Y. Bakircioglu, T. Szigeti, M. Óvári, G. Záray, Spectrochim. Acta Part B 70 (2012) 33–38.

[54] R.G.O. Araujo, B. Welz, M. Goreti, R. Vale, P. Smichowski, S.L.C. Ferreira, H. Becker-Ross, J. Anal. At. Spectrom. 25 (2010) 580–584.

[55] R. Araujo, F. Vignola, I. Castilho, D. Borges, B. Welz, M. Goreti Vale, P. Smichowski, S. Ferrerira, H. Becker-Ross, Spectrochim. Acta Part B 66 (2011) 378–382.

[56] I. Castilho, B. Welz, M. Goreti, R. Vale, J. de Andrade, P. Smichowski, A. Shaltout, L. Colares, E. Carasek, Talanta 88 (2012) 689–695.

[57] M. Savio, R.A. Olsina, L.D. Martínez, P. Smichowski, R.A. Gil, Microchem. J. 96 (2010) 243–246.

[58] B. Bocca, F. Petrucci, A. Alimonti, S. Caroli, Sci. Total Environ. 5 (2003) 563–568.

[59] B. Bocca, F. Petrucci, A. Alimonti, S. Caroli, J. Environ. Monit. 5 (2003) 563–568.

[60] A. Arana, A.L. Loureiro, H.M.J. Barbosa, R. Van Grieken, P. Artaxo, X-Ray Spectrom. 43 (2014) 228–237.

[61] J. Injuk, R.Van Grieken, Spectrochim. Acta Part B 50 (1995) 1787–1803.

[62] U.E.A. Fittschen, F. Meirer, C. Streli, P. Wobrauschek, J. Thiele, G. Falkenberg, G. Pepponi, Spectrochim. Acta Part B 63 (2008) 1489–1495.

[63] U.E.A. Fittschen, Spectrosc. Eur. 26 (2014) 10–14.

[64] E.H. El-Araby, M. Abd El-Wahab, H.M. Diab, T.M. El-Desouky, M. Mohsen, Appl. Radiat. Isot. 69 (2011) 1506–1511.

[65] J.C. Chow, J.G. Watson, Ion chromatography in elemental analysis of airborne particles, in: S. Landsberger, M. Creatchman (Eds.), Elemental Analysis of Airborne Particles, Gordon and Breach Science Publishers, Amsterdam, 1999.

[66] S. Canepari, E. Cardarelli, A. Giuliano, A. Pietrodangelo, Talanta 69 (2006) 581–587.

[67] M. Dos Santos, D. Gómez, L. Dawidowski, E. Gautier, P. Smichowski, Microchem. J. 91 (2009) 133–139.

[68] G.O. Da Rocha, P.C. Vasconcellos, S.G. Ávila, D.Z. Souza, E.A.O. Reis, P.V. Oliveira, O. Sanchez-Ccoyllo, J. Braz. Chem. Soc. 23 (2012) 1915–1924.

Chapter 9

Chromatographic Techniques for Organic Analytes

Pérola de Castro Vasconcellos,[1,*] Gisele Olímpio da Rocha,[2] Elina Bastos Caramão,[3,4] Maria Elisabete Machado[4] and Laiza Canielas Krause[3]

[1]*Institute of Chemistry, University of São Paulo — USP, São Paulo, State of São Paulo, Brazil;* [2]*Institute of Chemistry, Federal University of Bahia — UFBA, Salvador, Bahia, Brazil;* [3]*Nucleo de Estudos Coloidais, Universidade Tiradentes — UNIT, Aracaju, Sergipe, Brazil;* [4]*Institute of Chemistry, Federal University of Rio Grande do Sul — UFRGS, Porto Alegre, Rio Grande do Sul, Brazil*

**Corresponding author: E-mail: perola@iq.usp.br*

Chapter Outline

1. Introduction to Chromatography 268
- 1.1 General Fundamental Concepts 268
  - 1.1.1 Relative Retention 268
  - 1.1.2 Capacity or Retention Factor (k) 269
  - 1.1.3 Selectivity (α) 269
  - 1.1.4 Separation Efficiency: N, Height Equivalent Theoretical Plate 269
  - 1.1.5 Peak Capacity 271
  - 1.1.6 Resolution 271
- 1.2 Optimizing Resolution and Separation 272
- 1.3 Main Types of Chromatography 272
  - 1.3.1 Gas Chromatography 272
  - 1.3.2 High-Performance Liquid Chromatography 274
  - 1.3.3 Comprehensive Two-Dimensional Gas Chromatography 275

2. Sample Introduction 277
- 2.1 Split versus Splitless Liquid Injections 279
  - 2.1.1 Split Injection 280
  - 2.1.2 Splitless Injection 280
- 2.2 Liner Types 280
  - 2.2.1 Liners for Split Inlet 281
  - 2.2.2 Splitless Liners 281
- 2.3 Thermal Desorption Injection 282

3. Stationary Phases Commonly Employed in the Chromatographic Analysis of Air Pollutants 283
- 3.1 Polar versus Nonpolar Stationary Phases 285
- 3.2 Analyte-Specific Stationary Phases 287

Comprehensive Analytical Chemistry, Vol. 70. http://dx.doi.org/10.1016/bs.coac.2015.09.009

3.3 Orthogonality for GC × GC Separations 288
3.4 Developments in LC Stationary Phases 288
4. Chromatography Detectors Commonly Employed in Air Pollutant Analysis 289
4.1 General Requirements of Detectors for this Application 289
4.2 Flame Ionisation Detector 290
4.3 Electron Capture Detector 291
4.4 Nitrogen Phosphorus Detector 292
4.5 Photoionisation Detector 293
4.6 Mass Spectrometer (MS) Detector 294
5. Selected Applications of Chromatography in the Analysis of Air Pollutants 295
5.1 Persistent Organic Pollutants 297
5.2 Polycyclic Aromatic Hydrocarbons and Their Derivatives 298
5.3 Volatile Organic Compounds 301
6. Challenges and Future Outlook 304
References 304

1. INTRODUCTION TO CHROMATOGRAPHY

According to the International Union of Pure and Applied Chemistry, chromatography is defined as a physical separation method in which the components to be separated are distributed between two phases, one of which is stationary (stationary phase) while the other (the mobile phase), is moving in a defined direction [1]. Chromatography is a technique used to separate mixtures of compounds, allowing their individual identification or quantification.

The main classification of chromatographic methods is based on the physical state of the mobile phase. The classical division of liquid and gas chromatography (GC) is still used nowadays, but it is also considered that supercritical fluid chromatography is another kind of chromatographic method, in which the mobile phase is a supercritical fluid.

1.1 General Fundamental Concepts

Several volumes in the literature have described the theory of chromatography. Here only some topics are presented in order for the reader to better understand its objective. More comprehensive discussions can be found in the literature [2–12]. First, useful terms in chromatography are defined in this section.

1.1.1 Relative Retention

Relative retention (t'_R) is the ratio of the thermodynamic equilibrium constants associated with the retention of two compounds. Relative retention values with respect to a given standard are denoted by the symbol R.

t_R = retention time of a compound, represented by a chromatographic peak.

t_0 = retention time of a non-retained compound, that is, a compound that has no affinity for the stationary phase and elutes with the solvent front at a time t_0, which is also known as the 'hold-up time' or 'dead time'.

$t'_R = t_R - t_0$ = relative retention time, it represents the time that the solute interacts with the stationary phase.

1.1.2 Capacity or Retention Factor (k)

$k = t'_R/t_0$ = retention factor (or capacity factor), it is a means of measuring the extent of retention of an analyte on the chromatographic column. A high k value indicates that the sample is highly retained and that it has spent a significant amount of time interacting with the stationary phase.

1.1.3 Selectivity (α)

$\alpha = k_2/k_1$ = selectivity, it represents how two peaks can be separated by a chromatographic column, i.e. the ability of the system to provide different retention times for two specific compounds.

1.1.4 Separation Efficiency: N, Height Equivalent Theoretical Plate

A major goal in chromatography is to separate the analyte species in the minimal time. Hence, it is desirable to maintain narrow peak widths (minimize broadening) so that short columns (short elution times) can be used. Plate theory describes the equilibria in a chromatographic system involving the distribution of a solute between the stationary and mobile phases. For a better understanding, the column is divided into a number of imaginary theoretical plates (N) that correspond to equilibriums of sorption/desorption of the solute in both phases: stationary and mobile. A parameter that is used to describe broadening of a chromatographic peak is called chromatographic efficiency of a particular chromatographic system. This is significant because the separation of components of a mixture is achieved by increasing the number of plates in a column or decreasing the height equivalent theoretical plate (HETP or H). This model provides an equation where it is possible to see that chromatographic efficiency (N) is inversely related to the rate of band broadening per unit time:

$$\text{HETP} = L/N \tag{1}$$

where: L = column length, HETP = H = height equivalent of one theoretical plate and N = number of theoretical plates in the column.

The velocity theory, on the other hand, describes the migration of molecules throughout the column. This movement includes band shape, broadening, and diffusion of a solute. The velocity theory follows the Van Deemter equation, which is the most appropriate for prediction of dispersion in liquid chromatography (LC) columns. It takes into account the various pathways that a solute must travel through a column. Using the Van Deemter equation, it is possible to find the optimum velocity and a minimum HETP.

The Van Deemter equation relates the separation efficiency of a column to the linear mobile phase velocity by considering the physical, kinetic and thermodynamic properties of a separation [12]. This equation is represented by a hyperbolic function that predicts that there is an optimum velocity at which there will be the minimum variance per unit column length and, hence, it reaches the maximum efficiency.

The Van Deemter equation relates the resolving power (HETP, height equivalent to a theoretical plate or simply H) of a chromatographic column with various flow and kinetic parameters, which cause peak broadening, as shown in Eqn (2):

$$\mathrm{HETP} = \mathrm{A} + \frac{\mathrm{B}}{u} + (\mathrm{C_s} + \mathrm{C_m})\cdot u \tag{2}$$

where HETP = H = height equivalent of a theoretical plate, a measure of the resolving power of the column [m].
A term = eddy–diffusion parameter, related to channelling through a non-ideal packing and reflects the particle bed uniformity or how well packed the column is, and the particle size in packed columns (there is no A term for capillary columns).
B term = random molecular diffusion of the eluting particles in the longitudinal direction, resulting in dispersion, and reflects how long the diffusion zone is on the column.
C term = resistance to mass transfer coefficient of the analyte between mobile and stationary phases and reflects the diffusion within a column-packing particle or the rate of mass transfer in/out of pores and phases.
u = linear velocity [m s^{-1}]

The shape of the Van Deemter equation curve (Figure 1) is such that HETP achieves a minimum value at a particular flow velocity and when the resolving power of the column is maximized. During the optimization of the

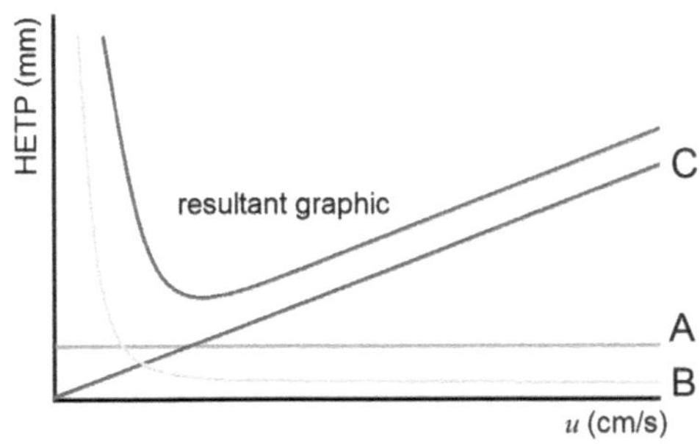

FIGURE 1 Graphic representation of the Van Deemter equation. HETP, height equivalent theoretical plate.

chromatographic process, the minimization of the values of A, B and C are sought in order to find the minimum HETP value that corresponds to a maximum value of N, that is, that enhances the process efficiency.

Since the terms A, B and C must be minimized for obtaining a maximum efficiency, they are strictly dependent on some parameters, such as:

- A is directly proportional to the diameter of the packing particles of the stationary phase;
- B is directly proportional to the diffusion coefficient of the mobile phase; and
- C is inversely proportional to the diffusion coefficient of the stationary phase and directly proportional to the diameter of the column and the film thickness of the stationary phase.

The minimization of these terms can be achieved either by reducing the pathway of the solute (utilising capillary columns, for example), by using mobile phases with low diffusion coefficients, and/or by using narrow columns with thin films of stationary phase. Capillary columns are tubular columns typically about 0.01 cm ID and 20 or more metres long, in which the liquid stationary phase is coated on and supported by the interior walls of the tube.

The number of theoretical plates (N) can be experimentally calculated by the Eqn (3):

$$N = 16\ (t_R/w_b)^2 \tag{3}$$

where t_R is retention time of the compound and w_b is the bandwidth at the base of the peak.

With a few considerations, this equation can be transformed into Eqn (4):

$$N = 5.54\ (t_R/w_h)^2 \tag{4}$$

where w_h is the bandwidth at the half height of the peak.

1.1.5 Peak Capacity

Another measure of column efficiency is the peak capacity, *n*. This is defined as the maximum number of separated bands that can be accommodated within the chromatogram. The peak capacity for isocratic separation, where the mobile phase composition is constant, can be calculated by the following equation:

$$n = 1 + \left(N^{1/2}/4\right) \cdot \ln\left(t_l/t_f\right) \tag{5}$$

Here, t_l and t_f refer to the retention times of the first and last bands, respectively. The peak capacity is a kind of 'universal' measure of separation power, which allows different chromatographic procedures to be compared.

1.1.6 Resolution

The relative separation of two bands is often referred to as their resolution, Rs. Resolution can be explained as the ratio between the difference between the peak centres of two adjacent peaks and their average baseline bandwidth and can be defined by Eqn (6).

$$\mathrm{Rs} = (\mathrm{t}_2 - \mathrm{t}_1/\mathrm{w}_{12}) = \Delta \mathrm{t}/\mathrm{w}_{12} \tag{6}$$

where t_1 and t_2 are the retention time of compound 1 and 2, respectively, and w_{12} is the average baseline bandwidth of both peaks: $w_{12} = (w_1 + w_2)/2$.

For quantitative analysis, it is desirable (but not always practical) to achieve a resolution of at least 1.5, since this corresponds to baseline separation for bands of similar size, i.e. return of the detector tracing to baseline between the two bands.

1.2 Optimizing Resolution and Separation

A general goal in chromatography is to achieve adequate resolution in the final separation. This requires understanding resolution as a function of a number of chromatographic conditions. Equation (6) is useful for measuring resolution, but not for indicating how to improve resolution. A new equation that relates resolution to separation conditions can be obtained through the combination of other expressions, by assuming that the two bands have equal bandwidths, w, which should be true for capillary columns:

$$\mathrm{Rs} = \frac{\sqrt{N}}{4} \cdot \frac{\alpha - 1}{\alpha} \cdot \frac{k}{k + 1} \tag{7}$$

where, k refers to the average retention factor of two bands. This is the fundamental equation of the resolution of chromatographic analysis.

According to Eqn (7), resolution depends on three separate parameters: α, N and k. If α is close to 1, the two compounds have very similar retention, and their separation will be difficult. If N is small, the bands will be quite wide, and again, good resolution may not be possible. Resolution is proportional to the term $\frac{k}{k+1}$ of Eqn (7), which corresponds to the fraction of the sample that is in the stationary phase. Small values of k mean that the sample is largely in the mobile phase, and under these conditions, little separation of the sample can occur ($\mathrm{Rs} \cong 0$). Figure 2 illustrates these parameters in a general chromatogram.

1.3 Main Types of Chromatography

1.3.1 Gas Chromatography

GC is a term used to describe the separation technique used to analyse volatile and semi-volatile substances in the gas phase. The components of a sample are vaporized in order to separate the analytes by distributing the sample between two phases: a stationary phase (solid or liquid) and a mobile phase (gas). The mobile phase is a chemically inert gas that carries the sample through the heated column. In GC, the interaction of the sample with the mobile phase can be disregarded. The stationary phase is either a solid adsorbent, in the case of

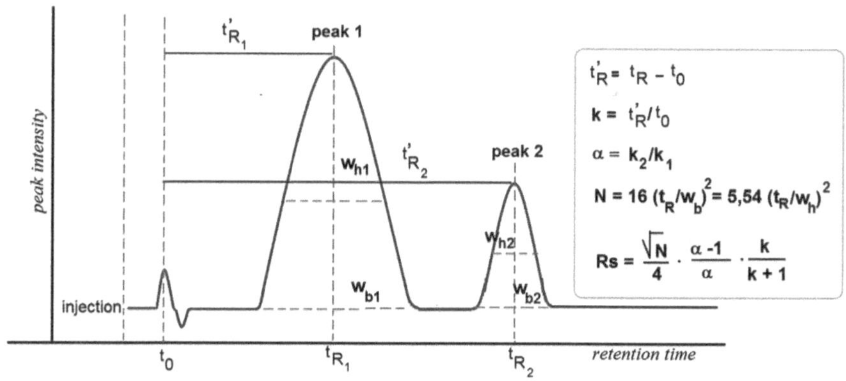

FIGURE 2 Some common parameters of a gas chromatogram.

gas–solid chromatography, or a liquid on an inert support, as it is in gas–liquid chromatography (GLC).

GLC is the method most commonly used to separate organic compounds, while the combination of GC and mass spectrometry (MS) is an invaluable tool in the identification of molecules. In GLC, the liquid stationary phase is adsorbed onto a solid inert packing or immobilized on the capillary tubing walls. In a capillary column, the tubing walls are coated with the stationary phase or an adsorbent layer, which is capable of supporting the liquid phase. The separation of the compounds is done by partitioning between both the phases.

The main parts of a gas chromatograph include a source of gas as the mobile phase, an inlet to deliver sample to a column, the column where separations occur, an oven as a thermostat for the column, a detector to register the presence of a compound in the column effluent, and data system to record and display the chromatogram. A typical gas chromatograph is presented in Figure 3.

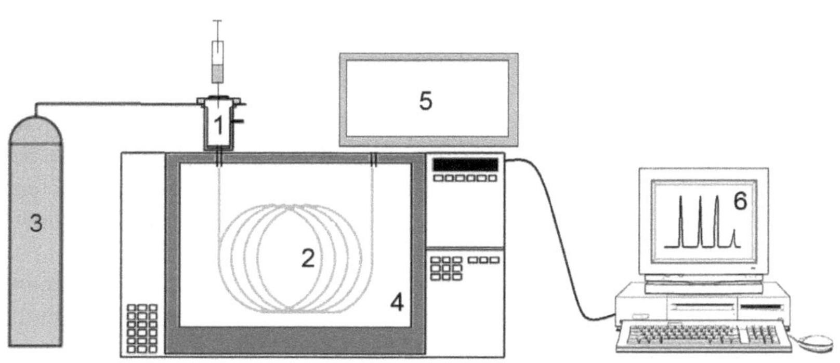

FIGURE 3 Schematic representation of a gas chromatograph: (1) injection port, (2) column, (3) carrier gas supply, (4) oven, (5) detector and (6) data processor.

Some of the best discussions of hardware can be found in publications released by instrument manufacturers [10,11]. The column may arguably be considered the key component of a gas chromatograph.

1.3.2 High-Performance Liquid Chromatography

High-performance liquid chromatography (HPLC) is an analytical technique used to separate and quantify each component in a mixture that must be soluble in the liquid phase. Compounds that cannot be volatilised or are thermolabile for GC analysis are the main compounds analysed by HPLC [13,14]. Today, compounds in trace concentrations as low as parts per trillion [ppt] may be easily analysed. HPLC can be, and has been, applied to samples such as pharmaceuticals, food, cosmetics, environmental matrices, forensic samples, and industrial chemicals [15]. HPLC can be used in both qualitative and quantitative applications (compound identification and quantification).

There are two main types of HPLC: normal phase HPLC (NP-HPLC) and reverse phase HPLC (RP-HPLC). If the stationary phase is more polar than the mobile phase, the separation is named as normal phase, while if the stationary phase is less polar than the mobile phase, the separation is considered as reverse phase. NP-HPLC is only rarely used now, while almost all HPLC separations are performed in reverse phase. In RP-HPLC, the retention time of a compound increases as its polarity decreases. The key to an effective and efficient separation is to combine the best separation between polar and nonpolar components in the least retention time possible without losses in resolution.

The goal is for all the compounds to elute in the shortest time possible, but ensuring the best separation of each individual peak. Typical columns for normal phase separation are alumina or silica. For RP-HPLC, typical columns are composed of alkyl, aliphatic or phenyl bonded phases. RP-HPLC is only ineffective for a few separation types. For instance, it cannot separate inorganic ions (which are separated by ion chromatography) and other highly hydrophobic compounds. Aside from these few exceptions, RP-LC can be used for the separation of almost all other compound varieties, such as aromatic hydrocarbons, amines, sugars, lipids, pharmaceutically active compounds, amino acids, peptides, proteins and also other molecules of biological origin. HPLC has been used for medical (e.g. detecting vitamin D levels in blood serum), legal (e.g. detecting performance enhancement drugs in urine), research (e.g. separating the components of a complex biological sample or of similar synthetic chemicals from each other) and manufacturing (e.g. during the production process of pharmaceutical and biological products) purposes [15].

Figure 4 shows a schematic representation of the basic instrumental components of a liquid chromatograph. The solvent inlet supplies the mobile phase, which is then pumped through the inline solvent filter and passed through the injection valve where it is mixed with the sample for injection. It then passes through another filter and then through the column where the

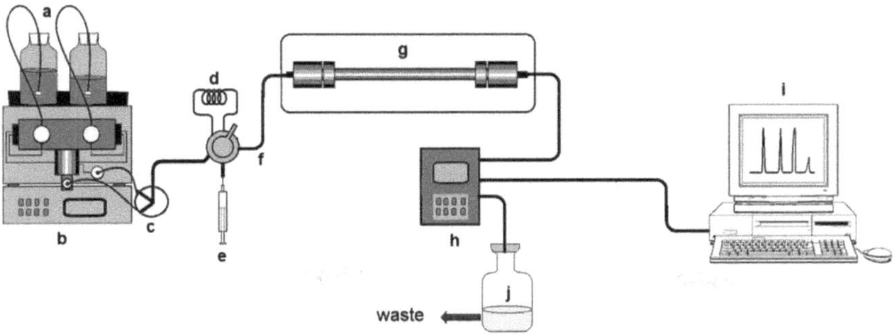

FIGURE 4 Schematic representation of a high-performance liquid chromatograph (HPLC): (a) solvents (mobile phases); (b) high pressure pump; (c) mixer for homogenizing mobile phases; (d) sample loop; (e) syringe with sample; (f) injector valve; (g) oven with the HPLC column; (h) detector; (i) data system (process and register) and (j) collection vial.

sample will be separated into its components. The detector detects the separated bands that correspond to the analytes and a computer will record and process this information. The effluent goes through a backpressure filter and then to waste [16].

The analyte retention time will vary depending on the interaction between the molecules being analysed and the stationary phase and the solvent, or solvents used. As the sample passes through the column, it partitions between the two phases at different rates, primarily due to different polarities of the analytes. Analytes that have the least amount of interaction with the stationary phase or the most amount of interaction with the mobile phase will exit the column faster [17].

1.3.3 Comprehensive Two-Dimensional Gas Chromatography

Comprehensive two-dimensional gas chromatography (GC × GC) is a relatively new technique, developed in 1991 by Liu and Phillips [18], and is a powerful analytical technique that has been applied to petrochemical samples, since its outset [19–25]. Nowadays it has a vast range of applications. GC × GC provides much more chromatographic information and separation capacity than the conventional GC separations [26–28]. The presence of ordered structures in the 2D separation space is a unique attribute of comprehensive techniques such as GC × GC and LC × LC that often allows unequivocal chemical compound classification and identification through a visual inspection of the contour plots [20].

For complex samples containing 150–250 compounds, for example, the separation obtained by 1D-GC with a single column is not sufficient for separation of all compounds [29]. Selective detectors, such as the nitrogen phosphorous detector (NPD) and electron capture detector (ECD), can also be a choice for 1D-GC, as they minimize matrix interferences and provide higher sensitivity for heteroatom-containing compounds. This type of detector may also be cost-effective when compared to mass spectrometer (MS) detectors, even though MS is more effective for qualitative analysis and may be employed in quantitative analysis by using the selective ion monitoring (SIM) mode attaining higher sensitivities [30,31]. Nevertheless, in all cases, the potential of the use of GC × GC with different detectors (ECD, NPD, MS) increases its application for a greater number of environmental samples. In GC × GC, two chromatographic columns of different polarities are coupled in series, typically one with conventional dimensions (1D, commonly 30 m length and internal diameter of 0.25–0.32 mm), and another one shorter in the 2D (the latter capable of generating fast-GC analysis, which thus requires a fast detector). This arrangement of columns is such that all the effluent coming from the first column is driven to the second through a modulator [32]. The modulator is the 'heart' of the technique, and has the main functions of cutting and refocusing narrow adjacent fractions of the first column eluate and rapidly injecting it into the second column [33]. In general, four main advantages of GC × GC over 1D-GC can be cited:

1. peak capacity is much higher, providing an improved separation of the analytes in a sample, firstly from each other, and secondly from the matrix constituents;
2. chemically related compounds present a spatial distribution as ordered structures, which greatly facilitates group-type analysis and the provisional classification of unknowns;
3. sensitivity is greatly enhanced due to the focusing obtained through modulation and the fast acquisition of the detectors which favours the detection of trace levels; and
4. two different sets of retention data for all constituents of a sample are provided, which yields an additional tool for their identification [29,34–36].

Basically, comprehensive 2D-GC exploits the same type of instrumentation utilized in monodimensional GC, with the novel introduction of the modulator within its apparatus. A simple schematic of a GC × GC system is shown in Figure 5. The apparatus is composed of the GC injector, carrier gas, one or two ovens, primary and secondary columns, modulator, detector and a computer.

Columns can be placed in the same or in two separate GC ovens. The latter is preferred as it allows for two independent temperature controls. In fact, the two columns are coated with different types of stationary phase, hence the limits of operating temperatures of the two columns very often vary from each other.

Cryogenic modulators are the most used modulators. A cryogenic modulator is composed of one or two cold-jets of a cooling gas (CO_2 or N_2) that is sprayed on the end of the 1D column (or in the beginning of the 2D column) at regular time periods (modulation periods). The analytes are instantaneously

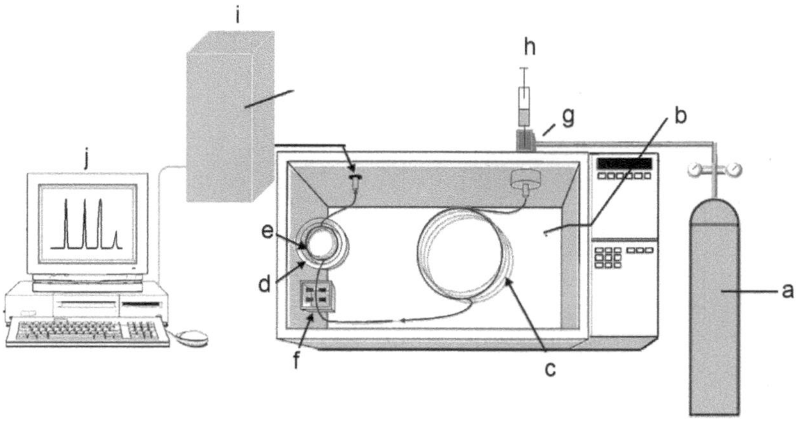

FIGURE 5 Schematic representation of a two-dimensional gas chromatograph: (a) carrier gas; (b) main oven – first dimension; (c) ID column; (d) secondary oven – second dimension; (e) 2D column; (f) modulator; (g) injector; (h) syringe; (i) detector; (j) data system (process and record).

frozen, re-concentrated and the cool jet is stopped (by closing a specific valve), the temperature is instantaneously increased, and the analytes are injected rapidly into the second column. The detector must have a high acquisition rate because the separation on the second column is completed in few seconds (normally the same time as the modulation period).

Figure 6 shows a schematic representation of the process of producing two-dimensional chromatograms. It is possible to obtain a two-dimensional diagram (colour diagram or apex diagram) or a three-dimensional diagram, as can be seen in Figure 6(e) and (f).

2. SAMPLE INTRODUCTION

The real chromatographic analysis starts with the introduction of the sample onto the column. The sample is vaporized and mixed with carrier gas prior to the separation. These processes should be achieved without reduction of the separation potential of the column, in the absence of thermal degradation, adsorption or rearrangement of sample components, without discrimination of sample components by boiling point, polarity or molecular weight, and with quantitative recovery for both trace analysis and for major compounds [37]. Also it is preferable that changes in the column operating conditions should not affect the sample introduction process. It is difficult to meet all of these requirements with a single inlet design for the wide range of sample and column types encountered in GC.

The absence of a universal inlet has resulted in the development of a number of specialized inlet systems required to manage the full range of gas chromatographic applications [38].

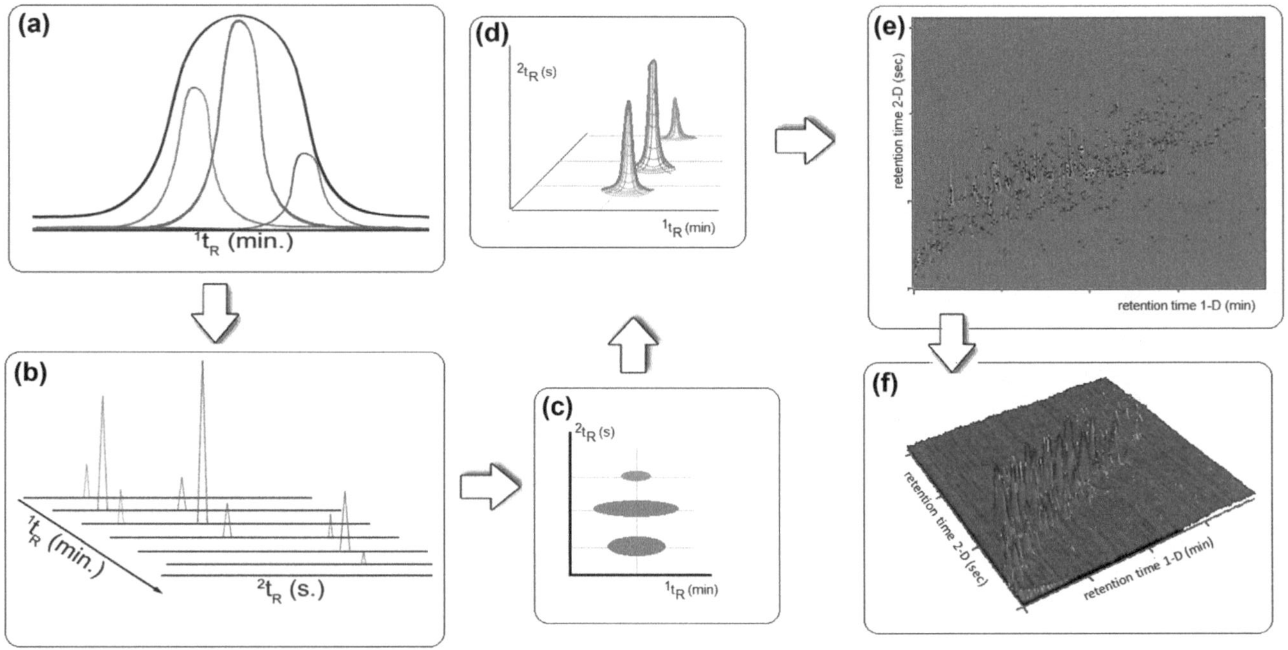

FIGURE 6 Schematic representation of the separation in two-dimensional gas chromatography (GC × GC). (a) Detail of three peaks overlapped in the 1D-GC; (b) modulation of these peaks; (c) two-dimensional representation of these peaks; (d) three-dimensional representation of these three peaks; (e) example of colour diagram (two dimensional) of a sample and (f) the same example in a three-dimensional colour diagram.

2.1 Split versus Splitless Liquid Injections

One of the most versatile and commonly used sample inlets in GC is the split/splitless inlet, schematically depicted in Figure 7. The injector contains the following elements: gas flow regulation (column flow and split operation), evaporation chamber – the glass liner, a septum, and a heated block. The central part is a vaporization chamber with internal walls covered by a glass liner. Carrier gas is continuously flushed through the liner and into the column. Liquid samples are vaporized in the chamber, which is connected to the chromatographic column. The vaporized sample can be divided into two parts, one part enters in the column and the other part is purged out through a split

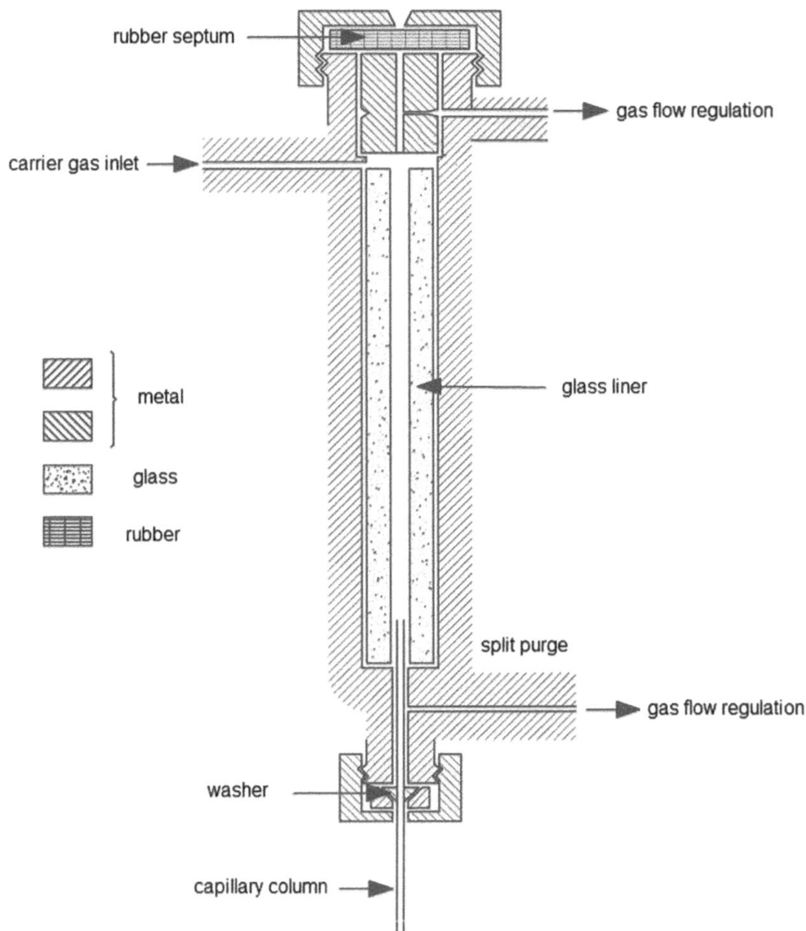

FIGURE 7 Schematic drawing of a vapourising split/splitless injector. *Adapted with permission from Ref. [39].*

vent (with gas flow regulation). The amount of sample entering the column is managed by changing the split ratio (split flow/column flow). Samples with a very low analyte concentration are transferred to the column almost in their entirety by keeping the gas flow regulation valve closed (the so-called splitless injection).

2.1.1 *Split Injection*

A split injection is used when the compounds of interest in sample are at relatively high concentration or when low limits of detection are not necessary to be achieved. As the name implies, the injection occurs by division, so that a controllable amount of sample is transferred onto the GC column. Split injections are accomplished by high flow rates through the inlet, with a minor flow (and sample) going to the GC column and some higher flow (and sample) going out through the split vent. The ratio, total inlet flow compared to the column flow, is referred to as the split ratio. Normal split ratios (ratio of vent-flow to column-flow) ranging from 10:1 to 200:1 are used, i.e., only ca. 10–0.5% of the sample enters the column [37]. Injection of lesser sample into the column is desirable, when possible, as this will usually greatly increase the column lifetime.

2.1.2 *Splitless Injection*

A splitless injection is used when the compounds of interest in the sample are present at lower levels. With this technique, the split valve is closed at the start of the injection and all flow passing through the inlet is directed onto the column during a programmed period of time, known as purge valve time. The split valve is then opened to flush out any remaining vaporized solvent. In a proper splitless injection, 99% of the compounds of interest will be transferred into the GC column [37].

2.2 Liner Types

The liner is a kind of empty glass pre-column where the mixing of the sample with the gas phase occurs as well as the instantaneous sample heating. The liner is used in the injection port to limit sample degradation and improve vaporization. The liner design can be altered or adjusted to optimize sample vaporization. A poor liner choice can be a major source of errors in the analysis. It is a crucial part of the whole chromatography process and needs to be fully understood to enable optimum separations.

Many inlet liners are available for use in GC, and they differ in geometric configuration/design, volume, base material (borosilicate, quartz or metal), deactivation, and the presence or absence of glass wool. Different styles of liners are used for different types of samples (liquid or gaseous) and different types of injections (split, splitless, on-column or direct injection). In this chapter, only liners for GCs equipped with split/splitless inlets are described

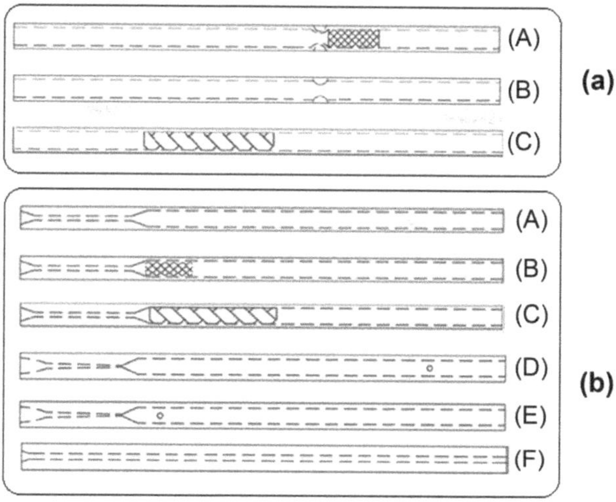

FIGURE 8 Common types of liners for gas chromatography: (a) liners for split injection: (A) straight inlet liner with glass wool, (B) without glass wool and (C) cyclo liner; (b) liners for splitless injection: (A) taper inlet liner with glass wool, (B) without glass wool, (C) cyclo liner, (D) with hole near top, (E) with hole near bottom and (F) straight/solid phase microextraction (SPME) inlet.

because these are more frequently used for environmental samples. Figure 8 shows several kinds of liners commonly used in GC. Some of them will be discussed below.

2.2.1 Liners for Split Inlet

Three liners are mostly suggested for use in a split injection, based on their ability to enhance vaporization and sample mixing. The first is the straight inlet split liner with glass wool (Figure 8(a-A)). This liner contains deactivated glass wool that is placed in the same location within every liner and fixed there by little dimples inside the liner. The wool enhances vaporization and sample mixing by increasing the surface area, while also trapping non-volatile impurities that can contaminate the GC column, and wiping the syringe needle during injection to increase repeatability.

If the sample interacts with wool (degradation or adsorption), the use of an empty liner (Figure 8(a-B)) or a cyclo-liner (Figure 8(a-C)) is recommended. The cyclo-liner does not contain wool. Instead, the bottom of the liner contains a corkscrewed glass section, which increases the interior surface area, assists with sample vaporization, and creates a vortex that favours sample mixing.

2.2.2 Splitless Liners

The main liners recommended for use in a splitless inlet are the gooseneck or single taper liner (Figure 8(b)). The taper at the bottom of the liner limits the interaction of the analytes with the metal inlet seal and helps direct or focus the sample to the head of the column. This liner is a good starting place for the majority of splitless injections. If compounds of interest degrade or adsorb on wool, a single taper liner without wool or a taper liner with a cyclo is recommended (Figure 8(b-B) and (b-C)) and ensures a highly inert liner for analysing compounds at trace levels.

Another type of splitless liner is designed with an internal 'press-fit' connection at the bottom of the liner. This seals the GC column to the liner so there is no contact between the sample and the metal seal at the bottom of the inlet. These liners are made in two configurations. One has a small hole drilled near the top of the liner (Figure 8(b-D)), while the other variation has a hole drilled near the bottom but above the liner/column seal (Figure 8(b-E)). If analytes are semi-volatile compounds, or if a tailing solvent peak could affect analysis, the best option is to use the liner with the hole drilled near the bottom. For aqueous injections or when compounds elute away from the solvent peak, the best choice is the configuration with the hole drilled near the top. They are ideal for trace samples.

For gaseous samples, the best inlet liner should have a small inner diameter (ID) to transfer the gaseous sample to the analytical column in the tightest sample band possible [40]. A straight liner with a 1.0 mm ID is recommended for these injections (Figure 8(b-F)). Small injection volumes of less than 0.5 μL are best used with a narrow bore and they are not suitable for high boiling samples.

Although liners with 4 mm internal diameter (ID) are primarily used for split injection of dirty samples, these liners are sometimes used for splitless injections. However, liners for splitless injection usually have a smaller ID (0.5–2 mm) than liners for split injection [41]. The reason for the use of these smaller ID liners is that the residence time in the injector liner is relatively long in splitless injections, and using a liner with a smaller surface area will reduce the likelihood of the breakdown of thermally labile compounds. These smaller ID liners require a slower injection speed.

It should be noted that the liners discussed in this section are only a small fraction of a great variety of different types of commercially available liners.

2.3 Thermal Desorption Injection

Apart from liquid injection, there are also other means to introduce analytes into the GC column, thermal desorption (TD) is an example. TD utilizes tubes or GC liners packed with sorbent materials, which are mainly used for airborne volatile analysis [41]. After sampling, the tubes are transferred to the GC and the analytes are desorbed in a special TD unit or in the GC injector (at high temperature). During desorption, the carrier gas flows through the tube in the opposite direction to the airflow used during sampling [40,43].

TD is also a valuable alternative to headspace techniques for the isolation of volatile compounds from non-volatile solid, semi-solid and occasionally liquid matrices [41–44].

TD methods are gradually superseding solvent extraction procedures for analysis of air samples collected on sorbent tubes/traps. Whole air monitoring (online or using containers) is already almost exclusively carried out using TD to extract the target organics from the sorbent-focusing traps and transferring them to the GC analytical system. TD is also the method of choice for most ambient air monitoring and atmospheric research studies due to the 1000-fold increase in sensitivity when compared with solvent extraction [44].

In TD, the samples are collected onto 'desorption tubes' which are heated in the desorption unit [45] to liberate the analytes. A high gas flow transfers the analytes to the GC injection device via a heated transfer line. The injection device is a programmable temperature vaporizer (PTV), which contains a liner packed with Tenax or similar sorbents.

The analytes are reconcentrated in this liner by a cold trap. Subsequently, the injector is rapidly heated in order to obtain a small injection band on the capillary column. Disadvantages of such a system are the limited sample capacity and the fairly complicated two-step desorption which involves high-flow conditions and interfacing via a heated transfer line. It will be obvious that analyte losses can easily occur. A preferred approach is to use direct thermal desorption-GC (DTD-GC). With this technique, the head of a PTV injector is opened, the liner containing the sample is introduced and the head is closed. The liner is purged with carrier gas and subsequently heated and the analysis is started [44]. The advantages of this approach are the use of only one PTV, the absence of a heated transfer line and the use of gas at a flow rate suitable to create optimum GC column conditions.

However, because of the (semi-) solid nature of TD sorbents, automation of the sample introduction is difficult. In 2002, de Koning et al. [45] designed a system which features fully automated liner exchange. To this end, a Focus XYZ sample preparation robot was equipped with a newly developed injector head to open and close the Optic 2 (ATAS GL, Veldhoven, the Netherlands) injection interface. In Figure 9, the injector head is shown in the open (left) and closed (right) position. The specially designed liners, capped with a standard crimp cap, are placed in a sample tray and are transported to the TD device. Both liner transport and liner exchange (which can be performed after each analysis) are automated and such systems are available commercially.

3. STATIONARY PHASES COMMONLY EMPLOYED IN THE CHROMATOGRAPHIC ANALYSIS OF AIR POLLUTANTS

For a chromatography instrument, one of the most important components is the chromatographic column with a specific stationary phase; liquid or solid. Separation of the sample components takes place by adsorption onto a solid or by dissolving in a nonvolatile liquid.

The stationary phase is attached to a support, in general, a solid inert material. The sample (gaseous, liquid or solid that may or may not be dissolved

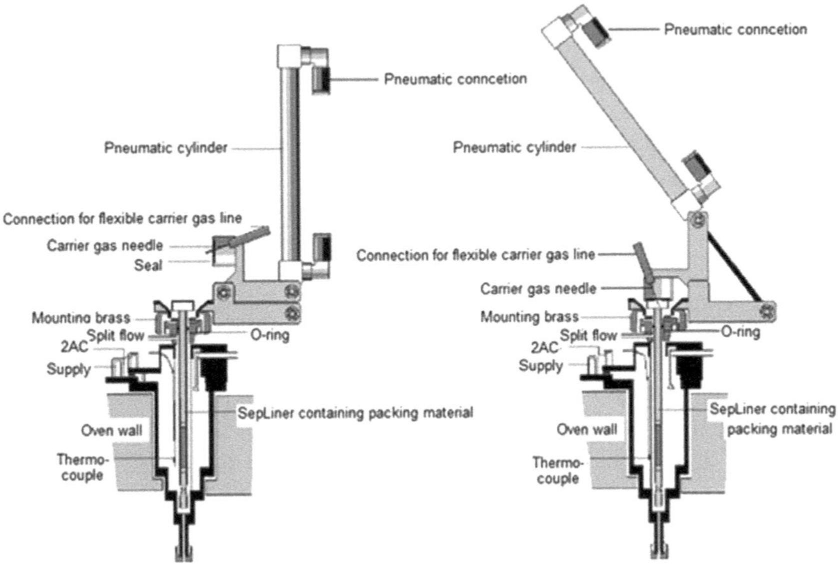

FIGURE 9 Direct thermal desorption automated liner-exchanging head. Left, open; right, closed. *Adapted with permission from Ref. [45].*

in solvents) is carried across the stationary phase by a mobile phase (gas or liquid). The sample components undergo a series of exchanges between the two phases due to the differences in their chemical and physical properties. These differences govern the rate of the movement, called migration, of the individual components.

In terms of classification of stationary phases, the chromatographic methods depend upon whether the phase is a solid or a liquid and if the technique is GC or LC [46].

In GC, the stationary phase is distinguished among solid, liquid and chemically bound. If a solid stationary phase is used, the technique is described as gas–solid adsorption chromatography, and if the stationary phase is liquid it is called gas–liquid partition chromatography. The latter is more commonly used, but in both cases the stationary phase is placed in a narrow column in an oven and the stationary phase is coated onto the inner surface of the column. The chemically bonded phases usually differ from the other two in their separation mechanisms which are based on partitioning, but also active (polar) groups on the surface may result in adsorption. Most authors consider this technique a separate method [4].

In LC, the stationary phase consists of solid particles packed in a column, which is traversed by the mobile phase. The types of LC are also named according to the interaction that occurs between the stationary phase and solutes present in the mobile phase. The four basic types of interactions are denoted: normal-phase, reverse-phase, ion exchange and gel filtration (also called exclusion or gel permeation) [47]. Some stationary phases are designed to interact with specific chemical groups and the technique is called affinity chromatography. The stationary phase also may be made with chiral groups linked to it. It is understood that no LC separations occur by means of only one type of interaction between the analytes and the stationary phase. There are always interactions besides the named, predominant types.

Different stationary phases combined with LC or GC may be used to characterize complex mixtures of compounds. Typical environmental applications for chromatography include the determination of levels of air pollutants, such as polycyclic aromatic hydrocarbons (PAHs). This class of compounds is the subject of a considerable number of studies because of their carcinogenic properties [48].

Traditionally, PAHs have been evaluated by U.S. Environmental Protection Agency (EPA) method 610 using nonpolar stationary phases operated at high temperatures [49]. Research focusing on the stationary phases for GC and LC analyses has been conducted for PAHs in order to improve selectivity and reduce the analysis time without compromising chromatographic resolution [48].

3.1 Polar versus Nonpolar Stationary Phases

Different types of analyses are classified based on the type of stationary phase employed and the mechanisms behind the separation in the column. Differences in polarity between the sample components and the bonding entities on the stationary phase result in preferential retention.

Phase selection in chromatography is based on the general chemical principle that 'likes dissolves like'. A nonpolar column is the recommended starting point for the analyses of nonpolar compounds. Likewise, polar columns are usually recommended for the separation of polar compounds.

For characterization of a stationary phase in GC, it is necessary to consider the polarity and selectivity. Polarity is sometimes considered the capacity of a stationary phase for dipole-type interactions but is more generally defined as the capacity of a stationary phase for all intermolecular interactions consisting of dispersion, dipole-type and complexation including hydrogen bonding [50]. The latter definition is quite sensible but its experimental measurement is not straightforward. No solute possesses the general quality of polarity but possesses a capacity for intermolecular interactions that is a property of its intrinsic structure and composition. Thus, any polarity ranking of stationary phases based on the retention of an individual substance are specific to that substance and substances with similar properties, but cannot be considered universal. A solute with a large dipole moment does not experience the same interactions with a hydrogen-bond base stationary phase, as would a solute, which is a strong hydrogen bond acid, and so forth. In each case, the polarity scale is massed in favour of the specific interaction dominated by the polar property of the selected solute [50].

It is important to recognize that in GC, the strengths of the individual solute–stationary phase interactions are very much a function of the solubility of that solute in that stationary phase. These forces attain their full potential only if the solute is fully soluble in the stationary phase. Polar stationary phases are those containing functional groups such as –CN, –CO, –C–O and –OH. Most hydrocarbon-type stationary phases (e.g. squalane and the Apiezon greases) are nonpolar. The low-polarity stationary phases: 100% polydimethylsiloxane and 95% dimethyl-5% diphenylpolysiloxanes are extremely useful and presently enjoy the broadest application. As the percentage of diphenylsiloxane subunits increases with respect to the methylsiloxane, the stationary phase becomes more polar. Phases containing 35–50% diphenyl substitution are considered 'midpolarity' phases. Greater polarity can be conferred by substituting cyanopropyl for up to 50–100% of the methyl groups [51].

In general, the name given to the specific application of LC is determined by the nature of the mobile and stationary phases that are used. Thus, in normal phase LC, the stationary phase is made of the more polar material. When the mobile phase is more polar, the operational mode is called reverse phase LC. Normal and reverse phase partition chromatography involve the use of stationary phases composed of a liquid film coated on or chemically bonded to an inert substrate and a mobile phase of opposite polarity. Chemical bonding precludes the possibility of the liquid film being 'stripped' from the substrate by the mobile phase. Thus, many stationary phases are prepared from silica, the surface of which has been modified by reacting it with a class of compounds called organosilanes.

In reverse phase liquid–solid adsorption chromatography, a polar mobile phase like water or ethylene glycol may be used with a nonpolar stationary phase like finely divided polyethylene or polypropylene powders. In normal phase adsorption chromatography, an aliphatic hydrocarbon may be used for the mobile phase in combination with a polar stationary phase like silica, glass or alumina.

The most commonly used mode of this technique is the reverse mode (RP-LC, reverse phase liquid chromatography) due to its numerous advantages such as use of mobile phases of lower cost and lower toxicity; rapid equilibration of the column after the change of mobile phase; possibility of conducting an elution mode separation gradient, resulting in faster and better separation in the analysis; good reproducibility of retention times; wide field of application because of the possibility of separating compounds with different polarities, molecular weights and functionalities, among others [52].

In RP-LC, stationary phases mostly comprise nonpolar alkyl hydrocarbons such as C-8 or C-18 chains bound to silica or other inert supports. C_{18} columns actually can handle more than 60% of the LC applications [53].

3.2 Analyte-Specific Stationary Phases

PAH-columns are an example of an analyte-specific stationary phase. PAHs are compounds containing two or more fused aromatic rings and there are several hundreds of these compounds. They are typically ubiquitous by-products of incomplete combustion or pyrolysis of carbonaceous and organic matter (e.g. tobacco, coal, charcoal, diesel, gas and wood) [54].

The potential carcinogenic and mutagenic properties of PAHs are well known [55]. Indeed, they are the key components of the 189 hazardous air pollutants in the U.S. Clean Air Act Amendments [54,55]. Furthermore, through long-range transport of polluted air masses, PAHs, particularly in the fine particulate matter fraction, can be transported to receptor sites far from their original sources [56].

Chromatographic methods for PAH analyses in environmental samples have been developed and evaluated extensively over the past few decades, as reviewed in the literature [57,58]. LC or GC are prominent techniques for PAH-related test methods of the EPA. In contemporary analysis of these complex matrices, GC, rather than LC, is often the preferred approach for separation, identification, and quantification of PAHs, largely because GC generally affords greater selectivity, resolution, and sensitivity than LC [57−59].

The most critical aspect of GC method development is selection of an appropriate stationary phase for a specific PAH separation problem [58]. Methyl- and phenyl-substituted polysiloxanes are the most widely used capillary column stationary phases for separation of PAHs in environmental samples.

Methylpolysiloxane phases containing 0%, 5%, 50%, and even 65% phenyl substitution are commercially available and give relatively low background from column bleed, even at high temperatures (>300 °C) and with non-selective detectors. The standard method for determination of PAHs in ambient air [60] uses a 5% phenyl methylpolysiloxane column. However, this generally does not provide results for a very wide range of PAH isomers.

In fact, one of the difficulties when analysing PAHs is achieving separation of isomeric pairs of which there are many of concern, necessitating adequate resolution, even when using GC. Three important pairs of PAHs that typically coelute on traditional 5% phenyl methylpolysiloxane columns are chrysene and triphenylene; benzo[b]fluoranthene and benzo[j]fluoranthene; and dibenz [a,c]- and dibenz[a,h]anthracene [58,61].

Columns containing liquid-crystalline stationary phases have shape selectivity aspects that are well suited for the separation of PAH isomers [58] that are otherwise not quantifiable by GC equipped with conventional capillary columns coated with phenyl methylpolysiloxane. Liquid-crystalline columns have been part of the NIST analytical scheme for the determination of PAHs in environmental standard reference materials (SRMs) [62]. However, this type of column has some limitations, including variations in selectivity, changes in the order of elution of PAHs among different columns and a limited temperature range [58,63].

HPLC in reverse mode with octadecylsilica (C_{18}) stationary phase [64,65] also provides good selectivity for separation of PAHs in environmental samples. There are a large number of papers in the literature reporting the analysis of PAHs employing C_{18} columns, including the determination in air particulate samples [66–69], diesel soot [70,71], biomass emissions [72,73], burning wood [74,76] and diesel exhaust particulates [77–79].

The use of C_{18} and HPLC led to the development of more sensitive methods allowing for detection limits and quantification lower than 1 ppb for isomers to be obtained, particularly alkyl-substituted PAHs with relatively high boiling points.

3.3 Orthogonality for GC × GC Separations

Orthogonality of a multi-dimensional system is obtained when the separation mechanisms of the two columns are completely independent. Generally, the first dimension (1D) stationary phase is nonpolar or of low polarity, where the separation process occurs by differences in boiling point of the analytes; whilst the secondary (2D) column has a polar or medium polarity stationary phase (polarity separation). This factor is of great importance because it determines the magnitude of the separation in two-dimensional space, without losses in the chromatographic resolution. The efficiency of the separation depends on the type of sample and columns employed in the first and second dimension and not necessarily on the orthogonality of the system. In practice, there is a direct relationship between orthogonality and increased peak capacity [80–82].

The combination of two chromatographic columns with stationary phases containing independent separation mechanisms confers increased selectivity and provides the separation of very similar compounds. Thus, it is possible to minimize the peak overlap of compounds (especially when less-intense signals for compounds of interest overlap with the major components of the mixture) [83].

3.4 Developments in LC Stationary Phases

By definition, LC encompasses all chromatographic techniques in which the mobile phase is a liquid. Thus, thin-layer chromatography, size-exclusion chromatography, ion-exchange chromatography, to name only a few techniques, all belong to this class, independent of the nature of the stationary phase or the equipment used. However, this section is devoted to stationary phases for HPLC.

The stationary phases most commonly used in HPLC are C_{18} and C_8, chemically bonded to chromatographic silica. These RP stationary phases, due to the inversion of selectivity with the introduction of organic groups onto a silica support, represent approximately 85% of all HPLC applications.

There are two types of C_{18} chromatographic sorbents based on the bonding chemistry, one is monomeric, the other polymeric. The monomeric type is usually prepared by monofunctional silane, whereas the polymeric materials are prepared via polycondensation of multi-functional silanes [84]. The monomeric phases are preferred due to their better mass-transfer capability. But, polymeric sorbents offer enhanced selectivity towards molecularly rigid solutes, so they are more favoured for separation of conformational isomers, such as PAHs [85,86].

However, organosilanes with reactive groups have substantially enriched the family of HPLC stationary phases via further derivatisation with ligands of suitable size. Meanwhile, efforts to explore new stationary phases have continued. New stationary phases are specifically designed for separation of a diverse range of solutes. Ligands with multi-functional groups, such as cyclodextrin [87,88], calixarene [89–91], crown ethers [92–94], ionic liquids (ILs) [95–97] and some embedded polar groups [98–101], have been bonded to silica. With these new stationary phases, separation of desired solutes from complex matrices can be achieved due to the involvement of multiple interactions. Over the past decade, separation science witnessed an ever-accelerating expansion of HPLC stationary phases induced by studies of new materials. The research highlights of novel stationary phases are promising with respect to useful ligands and/or substrates, amongst which carbonaceous nanomaterials [102–105], ILs [95–97], and metal-organic frameworks [106–108] have attracted great interest due to their unique nature, and their far-reaching influence within separation science. However, these materials generally have modest pore surface areas and wide pore size distributions, thus offering a limited molecular sieving effect on the separation of some molecules, particularly, isomeric small molecules.

Therefore, the development of HPLC stationary phases with novel materials has long attracted extensive attention from analytical and separation scientists in both academia and industry.

4. CHROMATOGRAPHY DETECTORS COMMONLY EMPLOYED IN AIR POLLUTANT ANALYSIS

4.1 General Requirements of Detectors for this Application

The diversity of compound types, sources and processes that regulates the physical and chemical properties of the compounds makes the analysis of atmospheric samples a difficult task. Still today, there is no consensus about the concentration levels of many compounds that can be harmful to human health and to other animals and plants. Therefore, the exact identification of pollutants and their concentrations are important for the study of atmospheric chemistry, for the observation of health consequences and for the determination of public policies.

As has already been mentioned in this chapter, GC is one of the most important and versatile techniques for the separation of mixtures of gases or volatile pollutants, and particulate bound pollutants. Since the inception of GC, over 60 detectors have been developed. Some are designed to respond to most compounds, while others are designed to be selective for particular types of substances.

Desirable detector properties are fast response time, low noise, low cost, same response for all masses, sensitivity and long-term stability. This list of properties makes it clear that is not possible to find all of these characteristics in only one detector. Most of the time increasing the sensitivity can improve the analysis due to the lower detection limit.

Here the most widely used detectors, their application and characteristics are discussed.

4.2 Flame Ionisation Detector

The flame ionisation detector (FID) was introduced in 1958 and since then it has become the most commonly used detector for GC. This is because the FID responds to all organic compounds with good sensitivity. The detector response is not affected by small changes in flow rate, pressure or temperature. Common carrier gas impurities like CO_2 and water also do not disturb the detector, but trace hydrocarbon levels in the gases can affect the baseline stability [109].

The operation of the FID is based on the phenomenon of the formation of charged particles in a hydrogen burner flame. The column eluate is burned in a mixture of H_2 and air. Most organic compounds form ions in a flame (except carbonyls and carboxyls), producing radicals $CH\cdot$, and then CHO^+ in the flame. The electrons flow from the anode to the cathode where they neutralize the CHO^+ ions present in the flame. The ionisation process results in an increase in the electric current between the burner jet (first electrode) and a second electrode located above the flame.

The number of ions collected depends on the number of carbon atoms in the sample and on the oxidation state of the carbon, as completely oxidized atoms do not ionise. Thus, the FID is a mass-sensitive detector, responding to the number of carbon atoms entering the detector per unit time.

The FID is sensitive to most hydrocarbons and is not sensitive to substances like H_2, He, O_2, CO_2, H_2O, NH_3, NO, H_2S and SiF_4.

The application of FID is wide in atmospheric chemistry. For example, determination of the compounds which can be used as molecular markers of different types of emissions [2]. FID was used to detect *n*-alkanes and PAHs in aerosol samples [110]. The distribution of the homologous series of *n*-alkanes ranged from C_{16} to C_{35} representing the contribution of anthropogenic and biogenic emissions. Figure 10 presents the distribution of *n*-alkanes detected in an atmospheric sample collected in an urban area of São Paulo city [110]. The compounds with higher concentrations are shown with the hump of the unresolved complex mixture (UCM).

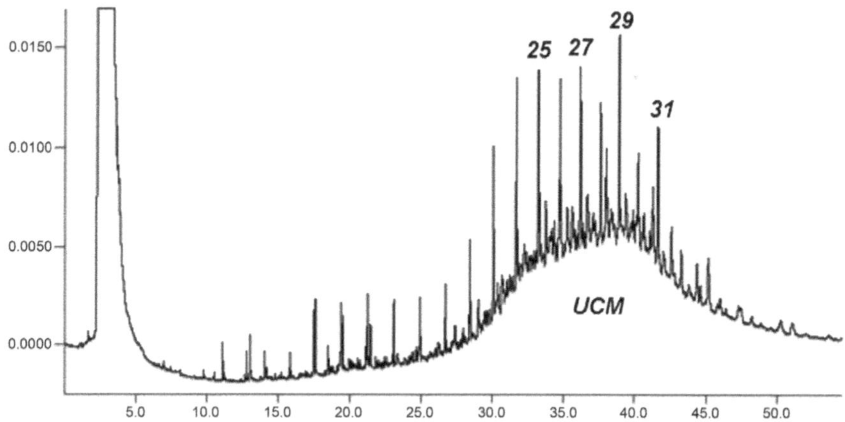

FIGURE 10 Flame ionisation detector chromatogram of *n*-alkanes showing the odd–even distribution of homologues identified in South American atmospheric samples [110]. The odd homologues from C_{25} to C_{31} are in evidence. UCM, unresolved complex mixture.

4.3 Electron Capture Detector

The ECD is highly sensitive for compounds that contain electronegative atoms. It is especially sensitive to molecules that contain halogens, conjugated carbonyls, nitriles, nitro and organometallic compounds, but it is insensitive to hydrocarbons, alcohols and ketones. The humidity decreases the sensitivity.

Hydrogen or nitrogen can be used as the carrier gas. These gases have low excitation energies compared to argon. Only compounds having high electron affinity are ionised by capturing electrons. The cathode consists of a metal impregnated with a β-emitting element, usually tritium or nickel-63. Tritium gives greater sensitivity than nickel, but it has a temperature limit of 220 °C because of tritium losses at high temperatures. Nickel can be used at temperatures up to 350 °C. In addition, nickel is easier to clean. Radioactive sources acquire a surface film that decreases the beta-emission intensity, decreasing the sensitivity [111]. The electrons are attracted to an anode establishing a current. When the analyte molecules, with high electron affinity, are introduced into the detector, they capture some electrons to create a large negative ion. To maintain a constant electric current, the detector responds by varying the pulse frequency of electrical potential between the anode and the cathode.

Many applications of this detector are described in the literature. Some recent studies are presented, which show the determination of important secondary pollutants: peroxyacetyl nitrate (PAN) and its homologues, and peroxypropionyl nitrate (PPN). These atmospheric species are formed from photochemical reactions of volatile organic compounds (VOCs) in the presence of NO_X. It is suggested that PAN acts as a temporary reservoir for NO_X through long-range transport and subsequent release of NO_X, especially in the lower troposphere, where it may control the production of O_3 in the remote marine troposphere. In addition, it has been recognized as being a lachrymator, and is mutagenic, and a phytotoxin [112,113]. Because GC-ECD is specific and selective, it is presented as a good method for the determination of PAN and PPN.

The GC-ECD method is sensitive to oxygen, therefore, a large oxygen peak was observed in the quantification of PAN and CCl_4. After optimization of the GC-ECD conditions, an interesting method was developed to better determine these photochemical pollutants [113].

The ECD is also largely used in the determination of pesticides in environmental samples. Persistent organic pollutants (POPs) are characterized by low water solubility and high lipid solubility, resulting in bioaccumulation in the tissues of living organisms. Many analytical methods were developed to measure these POPs and GC simultaneously coupled to MS and electron capture detection has been used [114].

Another import use of ECD is to detect the nitrated PAHs (nitro-PAHs) in ultrafine aerosols. These compounds are extremely important since some of them are carcinogenic and mutagenic [115]. After clean up, the extracts were derivatised where the original nitro-PAHs were fluorinated through specific chemical reactions. In this way, 1-nitronaphthalene, 2-nitrofluorene, 3-nitrofluoranthene, 1-nitropyrene and 6-nitrochrysene were successfully analysed by GC with electron capture detection [116].

4.4 Nitrogen Phosphorus Detector

Thermionic detectors for the analysis of phosphorus- and nitrogen-containing compounds were proposed by Karmen and Guifridda (1964) [117].

The NPD can be seen as a modified FID with a bead of an alkali metal salt mounted between the flame tip and the collector electrode with a flame temperature sufficient to vaporise the alkali metal salt and generate a stable population of alkali metal ions necessary for the thermionic process [118]. This detector employs a mixture of air and H_2 gas, but the gas composition will not by itself support a flame. The presence of an alkali metal salt bead or ceramic material, located just above the detector jet tip and before the collector electrode, suggests that the gases form a low-temperature plasma in the detector. The metal salt is formed at a ceramic bead on an electrical wire, the voltage of which can be separately controlled to maximize the signal and reduce the background response. The selective response comes when the degradation products of nitrogen and phosphorus compounds become negatively charged by electron-extraction processes. In comparison with FID, the sensitivity of this type of detector is 50–500 times higher for nitrogen- and phosphorus-containing compounds. It is largely used in the analysis of pesticides and drugs found in environmental samples.

Of extreme importance in atmospheric chemistry is the investigation of atmospheric nanoparticles due to their potential to affect human health, therefore, scientists investigate their toxicity, emission source, generation mechanism and environmental fate. Ochiai and collaborators [119] reported the use of TD-GC × GC and HRTOF-MS to analyse atmospheric nanoparticles collected in roadside dust with simultaneous detection using NPD and quadrupole mass spectrometry (qMS). They tentatively identified several nitrogen-containing compounds like benzonitrile, quinolone, nicotine and many others. The resulting mass spectra obtained at the elution times identified by NPD were used for the NIST library search.

Using NPD and MS detectors, organophosphorus pesticides were studied at trace levels in the atmosphere [119]. These pesticides are among the most widespread in the environment. They present a large degree of persistence which causes various health and safety problems. Concerning the analytical method, these compounds are normally analysed GC coupled to a selective detection method such as ECD or NPD.

4.5 Photoionisation Detector

The photoionisation detector (PID) is recognised as a sensitive detector with application particularly in environmental analyses and is largely used in the determination of VOCs in samples collected in forest areas, for example. An ultraviolet lamp emits high-energy photons to ionise many organic compounds as they elute from the column. The excitation of the analyte molecules takes place in a small ionisation chamber attached to the window of a gas discharge lamp, where the molecules are ionised. The lamp contains gases such as krypton or argon excited by a direct current at a potential of about 1000 V, or a radio frequency (RF) of 100 KHz. Radiation emitted by the krypton lamp has 10.2 eV energy corresponding to a wavelength of 121 nm, whilst argon emission is at 11.7 eV. The window is made of magnesium fluoride or lithium fluoride, transparent to the emissions. The ionisation chamber contains a repeller electrode at a potential of 100–200 V, which directs the ions to a shielded collector electrode [118].

The ions are collected and amplified to form the signal. The discharge ionisation detector is a related type that uses a spark to generate high-energy photons.

The most used lamp, krypton, has sufficient energy to ionise alkenes and aromatics and some alkanes, alcohols, esters and aldehydes. It can be used to selectively detect aromatic compounds in the presence of alkanes and other organic materials.

One interesting application of the PID is presented in the work developed by Harley et al. [120] where the air coming out from an enclosure system is collected onto cartridges and are analysed to study the emission of isoprene by tropical plants. The enclosure system contains a chamber (cuvette) with a glass cover, where part of the plant is inserted, and the temperature and the incident radiation are monitored by sensors. In this study, for initial screening of VOC emissions a hand-held PID was used. The emission inferred by PID screening was confirmed using other techniques.

4.6 Mass Spectrometer (MS) Detector

In 1913, J.J. Thompson demonstrated that neon consists of different isotopic species having atomic weights of 20 and 22 g mol^{-1}. Before him, Goldstein (1886) discovered positive entities and Weins (1898) demonstrated that positively charged ions could be deflected by electrical and magnetic fields [121].

A mass spectrometer is an instrument that measures the mass-to-charge ratio (m/z) of gas-phase ions and provides a measure of the abundance of each ionic species. The measurement is calibrated against ions of known m/z. In GC-MS, the charge is almost always 1 so that the calibrated scale is in Daltons or atomic mass units. All mass spectrometers operate by separating gas-phase ions in a low-pressure environment by the interaction of magnetic or electrical fields on the charged particles, as discussed in Chapter 10.

Analyte molecules must first be ionised in order to be attracted or repelled by the applied magnetic or electrical fields. There are numerous ionisation techniques and electron ionisation (EI) is the most common and most simple one.

The MS is widely used because of its high sensitivity, selectivity and time resolution. Ionisation is usually accomplished by the impact of a highly energetic (70 eV) electron beam. An ionising electron excites the parent molecule, ejecting one electron and generating the molecular ion. This species is not stable and rapidly decomposes to more stable fragments. The fragment with highest abundance is called the base peak and the data system plots it as 100% of the spectrum scale. Other peaks are plotted relative to the base peak, and the result is a typical mass spectrum.

Data can be plotted as a total scan (TIC – total ion chromatogram) or as a small number of individual ions (SIM – selected ion monitoring). A TIC is used to identify unknown compounds when a specified mass range is scanned. All the peaks in this range are reported, so the mass spectra can be retrieved from the computer and be used to identify each peak. The computer database compares the unknown mass spectrum with reference spectra in its library files. Otherwise, in SIM, only a small number of ions are monitored, and it cannot be used for qualitative analyses [122].

As mentioned, the mass analyser separates ions according to m/z. The most common analysers for GC are the quadrupole and ion-trap types.

The quadrupole analyser consists of four cylindrical rods oriented in a square arrangement. RF and direct current potentials are applied to the rods enabling ions with specific m/z to have a stable trajectory and pass through to the detector. By increasing the RF and DC potentials, ions of increasing m/z will pass through the analyser and be detected. Two advantages of the quadrupole analyser are its fast scanning rate and lower cost.

In the ion-trap analyser, ions can be confined by electric and magnetic fields. Ions of specific m/z value circulate in stable orbits within the analyser. A relatively high pressure of helium is used to stabilize the ion trajectories. As the RF is increased, ions of lower m/z values are destabilised and pass into the detector. Ions are typically introduced from the ion source in a pulsed fashion and the RF frequency is quickly scanned to produce the spectrum. These analysers are inexpensive and very sensitive.

The time-of-flight (TOF) analysers separate ions according to their migration time down a field-free region after acceleration. Since the TOF mass filter is an integrating rather than a scanning detector, the acquisition rate is limited only by the ion pulse frequency and the spectrum storage speed [123].

GC-MS systems are important for most analytical laboratories, as they play a major role in all environmental, food, petrochemical, chemical and pharmaceutical analyses. In atmospheric chemistry, GC-MS has been used in the analysis of hundreds of compounds. It is used for qualitative identification of unknown compounds and to give accurate quantitative determination.

Of special importance is the determination of compounds emitted by biogenic sources. Biogenic non-methane VOCs, such as isoprene and terpenes, including a large number of alkanes, alkenes, esters, carbonyls and acids are found in the atmosphere ranging from low ppb to several ppb in concentration. Some species can be very reactive under atmospheric conditions and can influence the oxidative capacity of the atmosphere. In urban areas, they contribute to the formation of secondary air pollutants and greenhouse acids, carbonyls, ozone, carbon monoxide and methane. To analyse these VOCs, TD is required that allows for the rapid desorption of the compound from the sampling cartridge (refer Chapter 4). After separation on the chromatographic column, the compounds are detected by MS.

Figure 11 presents the mixing ratios of the most abundant terpenoid and oxygenated species found in samples collected in the Amazon forest [124].

5. SELECTED APPLICATIONS OF CHROMATOGRAPHY IN THE ANALYSIS OF AIR POLLUTANTS

There are thousands of known organic compounds in the atmosphere distributed in the gaseous, particle and aqueous phases. In spite of this, studies from literature so far only were able to identify around 10–40% of the emitted organic mass, depending on the source [125]. From those already studied, about 20–80% of organic matter found in airborne particulate matter and vapour emissions are a complex mixture of nonpolar (normal, branched, mono-aromatic and/or polyaromatic hydrocarbons), polar (alcohols, carboxyls, esters and carbonyls), halogenated (pesticides, dioxin-like and perfluorinated) substances as well as compounds of sulphur, nitrogen, phosphorus, and metal-organics. These organic compounds possess a wide range of solubilities, re-activities and physical properties contributing to the continuing challenge of

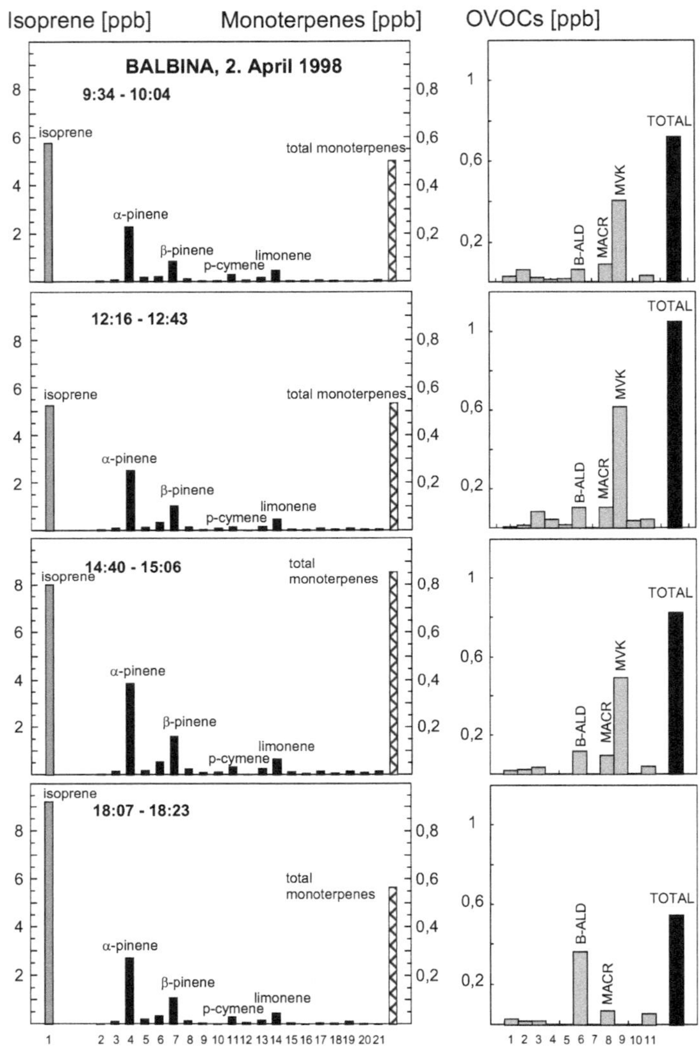

FIGURE 11 Mixing ratios of terpenoid and oxygenated species found in samples collected in the Amazon forest. The names of the most abundant species are presented. On the right side: benzaldehyde (B-ALD), methacrolein (MACR) and methylvinylketone (MVK). *Copyright (2011) from Elsevier [124].*

the identification of them. Moreover, added analytical complexity is attributed to the trace-level at which most organic compounds are found in atmosphere, as well as difficulties in sampling [126,127].

Depending on physicochemical characteristics, the organic compound content of atmospheric samples is mainly analysed either by mono-dimensional GC (mostly using GC-MS, or GC-ECD) and LC (coupled to MS, fluorescence detector or diode-array detector), or two-dimensional GC × GC coupled to a mass spectrometer detector. Some case studies reported in literature are presented in this section.

5.1 Persistent Organic Pollutants

Persistent organic pollutants (POPs) represent a group of anthropogenic organic chemicals of particular interest due to concern that they are resistant to degradation in the environment. Generally, POPs resist photolytic, biological or chemical degradation, in varying degrees, depending on their chemical structure. Since POPs possess low water solubility and high lipophilicity, they are subject to bioaccumulation in human and animal fatty tissues, and then biomagnify along food chains. Indeed, POPs are associated with chronic ecotoxicological effects [128–130].

Considering the major environmental and health aspects associated with POPs, considerable international efforts are being sought to control their worldwide distribution in the last decades. The Stockholm Convention on POPs classifies the following compounds as persistent: organochlorine pesticides (OCPs), polychlorinated biphenyls (PCBs), polybrominated diphenyl ethers (PBDEs) and polychlorinated dibenzo-*p*-dioxins and furans (PCDD/Fs), with more compounds being proposed for listing. Under this treaty, parties and signatories should monitor these chemicals and identify their source, as well as develop a national inventory [131,132]. So far, a number of monitoring studies on atmospheric POPs on local, regional and global scale have been conducted [129,133]. However, tracing past use and atmospheric fate of these compounds are not an easy task [133,134].

POPs occur in typically very low atmospheric concentrations, such as $\sim 1\ pg\ m^{-3}$ (1 ppq, parts per quatrillion) or even lower [133]. Considering such low ambient air concentration levels together with the large number and variety of POPs, there are considerable sampling and analytical challenges regarding their determination. In this way, it becomes highly necessary to sample air volumes that are large enough to overcome analytical limits of detection (LODs). Depending on the sampling site, it is advisable to sample air volumes of $\sim 50-100\ m^3$ for vapour phase and $\sim 500-1000\ m^3$ for particulate matter. POPs concentrations are variable both spatially and temporarily. Due to their low levels in air, they necessarily require elaborate sampling followed by high-performance, cost-effective analytical methods with skilled analysts. Real-time analysis of POPs is not yet feasible with existing instrument sensitivity [129,134].

Due to the large number of POP congeners and isomers, it is highly recommended to use high-resolution, highly sensitive and comprehensive chromatography systems. Considering most of the POPs are halogenated, one good option could be GC-ECD. But this detector, in spite the fact it is very sensitive for detection of halogenated species, does not unequivocally identify the molecular structures of POPs. Another good alternative would be GC coupled to a high-resolution mass spectrometer detector (GC-HRMS) or GC-HR-MS/MS, either in electronic ionization (EI) or negative ion chemical ionisation (NICI) modes [135]. Recently, comprehensive GC × GC, which is a high-resolution technique, has been used together with a high-resolution mass spectrometer for the determination of POPs [127,136,137]. Two-dimensional GC potentially allows global and selective detection of analytes in atmospheric samples as they are, without any clean-up steps prior to analysis.

Hashimoto et al. [127] successfully detected more than 200 POPs (16 PCDDs, 19 PCDFs, 62 PCBs, 39 PBDEs, 21 chlorinated PAHs, 11 brominated PAHs and 43 other POPs) in fly ash certified reference material (National Institute for Environmental Studies − NIES CRM No. 17) by using two-dimensional gas chromatography−tandem mass spectrometry (GC × GC−MS/MS). These POPs were measured in a neutral loss scan (NLS) mode with the GC × GC-MS/MS in order to optimize the conditions and parameters of the instrument, achieving then high levels of separation and sensitivity during analysis. Figure 12 shows the two-dimensional TIC of fly ash extract measured with ^{35}Cl-NLS and conventional modes. Many peaks in Figure 12 assumed to be organochlorine were detected in the ^{35}Cl-NLS, which were distributed between peaks estimated as non-substituted alkanes and PAHs. The total number of detected peaks in this mode was more than 1000.

5.2 Polycyclic Aromatic Hydrocarbons and Their Derivatives

PAHs are ubiquitous environmental contaminants, mainly formed during incomplete combustion or pyrolysis (such as biomass burning, forest fires, industrial fumes, fossil fuel burning and hydrothermal processes) [110,138−141]. Their chemical structures contain two or more fused benzene rings arranged in linear, angular or in cluster dispositions, forming a varied group of more than 100 molecules.

PAHs are widely recognized not only by their carcinogenic and mutagenic potential but also for their potential persistence and bioaccumulation [136,139]. With regard to toxicity, the parent PAHs may undergo photochemical reactions with NO_3, OH and/or O_3 yielding even more toxic PAH derivatives, such as nitro-PAHs and oxy-PAHs (quinones and hydroxy-PAHs) [55,136,137,139,142]. Oxy-PAHs and nitro-PAHs are secondarily formed in the atmosphere, therefore they are used as secondary process tracers.

The most used instruments for PAH determination are high-performance liquid chromatography coupled to fluorescence detection (HPLC-FLD) and gas chromatography coupled to mass spectrometry (GC-MS). Both methods present advantages and disadvantages. HPLC-FLD presents higher sensitivity than GC-MS and it is suggested to be used in samples collected with low total air volumes, such as cascade impactors [138,143]. On the other hand, the use of GC-MS with selective ion monitoring (SIM) or NICI provides structural information and also a reduction of matrix effects, which could happen with

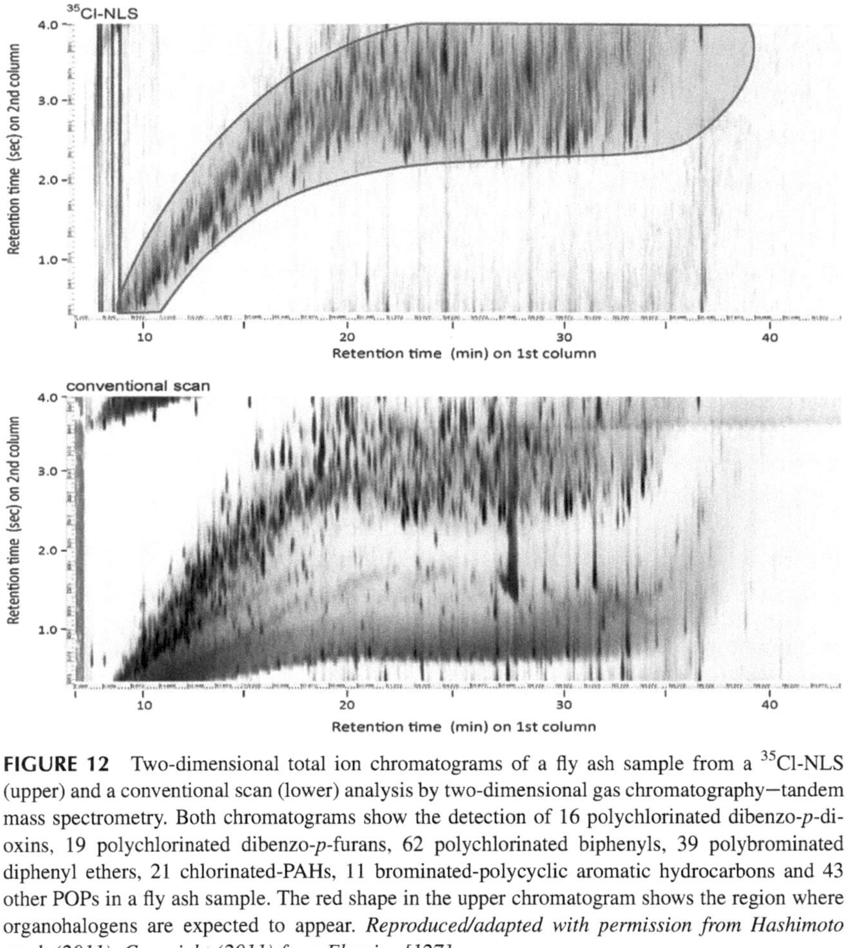

FIGURE 12 Two-dimensional total ion chromatograms of a fly ash sample from a ^{35}Cl-NLS (upper) and a conventional scan (lower) analysis by two-dimensional gas chromatography—tandem mass spectrometry. Both chromatograms show the detection of 16 polychlorinated dibenzo-*p*-dioxins, 19 polychlorinated dibenzo-*p*-furans, 62 polychlorinated biphenyls, 39 polybrominated diphenyl ethers, 21 chlorinated-PAHs, 11 brominated-polycyclic aromatic hydrocarbons and 43 other POPs in a fly ash sample. The red shape in the upper chromatogram shows the region where organohalogens are expected to appear. *Reproduced/adapted with permission from Hashimoto et al. (2011). Copyright (2011) from Elsevier [127].*

HPLC-FLD to some extent [139,143]. Additionally, by using similar sample extraction procedures and gas chromatographic programmings, it may be possible to determine PAHs and some of their nitro- and/or oxy-derivatives present in Hi-Vol collected samples. Alternatively, when the goal is to obtain an exploratory and more comprehensive determination of PAHs and PAH-derivatives, GC × GC-MS/MS or GC × GC-TOF/MS could be used to avoid the UCM normally present in GC-MS chromatograms [126,136,137]. The main disadvantage of GC × GC is that it is sometimes time-consuming due to data-processing requirements for the large quantity of information acquired in one single chromatogram.

In recent studies published by Barrado et al. [144,145], 13 PAHs, 5 nitro-PAHs and 2 hydroxy-PAHs were determined in vapour phase and particulate matter ($PM_{2.5}$ and PM_{10}) collected in a suburban area of Madrid, Spain. This determination was done using liquid chromatography coupled to a fluorescence detector (HPLC-FLD). The LODs were about 1; 5; and 5.0–10 ng mL^{-1} for PAHs, hydroxy-PAHs and nitro-PAHs, respectively [144].

In this study, gaseous phase samples from the coldest months contained 18,900 ± 2140 pg m^{-3} and 150 ± 97 pg m^{-3} of PAHs and nitro-PAHs, respectively. In turn, samples from the warmest months contained lower levels of both PAHs (9293 ± 1178 pg m^{-3}) and nitro-PAHs (97 ± 13 pg m^{-3}). In addition, total vapour phase PAH concentrations were one order of magnitude higher than in particulate matter, whereas the gaseous nitro-PAH concentrations were comparable to their aerosol concentrations [145].

In a study done by Albinet et al. [146], an analytical method was developed using GC-MS with NICI for the determination of 17 nitro-PAHs and 9 oxy-PAHs present at low concentrations in the atmosphere. Applications of this method were done on SRM 1649a urban dust reference material and on ambient samples collected in the French Alps. Examples from chromatographic separations for both oxy-PAHs and nitro-PAHs in a particulate ambient air sample are found in Figures 13 and 14, respectively. The LODs ranged from 0.01 to 2.60 pg and 0.03–0.07 pg for oxy-PAHs (9-fluorenone and 1,4-naphthoquinone) and for nitro-PAHs (3-nitrophenanthrene and 6-nitrobenzo[a]pyrene), respectively.

Nitro-PAH concentrations in a rural site during winter were about 0.2–100 pg m^{-3} in particulate phase and <LOD to 20 pg m^{-3} in the gas phase. In turn, oxy-PAH levels were approximately 10–100 times higher, ranging

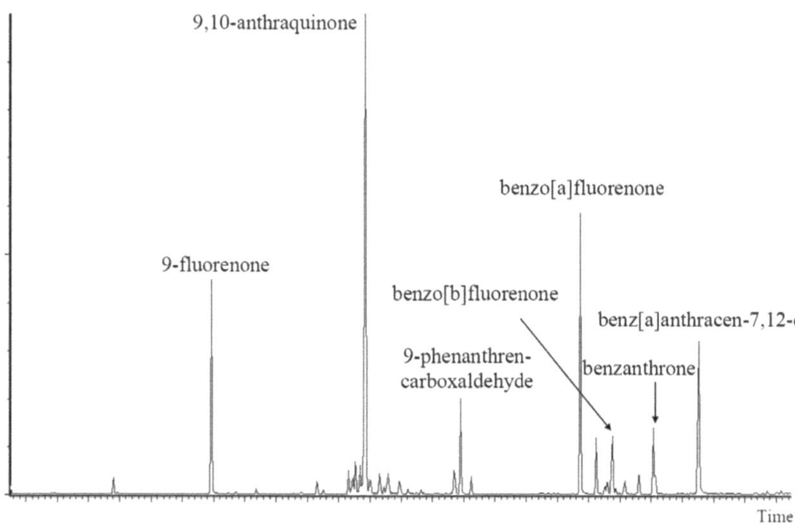

FIGURE 13 Reconstituted gas chromatography coupled to mass spectrometry total ion chromatogram for oxy-polycyclic aromatic hydrocarbons in a particulate ambient air sample. *Reproduced/adapted with permission from Albinet et al. [144]. Copyright (2006) from Elsevier.*

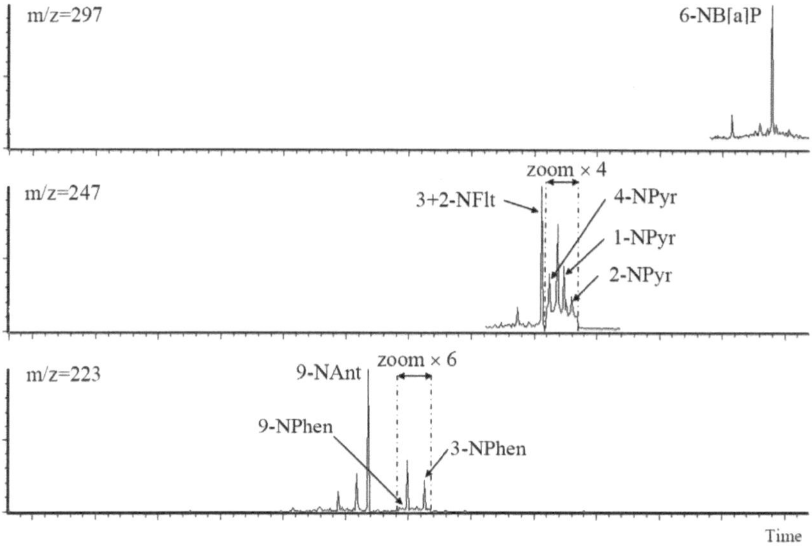

FIGURE 14 SIM chromatograms using molecular ions in a particulate ambient air sample. m/z = 223: 9-NAnt: 9-nitroanthracene; 9-NPhen: 9-nitrophenanthrene; 3-NPhen: 3-nitrophenanthrene; m/z = 247: 3+2-NFlt: 3+2-nitrofluoranthene; 4-NPyr: 4-nitropyrene; 1-NPyr: 1-nitropyrene; 2-NPyr: 2-nitropyrene; m/z = 297: 6-NB[a]P: 6-nitrobenzo[a]pyrene. *Reproduced/adapted with permission from Albinet et al. [144]. Copyright (2006) from Elsevier.*

from 0.1 to 2.0 ng m^{-3} and <LOD to 1.4 ng m^{-3} for aerosols and vapour phase, respectively.

5.3 Volatile Organic Compounds

VOCs are reactive substances in the atmosphere, which have a strong impact on atmospheric oxidative capacity [147,148]. VOCs in the atmosphere, which are emitted from many anthropogenic and biogenic sources, are numerous, diverse and ubiquitous. Anthropogenic VOCs (generally represented by some alkanes, alkenes, monoaromatics, polyaromatics and halocarbons) are very important mainly nearby urban or industrial regions. However, emissions of biogenic volatile organic compounds (BVOCs) constitute about 90% of global VOC emissions, estimated to be ~1150 Tg C annually [42]. BVOCs include not only isoprenoid substances, such as isoprene and terpenes, but also a large number of other species (alkanes, alkenes, alcohols, esters, carbonyls and acids), ranging from a few to several ppb [124]. BVOCs are oxidized in the atmosphere in the presence of NO_x, yielding tropospheric ozone, which is a strong oxidant and harmful to living organisms. Furthermore, oxidation products of BVOCs contribute to the formation of secondary organic aerosols through their condensation onto existing particles or the formation of new particles [42].

Many attempts to measure VOCs in the atmosphere are described in the literature. Most studies employ GC-FID or GC-MS as the main analytical techniques but recently there have been studies using GC × GC-MS. For instance, the study done by Maleknia et al. [149] applies GC-MS, proton-transfer reaction mass spectrometry (PTR-MS) and direct analysis in real-time (DART) MS to study VOC emissions from a series of eucalypts at temperatures ranging from ambient to combustion. Figure 15 shows the results of GC-MS analyses of VOCs emitted from fresh and senescent (brown) leaves of *Eucalyptus grandis* when heated at 40 °C for 5 min.

Some compounds (ethylvinylketone, diethylketone, 2-ethylfuran, 1,1-dietheylcyclopropane, hexanal, 2-hexenal, 3-hexene-1-ol and 1-hexenol) with short retention times (1.3−4.1 min) are present only in fresh leaves. In addition, some terpenes (α-pinene, β-pinene, α-phellandrene, eucalyptol, γ-terpinene) are emitted even from senescent leaves. In the same study, PTR-MS enabled trace level detection of VOCs from a complex forest atmosphere, and DART TOF-MS provided high mass accuracy and was unique in analysing plant tissues in open air.

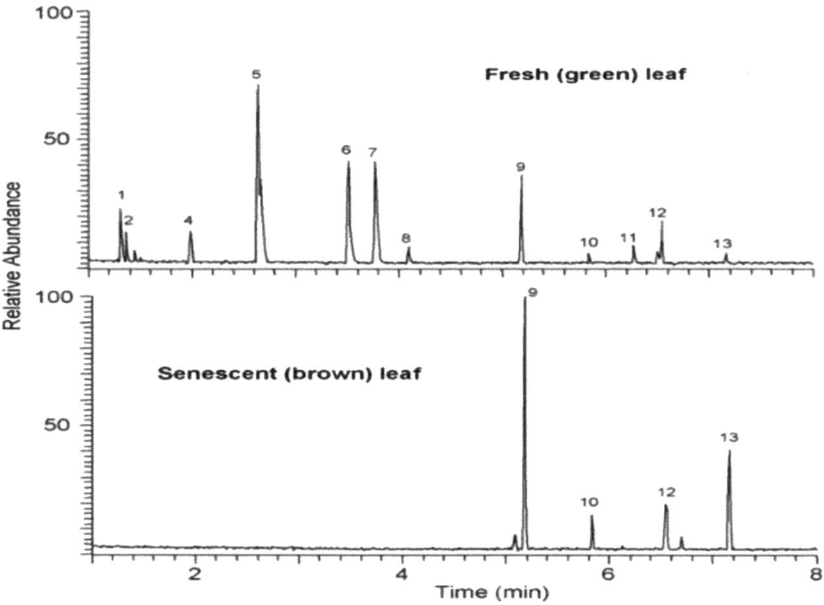

FIGURE 15 Gas chromatography coupled to mass spectrometry chromatogram of volatile organic compounds after 5 min incubation of leaves of *Eucalyptus grandis* in sealed headspace vials at 40 °C. The upper chromatogram refers to a fresh (green) leaf and the lower chromatogram is from a senescent (brown) leaf. Peaks are: #1: ethylvinylketone, #2: diethylketone, #3: 2-ethylfuran, #4: 1,1-diethylcyclopropane, #5: hexanal, #6: 2-hexenal, #7: 3-hexene-1-ol, #8: 1-hexenol, #9: α-pinene, #10: β-pinene, #11: α-phellandrene, #12: eucalyptol, #13: γ-terpinene. *Reproduced/adapted with permission from Maleknia et al. [149]. Copyright (2009) from Elsevier.*

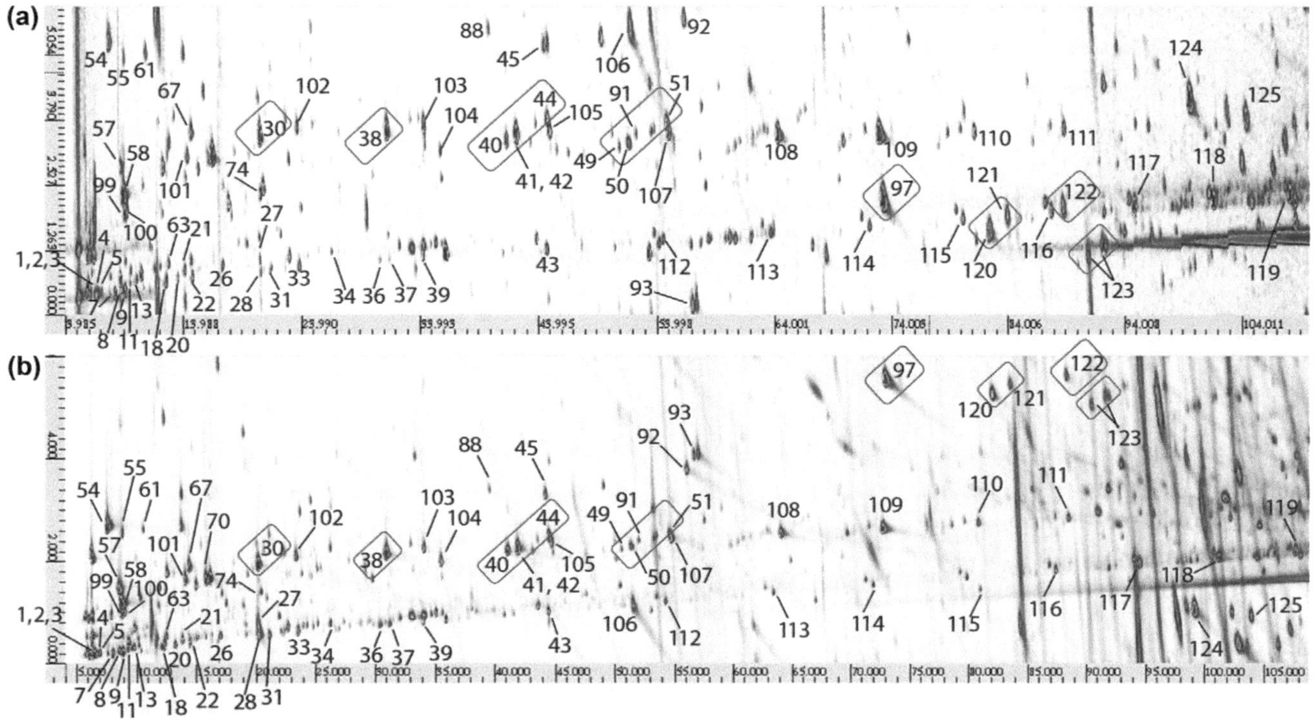

FIGURE 16 Two-dimensional gas chromatography with quadrupole mass spectrometry (a) and two-dimensional gas chromatography with flame ionisation detector (b) chromatograms of a Beijing air sample. In the chromatogram there are about 125 volatile organic compounds. See peaks names in [150]. *Reproduced/adapted with permission from Wang et al. [150]. Copyright (2014) from Elsevier.*

Another example is the study by Wang et al. [150], which used comprehensive GC × GC with qMS (GC × GC-qMS) for the identification of 125 VOCs in a Beijing urban air sample. The list of detected VOCs includes C1–C12 organic compounds, such as 31 alkanes, 16 alkenes, 2 terpenes, 12 monoaromatics, 7 polyaromatics, 30 halocarbons, 26 oxygenates and 1 sulphurated hydrocarbon (Figure 16). In this study a common qMS was combined with GC × GC with the purpose of the identification of VOCs in air samples and qualitative analysis by GC × GC-FID.

6. CHALLENGES AND FUTURE OUTLOOK

Modern GC instrumentation allows for the chemical constituents of a sample to be determined with precision, covering components present in trace amounts. The need to report an increasing number of components in a sample may require more than a single column analytical separation. A multi-column system can provide increased separation, improving the capability to measure more analytes.

It is clear that the interest in GC × GC will continue to increase. Developments in the MS field may lead to a reduced necessity for complete sample separation, prior to the ion source. The superior separation efficiency and sensitivity will lead to a better understanding of the composition of aerosols, their sources and the earth's climate system.

Many studies show that further evolution in the MS field will be welcome. GC-MS instrumentation employing hybrid technology, combining routine qualitative and quantitative analysis into one instrument, may save instrument costs and will increase the amount of information obtained from a single analysis. This is particularly important for the analysis of VOCs.

Many detector modes are available, and one of the most interesting is TOF-MS. The development of automated identification procedures will also be important to shorten the analysis time.

The development of new instruments that can be used *in situ* will be important because many compounds present in atmospheric samples are reactive which can lead to errors in the results.

REFERENCES

[1] IUPAC, International Union of Pure and Applied Chemistry, Commission on Analytical Nomenclature, Recommendations on nomenclature for chromatography, Pure Appl. Chem. 37 (1974) 445–462.

[2] B. Kolb, L.S. Ettre, Static Headspace-Gas Chromatography: Theory and Practice, Wiley-VCH, 1997.

[3] F.G. Kitson, B.S. Larsen, C.N. McEwen (Eds.), Gas Chromatography and Mass Spectrometry, Academic Press, 1996.

[4] W. Jennings, E. Mittlefehldt, P. Stremple, Analytical Gas Chromatography, second ed., Elsevier Inc, 1997.

[5] R.L. Grob, E.F. Barry, Modern Practice of Gas Chromatography, fourth ed., John Wiley & Sons, 2004.
[6] K. Dettmer-Wilde, W. Engewald, Practical Gas Chromatography: A Comprehensive Reference, Springer-Verlag Berlin Heidelberg, 2014.
[7] B. Salih, Ö. Çelikbıçak, Gas Chromatography – Biochemicals, Narcotics and Essential Oils, Intech Books, 2012.
[8] W.M.A. Niessen, Current Practice of Gas Chromatography-Mass Spectrometry, in: Chromatographic Science Series, CRC Press, 2001.
[9] H.F. Linskens, J.F. Jackson, Gas Chromatography/Mass Spectrometry, Springer-Verlag Berlin Heidelberg, 1986.
[10] K.J. Hyver, P. Sandra, High Resolution Gas Chromatography, third ed., Hewlett-Packard Corporation, HP Part No. 5950-3562, 1989.
[11] H.M. McNair, J.M. Miller, Basic Gas Chromatography, second ed., A John Wiley & Sons, Inc, 2011.
[12] F. Gritti, G. Guiochon, J. Chromatogr. A 1302 (2013) 1–13.
[13] D.A. Skoog, J.J. Leary, Principles of Instrumental Analysis, fourth ed., Saunders College, 1992.
[14] R. Ciola, Fundamentals of High Performance Liquid Chromatography, Edgard Blücher Ltda, 1998.
[15] A. Fallon, R.F.G. Booth, L.D. Bell, Applications of HPLC, in: Biochemistry, third ed., Elsevier Science Publishers, 1993.
[16] L.R. Snyder, J.J. Kirkland, Introduction to Modern Liquid Chromatography, second ed., John Wiley and Sons, 1979.
[17] D. Corradini, E. Eksteen, R. Eksteen, P. Schoenmakers, N. Miller, Handbook of HPLC, CRC Press, 2011.
[18] Z.Y. Liu, J. Phillips, J. Chromatogr. Sci. 29 (1991) 227–231.
[19] C. von Muhlen, E.C. Oliveira, P.D. Morrison, C.A. Zini, E.B. Caramao, P.J. Marriott, J. Sep. Sci. 30 (2007) 3223–3232.
[20] M.K. Jennerwein, M. Eschner, T. Groeger, T. Wilharm, R. Zimmermann, Energy Fuels 28 (2014) 5670–5681.
[21] C.J. Venkatramani, J.B. Phillips, J. Microcol. Sep. 5 (1993) 511–516.
[22] J. Blomberg, P.J. Schoenmakers, J. Beens, R. Tijssen, J. High Resol. Chromatogr. 20 (1997) 539–544.
[23] J. Beens, R. Tijssen, J. High Resol. Chromatogr. 20 (1997) 131–137.
[24] R.B. Gaines, G.S. Frysinger, Environ. Sci. Technol. 33 (1999) 2106–2112.
[25] C. Vendeuvre, R.R. Guerrero, F. Bertoncini, L. Duval, D. Thiebaut, M.C. Hennion, J. Chromatogr. A 1086 (2005) 21–28.
[26] D.A. Kidwell, L.A. Riggs, Forensic Sci. Int. 145 (2004) 85–96.
[27] P.J. Marriott, R. Shellie, Trends Anal. Chem. 21 (2002) 573–583.
[28] P.J. Marriott, P.D. Morrison, R. Shellie, M.S. Dunn, E. Sari, D. Ryan, LCGC Eur. 16 (2003) 23–31.
[29] J.B. Phillips, J. Beens, J. Chromatogr. A 856 (1999) 331–347.
[30] M.L. Feo, E. Eljarrat, D. Barcelo, Trends Anal. Chem. 29 (2010) 692–705.
[31] M. Tankiewicz, J. Fenik, M. Biziuk, Trends Anal. Chem. 29 (2010) 1050–1063.
[32] M.P. Pedroso, L.A.F. Godoy, C.H.V. Fidélis, E.C. Ferreira, R.J. Poppi, F. Augusto, Quim. Nova 32 (2009) 421–430.
[33] T. Górecki, J. Harynuk, O. Panic, J. Sep. Sci. 27 (2004) 359–379.
[34] L. Mondello, P.Q. Tranchida, P. Dugo, G. Dugo, Mass Spectrom. Rev. 27 (2008) 101–124.

[35] J. Dallüge, J. Beens, U.A.Th. Brinkman, J. Chromatogr. A 1000 (2003) 69–108.
[36] J.M. da Silva, C.A. Zini, E.B. Caramão, J. Chromatogr. A 1218 (2011) 3166–3172.
[37] P.J. Marriott, Gas chromatography, in: Chromatography, sixth ed., J. Chromatogr. Libr., Elsevier, London, 2004.
[38] C.F. Poole, The Essence of Chromatography, Elsevier, 2003.
[39] F.R. Aquino Neto, Scientia Chromatographica 4 (2010) 51–67.
[40] K. Dettmer-Wilde, W. Engewald (Eds.), Practical Gas Chromatography, Springer, 2014.
[41] O.D. Sparkman, Z. Penton, F.G. Kitson, Gas Chromatography and Mass Spectrometry: A Practical Guide, Academic Press Elsevier, 2011.
[42] S. Koning, H.G. Janssen, U.A.T. Brinkman, Chromatographia 69 (2009) 33–78.
[43] E. Woolfenden, J. Chromatogr. A 1217 (2010) 2674–2684.
[44] E. Woolfenden, J. Chromatogr. A 1217 (2010) 2685–2694.
[45] J.A. Koning, P. Blokke, P. Jünge, G. Alkerna, U.A.T. Brinkrnan, Chromatographia 56 (2002) 185–190.
[46] C.F. Poole, S.K. Poole, Chromatography Today, Elsevier, 2012.
[47] P. Campíns-Falcó, R. Herráez-Hernández, Y. Moliner-Martínez, Liquid chromatography: multidimensional chromatography, in: Reference Module in Chemistry, Molecular Sciences and Chemical Engineering, Elsevier, 2015.
[48] X. Li, Y. Yang, X. Xu, C. Xu, J. Hong, J. Cleaner Prod. 67 (2015) 79–87.
[49] J.F. Lawrence, Liquid Chromatography in Environmental Analysis, Springer Science & Business Media, 2013.
[50] M.H. Abraham, C.F. Poole, S.K. Poole, J. Chromatogr. A 842 (1999) 79–114.
[51] P. Jandera, Anal. Chim. Acta 692 (2011) 1–25.
[52] D.V. McCalley, J. Chromatogr. A 1217 (2010) 858–880.
[53] C. Galea, D. Mangelings, Y.V. Heyden, Anal. Chem. Acta 886 (2015) 1–15.
[54] J.E. Szulejko, K.-H. Kim, R.J.C. Brown, M.-S. Bae, Trend Anal. Chem. 61 (2014) 40–48.
[55] S.K. Pandey, K.H. Kim, R.J.C. Brown, Trend Anal. Chem. 30 (2011) 1716–1739.
[56] M.-S. Bae, J.-S. Shin, K.-Y. Lee, K.-H. Lee, Y.J. Kim, Sci. Total Environ. 466–467 (2014) 56–66.
[57] K.-H. Lee, in: W. Kleiböhmer (Ed.), Environmental Analysis. Handbook of Analytical Separations, Elsevier Science, 2001.
[58] D.L. Poster, M.M. Schantz, L.C. Sander, S.A. Wise, Anal. Bioanal. Chem. 386 (2006) 859–881.
[59] F.J. Santos, M.T. Galceran, Trends Anal. Chem. 21 (2002) 672–685.
[60] US Environmental Protection Agency, EPA/625/R-96/010b, Cincinnati: Environmental Protection Agency (1999) In: http://www.epa.gov/ttn/amtic/airtox.html
[61] S.A. Wise, M.M. Schantz, B.A. Benner, M.J. Hays, S.B. Schiller, Anal. Chem. 67 (2005) 1171–1178.
[62] S.A. Wise, Polycyclic Aromat. Compd. 22 (2002) 197–230.
[63] L.C. Sander, M. Schneider, S.A. Wise, C. Woolley, J. Microcolumn Sep. 6 (1994) 115–125.
[64] T. Shibamoto, Chromatographic Analysis of Environmental and Food Toxicants, CRC Press, New York, 1998.
[65] R. Ohmacht, M. Kele, Z. Matus, Chromatographia 39 (1994) 668–672.
[66] W. Maher, F. Pellegrino, J. Furlonger, Microchem J. 39 (1989) 160–165.
[67] K. Slezakova, D. Castro, C. Delerue–Matos, M.C. Alvim–Ferraz, S. Moraes, M.C. Pereira, Atmos. Res. 127 (2013) 141–147.

[68] A. Hanedar, K. Alp, B. Kaynak, J. Baek, E. Avsar, M.T. Odman, Atmos. Res. 99 (2011) 391–399.
[69] A. Hanedar, K. Alp, B. Kaynak, E. Avsar, Sci. Total Environ. 488 (2014) 437–446.
[70] H.J. Gotze, J. Schneider, H.G. Herzog, Fresenius' J. Anal. Chem. 340 (1991) 27–30.
[71] M. Carrara, J.-C. Wolf, R. Niessner, Atmos. Environ. 44 (2010) 3878–3885.
[72] A.N. Gachanja, P.J. Worsfold, Anal. Proc. 29 (1992) 61–63.
[73] I. Vassura, E. Venturini, S. Marchetti, P. Piazzalunga, E. Bernardi, P. Fermo, F. Passarini, Atmos. Environ. 82 (2014) 218–225.
[74] D.J. Freeman, F.C.R. Cattell, Environ. Sci. Technol. 24 (1990) 1581–1585.
[75] J. Cristale, F.S. Silva, G.J. Zocolo, M.R.R. Marchi, Environ. Pollut. 169 (2012) 210–216.
[76] X.-Y. Yang, K. Igarashi, N. Tang, J.-M. Lin, W. Wang, T. Kameda, A. Toriba, K. Hayakawa, Mutat. Res.-Gen. Tox. Em 695 (2010) 29–34.
[77] H. Li, R. Westerholm, J. Chromatogr. A 664 (1994) 177–192.
[78] M. Vojtisek-Lom, J. Czerwinski, J. Leníček, M. Sekyra, J. Topinka, Atmos. Environ. 60 (2012) 253–261.
[79] I. Sadiktsis, J.H. Koegler, T. Benham, C. Bergvall, R. Westerholm, Fuel 115 (2014) 573–580.
[80] C. von Mühlen, C.A. Zini, E.B. Caramão, P.J. Marriott, J. Chromatogr. A 1105 (2006) 39–50.
[81] D. Ryan, P.J. Morrison, J. Chromatogr. A 1071 (2005) 47–53.
[82] G. Vivó-Truyols, P.J. Schoenmakers, J. Chromatogr. A 1120 (2006) 273–281.
[83] S. Yuan, Z. Zhou, J. Li, F. Wang, Appl. Energy 92 (2012) 854–859.
[84] C. Meyer, U. Skogsberg, N. Welsch, K. Albert, Anal. Bioanal. Chem. 382 (2005) 679–690.
[85] C.A. Rimmer, L.C. Sander, S.A. Wise, Anal. Bioanal. Chem. 382 (2005) 698–707.
[86] L.C. Sander, K.E. Sharpless, N.E. Craft, S.A. Wise, Anal. Chem. 66 (1994) 1667–1674.
[87] Z. Guo, Y. Jin, T. Liang, Y. Liu, Q. Xu, X. Liang, J. Chromatogr. A 1216 (2009) 257–263.
[88] Y. Xiao, S.-C. Ng, T.T.Y. Tan, Y. Wang, J. Chromatogr. A 1269 (2012) 52–68.
[89] W. Zhao, K. Hu, C. Wang, S. Liang, B. Niu, L. He, J. Chromatogr. A 1223 (2012) 72–78.
[90] B. Mokhtari, K. Pourabdollah, N. Dalali, J. Inclusion Phenom. Macrocyclic Chem. 69 (2011) 1–55.
[91] M. Sliwka-Kaszynska, M. Slebioda, J. Sep. Sci. 37 (2014) 543–550.
[92] L. He, J. Zhang, Y. Sun, J. Liu, X. Jiang, L. Qu, J. Chromatogr. A 1217 (2010) 5971–5977.
[93] L. He, M. Zhang, L. Liu, X. Jiang, P. Mao, L. Qu, J. Chromatogr. A 1270 (2012) 186–193.
[94] L. He, M. Zhang, W. Zhao, J. Liu, X. Jiang, M. Zhang, Talanta 89 (2012) 433–440.
[95] T.D. Ho, C. Zhang, L.W. Hantao, J.L. Anderson, Anal. Chem. 86 (2014) 262–285.
[96] V. Pino, A.M. Afonso, Anal. Chim. Acta 714 (2012) 20–37.
[97] M. Zhang, X. Liang, S. Jiang, H. Qiu, Trends Anal. Chem. 53 (2014) 60–72.
[98] J. Omamogho, E. Stack, A. Santalad, S. Srijaranai, J. Glennon, H. Yamen, Anal. Bioanal. Chem. 397 (2010) 2513–2524.
[99] J. Layne, J. Chromatogr. A 957 (2002) 149–164.
[100] X. Liu, A.V. Bordunov, C.A. Pohl, J. Chromatogr. A 1119 (2006) 128–134.
[101] C.R. Silva, I.C.S.F. Jardim, C. Airoldi, J. Chromatogr. A 987 (2003) 127–138.
[102] A.-H. Duan, S.-M. Xie, L.M. Yuan, Trends Anal. Chem. 30 (2011) 484–491.
[103] K. Scida, P.W. Stege, G. Haby, G.A. Messina, C.D. Garcia, Anal. Chim. Acta 691 (2011) 6–17.
[104] A.V. Herrera-Herrera, M.A. Gonzalez-Curbelo, J. Hernandez-Borges, M.A. Rodriguez-Delgado, Anal. Chim. Acta 734 (2012) 1–30.
[105] E.P. Nesterenko, P.N. Nesterenko, D. Connolly, X. He, P. Floris, E. Duffy, Analyst 138 (2013) 4229–4254.

[106] A. Speltini, D. Merli, A. Profumo, Anal. Chim. Acta 783 (2013) 1–16.

[107] Z.Y. Gu, C.X. Yang, N. Chang, X.P. Yan, Acc. Chem. Res. 45 (2012) 734–745.

[108] J.R. Li, J. Sculley, H.C. Zhou, Chem. Rev. 112 (2012) 869–932.

[109] L.A. Colón, L.J. Baird, in: R.L. Grob, E.F. Barry (Eds.), Modern Practice of Gas Chromatography, fourth ed., 2004.

[110] P.C. Vasconcellos, D.Z. Souza, S.G. Ávila, M.P. Araújo, E. Naoto, K.H. Nascimento, F.S. Cavalcante, M. Dos Santos, P. Smichowski, E. Behrentz, Atmos. Environ. 45 (2011) 5770–5777.

[111] G. Christian, Analytical Chemistry, John Wiley & Sons, 1986.

[112] O.C. Taylor, JAPCA 19 (1969) 347–351.

[113] G. Zhang, Y. Mu, J. Liu, A. Mellouki, J. Chromatogr. A 1266 (2012) 110–115.

[114] G. Aslan-Sungur, E.O. Gaga, S. Yenisoy-Karakas, J. Chromatogr. A 1325 (2014) 40–48.

[115] P. Ciccioli, A. Cecinato, E. Brancaleoni, M. Frattoni, P. Zacchei, A.H. Miguel, P.C. Vasconcellos, J. Geophys. Res. 101 (1996) 19567–19581.

[116] K.O. Garcia, E.C. Teixeira, D.M. Agudelo-Castaneda, M. Braga, P.G. Alabarse, F. Wiegand, R.M. Kautzmann, L.F.O. Silva, Sci. Total Environ. 479–480 (2014) 57–65.

[117] A. Karmen, L. Giufridda, Nature 201 (1964) 1204–1205.

[118] A. Braithwaite, F.J. Smith, Chromatographic Methods, fifth ed., Blackie Academic & Professional, 1996.

[119] N. Ochiai, T. Ieda, K. Sasamoto, A. Fushimi, S. Hasegawa, K. Tanabe, S. Kobayashi, J. Chromatogr. A 1150 (2007) 13–20.

[120] P. Harley, P.C. Vasconcellos, L. Vierling, C.C.S. Pinheiros, J. Greenberg, A. Guenther, L. Klinger, S.S. Almeida, D. Neil, T. Baker, O. Phillips, Y. Malhi, Glob. Change Biol. 10 (2004) 1–21.

[121] F.G. Kitson, B.S. Larsen, C.N. McEwen, Gas Chromatography and Mass Spectrometry. A Practical Guide, Academic press, 1996.

[122] H.M. Macnair, J.M. Miller, Basic Gas Chromatography, second ed., A John Wiley & Sons, Inc., Publication, 2009.

[123] J.A. Masucci, G.W. Caldwell, in: R.L. Grob, E.F. Barry (Eds.), Modern Practice of Gas Chromatography, fourth ed., 2004.

[124] J. Kesselmeier, U. Kuhn, M. Andrea, P. Ciccioli, E. Brancaleoni, M. Frattoni, M. Ganzeveld, A. Guenther, J. Greenberg, P. De Castro Vasconcellos, T. Tavares, P. Artaxo, Atmos. Environ. 34 (2000) 4063–4072.

[125] J.H. Seinfeld, S.N. Pandis, Atmospheric Chemistry and Physics: From Air Pollution to Climate Change, second ed., Wiley-Interscience, USA, 2006.

[126] M.S. Alam, C.E. West, A.G. Scarlett, S.J. Rowland, R.M. Harrison, Atmos. Environ. 65 (2013) 101–111.

[127] S. Hashimoto, Y. Takawaza, A. Fushimi, K. Tanabe, Y. Shibata, T. Ieda, N. Ochiai, H. Kanda, T. Ohura, Q. Tao, S.E. Reichenbach, J. Chromatogr. A 1218 (2011) 3799–3810.

[128] WHO, Health Risks of Persistent Organic Pollutants from Long-range Transboundary Air Pollution. Joint WHO/Convention Task Force on the Health Aspects of Air Pollution, World Health Organization, 2003.

[129] C. Bogdal, E. Abad, M. Abalos, B. van Bavel, J. Hagberg, M. Scheringer, H. Fiedler, Trends Anal. Chem. 46 (2013) 150–161.

[130] A. Bacaloni, S. Insogna, L. Zoccolillo, Remote Zones Air Quality. Persistent Organic Pollutants: Sources, Sampling and Analysis, InTech, 2011, ISBN 978-953-307-317-0.

[131] K. Pozo, T. Harner, F. Wania, D.C.G. Muir, K.C. Jones, L.A. Barrie, Environ. Sci. Technol. 40 (2006) 4867–4873.

[132] Stockholm Convention on Persistent Organic Compounds, chm.pops.int/default.aspx, (accessed 15.05.15).

[133] H. Hung, M. MacLeod, R. Guardans, M. Scheringer, R. Barra, T. Harner, G. Zhang, Atmos. Environ. 80 (2013) 591–598.

[134] S.-Y. Baek, J. Jurng, Y.-S. Chang, Atmos. Environ. 64 (2013) 40–46.

[135] F. Gueguen, P. Stille, M. Millet, Talanta 111 (2013) 140–146.

[136] C. Manzano, E. Hoh, S.L.M. Simonich, Environ. Sci. Technol. 46 (2012) 7677–7684.

[137] A. Fushimi, S. Hashimoto, T. Ieda, N. Ochiai, Y. Takazawa, Y. Fujitani, K. Tanabe, J. Chromatogr. A 1252 (2012) 164–170.

[138] C. Liaud, M. Millet, S. Le Calve, Talanta 131 (2015) 386–394.

[139] C.V. De Souza, S.M. Correa, Atmos. Environ. 103 (2015) 222–230.

[140] G.O. Da Rocha, W.A. Lopes, P.A.P. Pereira, P.C. Vasconcellos, F.S. Oliveira, L.S. Carvalho, L.S. Conceição, J.B. de Andrade, J. Braz. Chem. Soc. 20 (2009) 680–692.

[141] J. He, R. Balasubramanian, Anal. Lett. 42 (2009) 1603–1619.

[142] C.S. Casal, G. Arbilla, S.M. Correa, Atmos. Environ. 96 (2014) 107–116.

[143] A.L.N. Guarieiro, J.V.S. Santos, A. Eiguren-Fernandez, E.A. Torres, G.O. da Rocha, J.B. de Andrade, Fuel 116 (2014) 490–497.

[144] A. Albinet, E. Leoz-Garziandia, H. Budzinski, E. Villenave, J. Chromatogr. A 1121 (2006) 106–113.

[145] A.I. Barrado, R.M. Perez-Pastor, S. Garcia-Alonso, Talanta 101 (2012) 428–434.

[146] A.I. Barrado, S. Garcia, M.L. Sevillano, J.A. Rodriguez, E. Barrado, Chemosphere 93 (2013b) 1678–1684.

[147] C.N. Hewitt, Reactive Hydrocarbons in the Atmosphere, Academic Press, San Diego, 1999.

[148] J. Timkovsky, P. Gankema, R. Pierik, R. Holzinger, Environ. Sci. Processes Impacts 16 (2014) 2301–2312.

[149] S.D. Maleknia, T.L. Bell, M.A. Adams, Int. J. Mass Spectrom. 279 (2009) 126–133.

[150] Y. Wang, X. Xu, L. Yin, H. Cheng, T. Mao, K. Zhang, W. Lin, Z. Meng, J.A. Palasota, J. Chromatogr. A 1361 (2014) 229–239.

Chapter 10

Mass Spectrometry and Ion Mobility Spectrometry

Fatkhulla K. Tadjimukhamedov[1] and Patricia B.C. Forbes[2,*]
[1]*US Pharmacopeial Convention, Rockville, MD, USA;* [2]*Department of Chemistry, Faculty of Natural and Agricultural Sciences, University of Pretoria, Pretoria, South Africa*
**Corresponding author: E-mail: patricia.forbes@up.ac.za*

Chapter Outline

1. Introduction **311**
2. Ionisation Methods **313**
2.1 Electron Ionisation 313
2.2 Secondary Ionisation Methods 314
2.3 Laser Desorption/Ionisation 315
3. Methods of Sample Introduction **316**
3.1 Thermal Desorption 316
3.2 Membrane Introduction Mass Spectrometry 317
4. Instrumentation **318**
4.1 Mass Analysers 318
4.2 Gas Chromatography–Mass Spectrometry 319
4.3 Portable/Miniature Mass Spectrometers 320
4.4 Aerosol MS 321
4.5 Ion Mobility Spectrometry 323
4.6 Proton Transfer Reaction–Mass Spectrometry 325
5. Conclusions **326**
References **326**

1. INTRODUCTION

Mass spectrometry (MS) is an analytical technique that performs measurements of ionised molecules in the gas phase. Among other analytical techniques, MS provides the most specific information – the molecular mass of a compound – and, therefore, is often considered the gold standard in analytical chemistry [1]. A basic mass spectrometer consists of three main components: an ionisation source, a mass analyser and a charge-ion detector (Figure 1). Compounds from any of the three phases are converted into gas phase ions by means of various ionisation sources. Following ionisation, the gas phase ions are guided into the mass analyser region where they are separated based on their mass-to-charge

Comprehensive Analytical Chemistry, Vol. 70. http://dx.doi.org/10.1016/bs.coac.2015.10.001

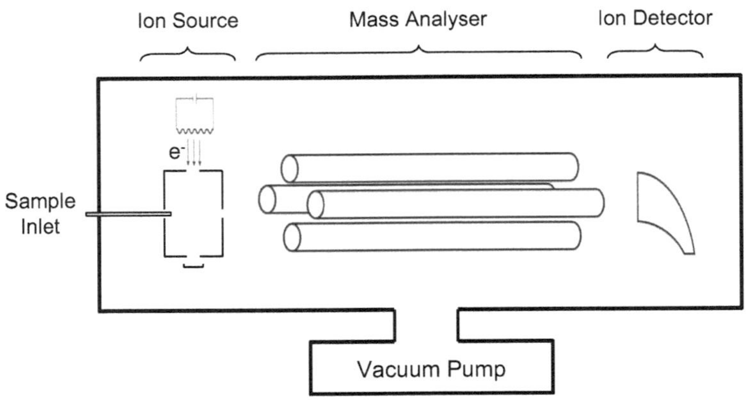

FIGURE 1 Schematic diagram showing the components of a mass spectrometer.

ratios (m/z). Today there are several types of mass spectrometers equipped with a variety of ion sources, and mass analysers are used for a variety of applications. The most popular, low-cost and widespread instruments are those that are based on the principles of separating ions under quadrupole electric fields. These include quadrupole mass filters and various ion traps [2].

A related technique called ion mobility spectrometry (IMS) also analyses gas phase ions but on the basis of their *shape-to-charge* ratios (Ω). Unlike MS, which requires moderate to high vacuum in the mass analyser region to operate, the IMS instruments require a buffer gas, often referred to as *drift gas*, which can range from a few torr up to ambient pressure in the analyser region. This eliminates the need for vacuum pumps that, in turn, facilitates portable instrumentation. There are several variations of both MS and IMS instruments developed over several decades, and some of the technologies relevant to monitoring air quality will be discussed in this chapter. Both techniques are especially well suited for detecting airborne pollutants, partly due to the ease of ion formation directly from volatile organic compounds (VOCs) using traditional ionisation sources. The use of more advanced ionisation techniques, such as laser desorption methods, has allowed the analyses of aerosols, which also play crucial role in monitoring particulate pollutants in ambient air.

The nature of VOCs and semi-volatile organic compounds (SVOCs) found in the atmosphere is extremely diverse. Natural emissions from decaying plants and animals or due to photochemical processes of plants, or gases emitted by animals' digestive and respiratory systems, constitute a significant portion of VOCs in the air [3]. At the same time, anthropogenic sources, such as burning fossil fuels, emissions and waste products from industrial processes, use of various industrial and household volatile solvents, halocarbons and hydrocarbons significantly contribute to the pollution of the atmosphere by organic pollutants. The latter has increased dramatically over the last century following the industrial revolution and an exponential increase in use of man-made materials that leave their imprint in the environment.

Moreover, aerosols formed from burning of trees and fossil fuels contribute to the pollution of air, especially to the accumulation of particulates. Therefore, it is imperative to monitor ambient air quality using analytical techniques that provide high specificity.

Although the bulk of this chapter is dedicated to MS, focussing on instrumentation used for monitoring ambient air quality, recent developments in novel and portable IMS instruments are also briefly covered, as are the advantages of each technique.

2. IONISATION METHODS

As the ionisation process is the crucial first step of an analysis with MS, several relevant techniques used in monitoring air quality are discussed in this section.

2.1 Electron Ionisation

Ionisation is a process of converting neutral molecules into gas phase ions, which may be either positively or negatively charged. Ionisation is necessary in order to analyse compounds with MS. Unlike neutral molecules, ions can be easily manipulated under both electric and magnetic fields: they can be trapped, fragmented, moved from one place of the instrument to another and subjected to chemical modifications with other neutral molecules or ions, so that even more specific information can be obtained. The classical ionisation technique for MS is known as an electron impact (or simply as electron ionisation, EI) ion source, where a gaseous stream of compounds of interest are subjected to a beam of high-energy electrons (e^-). Because the energy of the electrons (70 eV) is much higher than the ionisation potential of typical small organic molecules which averages around 10–11 eV, collision with these electrons induces extensive fragmentation of analyte molecules and the molecular ions are not always observed, as per the reaction:

$$M + e^-(70\,eV) \rightarrow M^{\cdot +} + \text{fragments} + 2e^-$$

where M is the neutral analyte molecule and e^- is an electron.

Although this fragmentation is sometimes considered detrimental due to the fact that molecular ions may not be directly observed to deduce the molecular weight of the analyte, the fragments and the fragmentation pattern, on the other hand, provide fingerprint-style robust structural information from which the identities of analytes can be deduced. This is especially useful when identifying compounds that are isomeric (those whose molecular ions share identical m/z values) and isobaric (compounds with similar m/z values that share the same nominal mass), because the fragment ions from different molecules tend to be different. As a result, spectral libraries of compounds have been established to support the identification of unknowns and have been the cornerstone of all gas chromatography–mass spectrometry (GC-MS) instrumental analyses, which will be covered in Section 4.2.

2.2 Secondary Ionisation Methods

In the analyses of complex mixtures, where the gas phase ions result from more than one analyte, the EI fragmentation spectra become difficult to interpret and the unequivocal identification of analytes becomes impossible. More importantly, quantitative assessment of each individual component in the mixtures is compromised. In such cases, the need for softer ionisation methods arise, and a secondary ionisation method called chemical ionisation (CI) [4] has been developed to mitigate excessive fragmentation. In this method, the EI chamber is pressurised with a reagent gas, typically methane, at about 0.5–1 torr pressure [5]. The electrons emitted from the EI source primarily hit the reagent gas producing the primary reagent ions:

$$CH_4 + e^- \rightarrow CH_4^{\cdot +} + 2e^-$$

Successive interactions of the $CH_4^{\cdot +}$ radical ions with other methane molecules lead to the formation of other ionic species, such as $CH_3^+, CH_2^{\cdot +}, CH_5^+$ and $C_2H_5^+$ all of which have been observed in the reagent gas plasma. These reagent ions in turn react with analyte molecules (XH) through a secondary ionisation step yielding primarily molecular ions of the analytes, predominantly via proton transfer [6]:

$$CH_5^+ + XH \rightarrow XH_2^+ + CH_4$$

$$C_2H_5^+ + XH \rightarrow XH_2^+ + C_2H_4$$

Or via hydride transfer [6]:

$$C_2H_5^+ + XH \rightarrow X^+ + C_2H_6$$

Fragmentation of molecular ions, although still present and sometimes significant, is minimised and, in most cases, abundant molecular ions of analytes are observed. Isobutane reagent gas is used to further mitigate the fragmentation of easily fragmented ions. However, because the ionisation still takes place at subatmospheric pressures, the CI has been used in applications where the analytes are directly introduced into the CI source interface (such as GC-MS) or through membrane introduction systems (refer to Section 3.2). An ambient pressure version of CI, aptly called atmospheric pressure chemical ionisation (APCI) [7], is a technique that is simple in its operation: it uses a corona discharge between a needle point, usually kept at several kilovolts of DC voltage, and a mass spectrometer inlet; ambient vapours of VOCs entering the APCI chamber are ionised under the strong DC electric field and are immediately sucked into the mass spectrometer inlet. Polar and nonpolar VOCs alike ionise well in an APCI source, which is one of the main advantages of the technique.

A variant of APCI that is widely used in the analyses of VOCs is the proton transfer reaction (PTR) source, which uses an electric discharge inside a hollow cathode tube to ionise water vapour to form hydronium ions, H_3O^+.

Analyte molecules (M) are introduced into the drift tube ionisation/reaction chamber as ambient air is continuously pumped through it, and undergo a gas phase proton exchange reaction with hydronium ions that lead to the formation of protonated analyte ions:

$$M + H_3O^+ \rightarrow MH^+ + H_2O$$

The proton affinity of water ($\sim$697 kJ mol^{-1}) [8] exceeds that of the major gases that constitute air (nitrogen, oxygen, hydrogen, carbon monoxide, carbon dioxide, noble gases, methane and ethane) [9], but it is far less than most VOCs and, therefore, undergoes facile PTRs with most VOCs. This process yields very soft ionisation with practically no fragmentation [10]. The protonated molecules are then easily analysed by MS, although the formation of cluster ions containing multiple water molecules may complicate the interpretation of the mass spectra. This effect is minimised by enhancing the kinetic energy of the reagent ions [11]. The sensitivity depends on the pressure in the drift tube and on the humidity of the air which is sampled [12].

Continuous radioactive ionisation sources may also be used to ionise VOCs, and the principles of ionisation are similar to that of PTR. The most commonly used radioactive sources are ^{241}Am and ^{63}Ni that emit energetic particles (alpha and beta particles, respectively) [13,14]. These particles undergo several steps of interactions with air and water vapour eventually leading to ionised water clusters similar to those observed in PTR technique. Radioactive sources are the preferred choice for many portable ion mobility spectrometers, partly because they do not require any additional power to operate, however, safety precautions must be taken into consideration when handling these sources to avoid radioactive contamination.

2.3 Laser Desorption/Ionisation

When it comes to the analyses of nonvolatile substances (solids and liquids alike), a challenge arises to convert these compounds into the gas phase before they can be ionised with EI, CI or PTR ionisation methods. Analysis and detection of the components of aerosol particles present a unique challenge in this regard. Laser desorption/ionisation utilises focused laser beams to simultaneously desorb and ionise the samples, primarily in the solid state, transferring them into gas phase ions. The laser beam vaporises the aerosol particles ionising the gas molecules simultaneously through a photoionisation mechanism:

$$M \xrightarrow{h\nu} M^* \xrightarrow{h\nu} M^{+\cdot} + e^-$$

Because the lasers are pulsed ionisation sources, they are most compatible with time-of-flight (TOF) mass analysers which are fast detectors. Resonant two-photon ionisation with an ultraviolet laser has been used as a soft ionisation source for polycyclic aromatic hydrocarbons (PAHs) coated on a desorption rod, where one photon excites the molecule to a resonant intermediate electronic state and a second photon ionises the molecule [15]. A limit of detection for gas phase naphthalene of 4 ppb was demonstrated when using a KrF excimer laser (248 nm) for the photoionisation of the analytes with time of flight mass spectrometry (TOFMS) analysis, as many aromatic hydrocarbons have high UV absorption cross sections in the 200−300 nm range, and they are efficiently ionised by two-photon processes [16].

3. METHODS OF SAMPLE INTRODUCTION

As in any other field of chemical analysis, efficient sampling and recovery is essential for the detection of ambient air pollutants, such as airborne particles and VOCs, with MS. As stated earlier, mass spectrometers only detect gas phase ions generated by ionisation sources. Accordingly, sampling and sample introduction methods must accommodate the needs of an ion source so that maximum possible detection sensitivity can be achieved. The methods of sample introduction vary depending of the type of mass spectrometer or instrument. Sampling and sample preparation methods are covered in detail in Part II of this book, where many references are made to the analyses of samples by mass spectrometric methods. In this section, two means of sample introduction to MS systems are described, whilst GC introduction is discussed in Section 4.2. It should be noted that it is also possible to introduce samples directly into a mass spectrometer via numerous direct open air surface sampling and ionisation methods, as reviewed by Monge et al. [17].

3.1 Thermal Desorption

Thermal desorption (TD) was discussed in Chapter 9, in terms of GC analyses of atmospheric analytes. TD provides a means of introducing analytes into a mass spectrometer following either sampling onto adsorbents or absorbents (as discussed in Chapter 4), or directly from aerosol particles. Typically GC separations are employed prior to MS analysis in these ambient air applications. Trace quantities of SVOCs in ambient atmospheric aerosols sampled onto aluminium foil have been determined utilising the latter technique, for example, where quinolones, PAHs and monocarboxylic acids were found [18]. Advantages of this means of sample introduction, as compared to traditional solvent extraction techniques, are the small sample quantities required, and the lack of complicated sample preparation steps, which facilitates rapid screening of the organic content of ambient particles [18]. TD has been employed in situ for hourly measurements of speciated organic compounds in atmospheric aerosols, where aerosols are collected into a TD cell by humidification and inertial impaction, with analysis by GC-MS or GC-FID (flame ionisation detection) [19]. Humidification increases adhesion and eliminates particle bounce following impaction, and the system was found to efficiently collect particles in the 0.06−2.5 μm size range.

3.2 Membrane Introduction Mass Spectrometry

Membrane introduction mass spectrometry (MIMS) utilises a porous polymer membrane introduction system to a mass spectrometer and is capable of continuous online monitoring. The permeation process involves absorption of an analyte into the membrane, diffusion through it according to Fick's diffusion laws, and then evaporation from the membrane surface into the vacuum [20]. The semipermeable membrane rejects the sample matrix and enriches certain analytes from the gas sample. The membrane can be chosen (for example, porous polypropylene) such that sufficient solvent penetrates into the ion source for CI to occur, where the vaporised solvent serves as the reagent gas [21]. Teflon has also been employed in membranes for gas monitoring applications [20]; however, the majority of applications for direct organic vapour sampling use silicone membranes [22].

The separated analytes are directly transferred as a mixture to the mass spectrometer, often using helium carrier gas [23]. Membrane introduction (MI) can be interfaced with quadrupole, ion trap or magnetic sector mass analysers, and this clean introduction system reduces source cleaning requirements [20]. MIMS has also been used for PTR-MS, which is useful for high humidity applications, as the membranes discriminate against water [24].

MIMS has been employed in the monitoring of VOCs in air at part per trillion levels, as well as for peroxyacetyl nitrate produced in atmospheric photochemical reactions; atmospheric SVOCs such as PAHs emitted from combustion processes and pesticides such as malathion and lindane; and organometallic dimethylmercury in air (as discussed in Refs [22,25]). The sensitivity of MIMS for SVOC analytes can be enhanced by thermal modulation of the membrane, however, this also increases the permeation of water through the membrane, which may be a problem in humid environments [23]. Alternatively SVOCs can be derivatised to form more volatile analyte species for MIMS determinations.

Membrane introduction is a useful sample introduction method for portable MS instruments, as it is cost-effective and may eliminate sample degradation. They provide a rugged interface between air and instrument vacuum and may be placed within or outside the ion source of the mass spectrometer [27]. Online permeation tubes have been employed to continuously infuse an isotopically labelled internal standard into a MIMS system to allow for quantitation without the need for offline calibration [23]. This was applied in the monitoring of BTEX (benzene, toluene, ethylbenzenes and xylenes), pinenes, naphthalene and 2-methoxyphenol in urban air plumes at parts per billion by volume levels [23]. Online continuous calibration accounts for any potential instrumental drift or membrane interface modifications which may occur over time [23]. MIMS provides a rapid and sensitive online analytical tool, which fills a niche between hand-held sensors and conventional GC-MS-based analytical methods.

4. INSTRUMENTATION

4.1 Mass Analysers

The central component of a mass spectrometer is the mass analyser. There are several types of mass analysers that all measure m/z values of ions based on different principles of operation and each have their advantages and limitations. The earliest mass spectrometers utilised magnets to deflect the beam of ions and thus measure their m/z values as a function of deflected angle. These include the early pioneers of MS: Sir J.J. Thomson's first parabolic mass spectrograph, F.W. Aston's first mass spectrograph that utilised both electrostatic and magnetic focussing fields and A.J. Dempster's first mass spectrometer [26]. Although later improvements with the magnetic sector instruments allowed for the achievement of high-resolution mass spectra, the use of heavy magnets with these instruments and the cost involved limited their use.

Later developments led to the discoveries of quadrupole and time of flight (TOF) mass analysers that did not rely on the use of magnets. Like the magnetic sector analyser, the quadrupole is a scanning instrument: a DC voltage and RF amplitude are both scanned to analyse ions of different m/z values. Ions with a stable trajectory at given DC and RF voltages (which correspond to a given m/z value) pass through the analyser, whilst other ions are lost due to their unstable trajectories. Due to the fact that only ions with a given m/z value can pass through at a given ratio of DC/RF voltages, the quadrupole is often referred to as a filter, and scanning the voltages results in the acquisition of a complete mass spectrum. Moreover, not only can ions be filtered, the quadrupole electrodes and the electric fields can be arranged in such a way that ions can be trapped within an analyser referred to as 3D trap, Paul trap or simply an ion trap. The quadrupole and ion trap analysers can be made fairly compact: most current bench-top and portable mass spectrometers are based on these two mass analysers.

Unlike any other mass analyser, TOF analysers are not scanning devices, but rather are pulsed analysers. The TOF analyser is relatively simple in its principle of operation: after initial acceleration, ions enter a field-free region within a flight tube and the time it takes for a packet of ions to reach the detector, located at the end of the flight tube, is a function of the m/z values. Thus, a mass spectrum is generated by converting this time to m/z, often with the help of calibrant ions with known m/z values. The overall analysis time for each pulsed ion packet takes less than a millisecond, compared to several hundred milliseconds or even seconds required for most scanning analysers. Therefore, the TOF analyser is best suited for applications where analysis speed is a crucial factor. The best example is coupling with high-speed gas chromatographs, where separation is fast: full mass spectral analysis of a chromatographic peak having a width of 100 ms can be achieved, whereas a scanning mass analyser, such as a quadrupole, may miss it or may not provide sufficient data points across the chromatographic peak for successful quantitation. TOF analysers are therefore the best choice for such applications, such as for GC × GC (refer to Chapter 9).

4.2 Gas Chromatography–Mass Spectrometry

Analyses of complex mixtures of airborne organic compounds are best achieved when individual compounds making up the mixture are first separated. This is achieved with GC, which can be coupled to various detectors. The most popular detectors for these applications are the FID, electron capture, photoionisation and thermal conductivity detectors, some of which were discussed in Chapter 9. Although a number of these detectors have better selectivity or sensitivity to some classes of compounds, the detector that provides the best specificity (i.e. the ability to identify any given compound) for GC is the mass spectrometer, because each signal in the mass spectrum belongs to some molecular ion or a fragment thereof. Since most organic compounds encountered in the atmosphere are in the form of mixtures with other pollutants, GC-MS is best suited for these applications due to the ease of ion formation directly from the gas phase of separated analytes and allows for the identification and quantification of a wide range of compounds even at trace concentrations. These include the analysis of PAHs and their oxygenated derivatives, such as that present in particulate matter (PM) collected onto filters at an urban background site in Augsburg, Germany [27].

Quadrupole, ion trap and TOF mass analysers have been coupled to GC for the analyses of organic compounds in ambient air. Whilst the quadrupole mass analyser is the most cost-effective detector and may be employed in selected ion monitoring mode for enhanced sensitivity in ultra-trace analysis, the ion trap analyser provides tandem MS/MS capabilities to further confirm the identities of compounds and increase selectivity. TOFMS provides higher mass resolution and mass accuracy along with the fastest analytical speed, which is essential for fast chromatographic separations with very narrow peaks [28].

In TOFMS instruments, good mass resolving power is achieved, even from continuous ion sources, by pulsed orthogonal sampling of the ions which are generated, which enhances spatial focussing. Mass resolution is further improved using a reflectron for energy focussing, which utilises a retarding field created by a series of ring electrodes with linearly increasing voltages. Ions with higher energy penetrate more deeply inside, which extends the time until they are reflected, resulting in ions of the same m/z with different initial energies reaching the detector at almost the same time. The instrument may also be configured such that ions pass twice along the TOF flight tube before reaching the detector, which further enhances resolving power [28]. In high-resolution TOFMS (HR-TOFMS), a reference compound is introduced into the ion source as a 'lock mass', which can provide mass accuracies of <5 ppm, although at low and high signal intensities deviations from this may occur.

At low concentrations, the analytical signal may be too weak, and at high concentrations the detector or recording device may become saturated [28]. GC-TOFMS has been utilised in the monitoring of a wide range of atmospheric pollutants, including pesticides, polychlorinated biphenyls and PAHs [28].

High-resolution mass spectrometry (HRMS) is required in a number of ambient air monitoring applications, where comprehensive isomer and congener characterisation is required. In these analyses, high-resolution GC (HRGC) is often used as a means of separation of analytes in HRGC-HRMS systems, which are typically used for the analysis of persistent organic pollutants (POPs) in air. The global atmospheric passive sampling network, for example, analyses passive sampler extracts (as described in Chapter 2) using HRMS for a number of POPs and new priority pollutants, such as short-chain chlorinated paraffins and polychlorinated naphthalenes [29]. HRGC-HRMS has also been employed in the determination of background air concentrations of polychlorinated dibenzo-*p*-dioxins, polychlorinated dibenzofurans and dioxin-like polychlorinated biphenyls by the National Dioxin Air Monitoring Network established by the United States Environmental Protection Agency [30]. In addition, HRGC-HRMS has been used in the comprehensive congener specific evaluation of atmospheric polychlorinated biphenyls and polychlorinated naphthalenes [31].

4.3 Portable/Miniature Mass Spectrometers

Portability of mass spectrometers allows direct in situ analyses of environmental samples. Instead of bringing the sample to the laboratory, one could bring the laboratory to the field and obtain real-time analyses and eliminate all the steps needed in between. Portable instruments provide insight into the geographic variations of atmospheric aerosols, as well as their transport characteristics and chemical dynamics [32]. Building a portable mass spectrometer, however, is not an easy task, primarily because mass spectrometers require vacuum pumps to sustain acceptable levels of low pressure inside the mass analyser. Trace atmospheric analysis requires the exclusion of the air matrix from the vacuum manifold, and/or enrichment of the analyte [33]. MIMS is useful in this regard, however, the choice of the membrane material will limit the range of analytes which can be determined. The use of an atmospheric inlet has been employed, but this requires significant pumping capacity, therefore, TD of analytes sampled onto a solid sorbent provides a useful alternative [33].

As stated earlier, quadrupole and ion trap electrodes are the best suited to miniaturisation, and several variations of miniature portable mass spectrometers have been built in the past [34]. In fact, the first reference to miniaturised mass spectrometers was made with regard to the use of an ion trap in the early 1970s, because the size of the ion trap and quadrupole analysers was very small relative to sector and TOF analysers at the time [35]. Quadrupoles and ion traps are also more tolerant to slightly lower vacuums (i.e. higher pressures inside the analyser), and the presence of some residual gas pressure is required to achieve a desired mass resolution and also to obtain tandem MS/MS fragments of compounds for further identification and structural confirmation [36]. Real-time direct air monitoring by means of a portable triple quadrupole mass spectrometer has been reported, where trichloroethylene and tetrachloroethylene were monitored utilising a low-pressure CI source with detection limits of 0.5 $\mu g\ m^{-3}$ [37].

The advent of compact ion traps such as cylindrical [38] and rectilinear [39] ion trap geometries allowed for further reduction in size of mass analysers, which led to the realisation of battery operated miniature mass spectrometers that can be taken directly into the field to perform in situ analyses. VOCs in air (including naphthalene) were sampled onto a solid-phase microextraction fibre prior to TD of the fibre utilising MIMS, with EI and analysis utilising a portable miniature cylindrical ion trap mass spectrometer [40]. Limits of detection (LODs) were in the low parts per billion range, and the method was found to be simple, rapid, robust, reproducible, accurate and sensitive. A miniature cylindrical ion trap mass spectrometer was later successfully used to detect, identify and quantify gaseous, neutral ammonia, where an ion/molecule reaction with halobenzene radical ions produced a product that could be detected above the instrumental m/z limit of 40 [41]. A hand-held, battery powered, rugged rectilinear ion trap MS system was developed, which allowed for continuous air monitoring and had tandem MS capabilities for compound identification [42]. This instrument provided a LOD of 50 ppb for toluene in air and was later modified with an atmospheric pressure interface to implement APCI for direct analysis of trace levels of benzene, toluene and ethylbenzene in air, where a LOD of 0.2 ppb was achieved for benzene without any sample pre-concentration [43]. A sorption trap inlet was employed with the hand-held rectilinear ion trap MS system as the interface between the atmosphere and the vacuum chamber of the mass spectrometer, where the analytes were pre-concentrated by sampling onto the trap and were then thermally desorbed into the instrument [33]. Nine toxic gases (including sulphur dioxide and formaldehyde) were sampled, detected, identified and quantified in less than 2 min, with detection limits ranging from 800 ppt to 3 ppm and a linear range of one to two orders of magnitude [33].

A problem with TOF analysers is that, unlike linear quadrupoles and quadrupole ion traps, there is the need for relatively low-operating pressures. This requires more powerful and costly pumps that are, at this time, the main factor limiting the degree of miniaturisation possible [36].

4.4 Aerosol MS

The analysis of single airborne particles using TOFMS to detect analyte ions generated by laser-induced ionisation (employing a UV laser) was demonstrated by Hinz et al. [44]. The particles were introduced into the mass spectrometer through a differentially pumped interface and were detected by light scattering using a He-Ne laser beam. This instrument allowed for the online MS analysis of ambient aerosol particles. A challenge with the development of these instruments is the efficient transport of particles through the system with minimal loss. Systems using these techniques in real time have been developed which determine the particle size, based on aerodynamic particle sizing, in addition to individual particle composition [45] and include portable systems [32].

Real-time aerosol analysis with high temporal resolution is useful in the elucidation of chemical changes occurring in individual particles and in understanding the chemical dynamics within the aerosol [32]. A HR-TOFMS system has been developed for aerosol analysis, which is portable and provides high time resolution (0.5 Hz) [46]. This instrument has two optical modes: a single reflection configuration for higher sensitivity (LOD $< 0.04\ \mu g\ m^{-3}$) but lower resolving power and a two reflectron configuration for higher resolving power but reduced sensitivity (LOD $< 0.4\ \mu g\ m^{-3}$). Such a system has been employed in the identification and quantification of organic aerosol in PM_1 (PM with an aerodynamic diameter of $<1\ \mu m$) from cooking and other sources at an urban background site in Barcelona [47], as well as in the determination of the chemical composition of the organic fraction of aerosols and their source apportionment in the metropolitan area of Paris [48].

An alternative means of ionisation to laser-induced methods is vacuum ultraviolet photoionisation (VUV) utilising a VUV krypton lamp as an ionisation source in an aerosol TOFMS (ATOFMS) system [49], which was found to perform well in detecting PAHs in aerosol particles.

Quadrupole mass spectrometers have been used in aerosol MS (AMS), however because they are scanning instruments and the ion burst from a single particle is so brief, only one mass-to-charge ratio can be analysed for each particle [32], which limits the analytical information which can be obtained from such an analysis. Commercial quadrupole AMS instruments are available, however, and can be used to quantify atmospheric mass concentrations of chemical species (in $\mu g\ m^{-3}$) upon calibration [50]. These instruments typically use flash thermal vaporisation under high vacuum followed by electron impact ionisation and not laser ionisation methods [46]. TOFMS in comparison allows for the entire mass spectrum to be obtained for a single particle and has enhanced sensitivity [32].

An instrument which utilises an ion trap mass spectrometer has been utilised for single-particle aerosol measurements, where fibre optics are used to detect particles entering the ion trap, and a Nd-YAG laser pulse desorbs and ionises the analytes from the particles [51]. An advantage of this system is that tandem mass analysis (MS/MS) is possible for individual particles, which may assist with analyte identification based on fragmentation patterns, although this limits sampling frequency [32].

AMS has been employed in both ground-based and airborne aerosol monitoring campaigns, where many components of aerosols have been measured, from volatiles to refractory elements, including sulphates, nitrates, carbonaceous material, sea salt and minerals [52]. ATOFMS has been used to determine chemical markers from the combined size and chemical compositional analysis of individual particles derived from biomass combustion [53], and AMS has been used to study organic aerosols in the northern hemisphere and to determine its evolution in the atmosphere [54]. Seasonal differences in urban organic aerosol composition have been studied by means of an ultra high resolution (mass resolution of >100,000) Orbitrap-MS system [55]. Here an electrospray ionisation system was used in positive and negative ionisation modes to determine structural information and tentative identification of selected compounds by MS/MS using argon as collision gas and 20–35 eV collision energy. AMS employed in conjunction with ultrasonic anemometry allowed for the correlation of chemical particle and wind data, providing a description of turbulent particle behaviour near the surface of the earth [56].

A real-time single-particle MS system was used to study regional and local contributions to fine and ultrafine urban PM, in order to characterise and identify sources [57]. Ambient air was drawn through a PM_{10} inlet extending 5 m above the ground and sampled at a flow rate of 30 L min^{-1}. Artefacts were however observed for SVOCs, and it was noted that the total volume and number of aerosol particles analysed were relatively small, leading to statistical uncertainty [57].

Laser microprobe mass spectrometry has also been utilised to determine the surface composition of aerosol particles, although quantitation may be difficult with this sensitive technique [58]. It is also a relatively slow analytical method, which precludes its use for real-time aerosol monitoring [32].

The analysis of single ultrafine aerosols (<100 nm diameter) by both offline and online techniques has been reviewed by Bzdek et al. [59]. Ultrafine aerosols have potential impacts on both human health and climate change, and they constitute the largest portion of ambient aerosol loading based on number of particles, and their concentrations and composition are very time and location dependent. In offline MS methods, the aerosol particles are first collected onto a filter or grid prior to analysis, whilst online measurements of nanoparticles is more challenging because the amount of material is very limited, therefore the signal intensities are low, limiting quantitation. A water-based growth tube can be utilised at the inlet to the MS instrument, in order to grow particles by condensation that would otherwise be too small for analysis [59].

4.5 Ion Mobility Spectrometry

IMS is an analytical technique that measures mobilities of ions, created at ambient pressure from radioactive materials, in the gas phase. When a plasma (a mixture of ions and gases) is placed in an electric field, ions 'drift' at a certain drift velocity, v_d, proportional to the strength of the imposed electric field (E) and is described by the following equation:

$$v_d = K \cdot E$$

The proportionality coefficient, K ($cm^2\ V^{-1}\ s^{-1}$) is called the *ion mobility coefficient*, or simply *ion mobility*. Ions that have different shapes, such as isomers and isobars, have different drift velocities in the gas phase as they move towards low potential. Consequently, ions travelling a distance (d) at different velocities arrive at different times, called *drift time* (t_d), which is usually measured in microseconds to milliseconds and can be calculated from the equation:

$$t_d = d/v_d$$

Unlike mass spectrometers, where ions move virtually without collisions with neutral gas molecules (except for when collisions are required to induce fragmentation), in IMS instruments ions drag against the *drift gas* and undergo collisions with the gas molecules. Therefore, the collisional cross sections of ions (Ω) effect their velocity and is a measure of the shape and size of an ion. Thus, the collision cross section is inversely related to ion mobility:

$$K = \frac{3}{16}\frac{ze}{N}\left(\frac{2\pi}{\mu k T_{eff}}\right)^{1/2}\frac{1}{\Omega}$$

where N is the number density (molecules cm^{-3}); μ is the reduced mass; T_{eff} is the effective temperature (K); k is the Boltzmann constant.

Isomeric and isobaric compounds that share identical or similar m/z values might not get resolved and differentiated with quadrupole or ion trap mass analysers, which typically provide a unit mass resolution. Utilising IMS and its capability to separate ions based on their shape-to-charge ratio can separate these ions in the gas phase, and stand-alone IMS instruments have been used to separate isomeric [60,61] as well as isobaric compounds.

Several variations of IMS analysers have been developed. Traditional drift tube IMS instruments use stacked ring electrodes that are separated with insulators and connected through a chain of resistors, which create an electric field along the drift tube (Figure 2). IMS can provide rapid analysis of samples spanning a range of vapour pressures and chemical classes; it has been used for measuring VOCs in the indoor air of space stations and has found military use (detection of explosives and chemical agents) in addition to stack emissions monitoring [62]. Particulates in workplace air collected on Teflon membrane filters have been thermally desorbed for sample introduction [63]. Limitations of IMS include the complex gas phase chemistry involved, which can lead to false negative screening due to competitive ionisation [63] as well as the cost, although on the positive side, spectral libraries have been established for IMS which assist with identification of unknown compounds [62]. In addition,

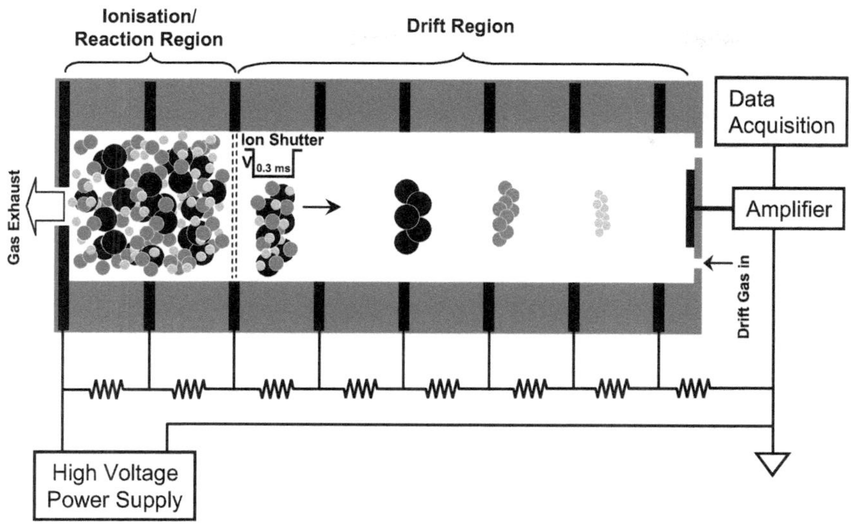

FIGURE 2 Schematic representation of a traditional ion mobility spectrometry drift tube.

atmospheric pressure operation (lending to portability), speed, operational simplicity and sensitivity (ng to pg range) are potential advantages [63].

4.6 Proton Transfer Reaction–Mass Spectrometry

PTR-MS was first used to study ion-neutral reactions [64], but has since been employed in VOC emissions monitoring, such as that arising from decaying biomatter, which was consequently found to be an important source of tropospheric acetone, methanol and ethanol [9]. The sensitivity of PTR-MS was found to be in the range of a few parts per trillion by volume when it was used in the online monitoring of the diurnal variations of VOCs in the troposphere [9]. The sample is typically introduced into the drift tube at a flow rate of ~11 mL min^{-1}, which maintains the drift tube pressure at around 2 mbar [8]. Aside from biogenic VOC (BVOC) monitoring of isoprene and pinene in ambient air, PTR-MS has been employed in monitoring traffic emissions (benzene, toluene, xylene and alkylbenzenes), with sampling times ranging from 3 to 10 s per mass [9]. Compounds ranging from VOCs, such as benzene, toluene and xylenes, to hydrocarbons (propene and isoprene), oxygenated compounds (including formaldehyde and acetone) and nitrogen-containing compounds (for example, acetonitrile) were continuously monitored in real time using PTR-MS near a ship channel [65]. The high resolution of the data set assisted in the identification of the anthropogenic sources, and the method was valuable in monitoring rapidly changing VOC plumes which passed over the site.

Atmospheric dimethyl sulfide and benzene were analysed by PTR-MS in another study [66] to monitor diurnal variations in concentrations at both a clean air and a traffic-impacted site at <5 min resolution with LODs in the tens of parts per trillion by volume. PTR-MS has the additional advantage of a large dynamic range [24], although instrumental carryover may be a concern [9]. PTR-MS cannot be used for the detection of certain hydrocarbons, such as alkanes and ethane [65], although a response time of ~1 s, and low detection limits (10–100 ppt) have been reported for a number of environmentally important VOCs [12]. Interferences are of concern for some VOCs; for example, a large number of BVOCs produce a signal at 69 amu, which requires the use of GC-PTR-MS methods or alternatively the reporting of the sum of concentrations of a compound class (such as C_2-benzenes) [12].

An ion trap MS instrument has been used as an alternative to the more commonly employed quadrupole mass analysers in PTR-MS instruments for online trace gas analysis of VOCs [67]. Collision-induced dissociation and ion–molecule reactions can occur in the ion trap, which allow for the identification of ions that are otherwise indistinguishable by PTR-MS. The advantages of the ion trap system were related to improved identification of VOCs and an almost 100% duty cycle, with detection limits between 0.05 and 0.3 ppbv for a range of oxygenated VOCs, aromatic compounds, isoprene, dimethyl sulphide, etc. [67]. It was noted that ion trap PTR-MS may find application in the analysis of highly polluted air masses, where identification of individual compounds is challenging, but when the higher detection limits of the ion trap system would not be a problem [67]. TOFMS has also been used as the mass analyser in PTR-MS systems, which has the advantage of high mass resolution. However, the ions need to be pulsed into the TOF mass spectrometer, therefore some of the ions are lost as a constant current of ions are generated in the PTR-MS instrument [11].

5. CONCLUSIONS

Sophisticated deconvolution software has provided an invaluable tool in modern MS instrumentation and with spectral libraries that are available today, one can unequivocally identify and quantify many compounds of relevance to ambient air monitoring. MS is therefore an invaluable tool in this field. Future developments in MS technologies which further enhance resolution and sensitivity are likely, including research into the ambient air monitoring applications of GC-Orbitrap instrumentation. An ongoing focus on portable and miniaturised MS instrumentation for on-site, real-time ambient air monitoring is also expected.

REFERENCES

[1] Z. Ouyang, R.J. Noll, R.G. Cooks, Anal. Chem. 81 (7) (2009) 2421–2425.
[2] R.E. March, J.F.J. Todd, Quadrupole Ion Trap Mass Spectrometry, second ed., J. Wiley, Hoboken, NJ, 2005.
[3] R.S. Blake, P.S. Monks, A.M. Ellis, Chem. Rev. 109 (3) (2009) 861–896.
[4] M.S.B. Munson, F.H. Field, JACS 88 (12) (1966a) 2621–2630.

[5] M.S.B. Munson, F.H. Field, JACS 88 (19) (1966b) 4337–4345.
[6] M.S.B. Munson, Anal. Chem. 43 (13) (1971) 28A–43A.
[7] E.C. Horning, D.I. Carroll, I. Dzidic, K.D. Haegele, M.G. Horning, R.N. Stillwell, J. Chromatogr. Sci. 12 (11) (1974) 725–729.
[8] C.N. Hewitt, S. Hayward, A. Tani, J. Environ. Monit. 5 (2003) 1–7.
[9] W. Lindinger, A. Hansel, A. Jordan, Int. J. Mass Spectrom. Ion Processes 173 (1998) 191–241.
[10] A. Lagg, J. Taucher, A. Hansel, W. Lindinger, Int. J. Mass Spectrom. Ion Processes 134 (1994) 55–66.
[11] J. de Gouw, C. Warneke, Mass Spectrom. Rev. 26 (2007) 223–257.
[12] J. de Gouw, C. Warneke, T. Karl, G. Eerdekens, C. van der Veen, R. Fall, Int. J. Mass Spectrom. 223-223 (2003) 365–382.
[13] C.-M. Yun, Y. Otani, H. Emi, Aerosol Sci. Technol. 26 (5) (1997) 389–397.
[14] E.C. Horning, M.G. Horning, D.I. Carroll, I. Dzidic, R.N. Stillwell, Anal. Chem. 45 (6) (1973) 936–943.
[15] R. Tembreull, D.M. Lubman, Anal. Chem. 58 (1986) 1299–1303.
[16] M.J. Castaldi, S.M. Senkan, J. Air Waste Manage. Assoc. 48 (1998) 77–81.
[17] M.E. Monge, G.A. Harris, P. Dwivedi, F.M. Fernández, Chem. Rev. 113 (2013) 2269–2308.
[18] A.H. Falkovich, Y. Rudich, Environ. Sci. Technol. 35 (2001) 2326–2333.
[19] B.J. Williams, A.H. Goldstein, N.M. Kreisberg, S.V. Hering, Aerosol Sci. Technol. 40 (2006) 627–638.
[20] T. Kotiaho, F.R. Lauritsen, T.K. Choudhury, R.G. Cooks, G.T. Tsao, Anal. Chem. 63 (18) (1991) 875–883.
[21] F.R. Lauritsen, T. Kotiaho, T.K. Choudhury, R.G. Cooks, Anal. Chem. 64 (1992) 1205–1211.
[22] R.A. Ketola, T. Kotiaho, M.E. Cisper, T.M. Allen, J. Mass Spectrom. 37 (2002) 457–476.
[23] J.M. Etzkorn, N.G. Davey, A.J. Thompson, A.S. Creba, C.W. LeBlanc, C.D. Simpson, E.T. Krogh, C.G. Gill, J. Chromatogr. Sci. 47 (2009) 57–66.
[24] M. Alexander, E. Boscaini, W. Lindinger, T.D. Märk, Int. J. Mass Spectrom. 223–224 (2003) 763–770.
[25] N. Srinivasan, R.C. Johnson, N. Kasthurikrishnan, P. Wong, R.G. Cooks, Anal. Chim. Acta 350 (1997) 257–271.
[26] R.W. Kiser, Introduction to Mass Spectrometry and Its Applications, Prentice-Hall, Englewood Cliffs, NJ, 1965.
[27] M. Sklorz, J.-J. Briedé, J. Schnelle-Kreis, Y. Liu, J. Cyrys, T.M. de Kok, R. Zimmermann, J. Toxicol Environ. Health Part A 70 (2007) 1866–1869.
[28] T. Cajka, in: I. Ferrer, E. Thurman (Eds.), Advanced Techniques in Gas Chromatography-Mass Spectrometry (GC-MS-MS and GC-TOF-MS) for Environmental Chemistry, Comprehensive Analytical Chemistry, vol. 61, Elsevier, 2013, ISBN 978-0-444-62623-3, pp. 271–302 (Chapter 12).
[29] T. Harner, J. Schuster, E. Sverko, L. Jantunen, S.C. Lee, K. Pozo, Organohalogen Compd. 76 (2014) 1309–1312.
[30] M. Lorber, J. Ferrario, C. Byrne, Atmos. Environ. 77 (2013) 311–317.
[31] J.N. Hogarh, N. Seike, Y. Kobara, A. Habib, J.-J. Nam, J.-S. Lee, Q. Li, X. Liu, J. Li, G. Zhang, S. Masunaga, Chemosphere 86 (2012) 718–726.
[32] E. Gard, J.E. Mayer, B.D. Morrical, T. Dienes, D.P. Fergenson, K.A. Prather, Anal. Chem. 69 (1997) 4083–4091.
[33] A. Keil, H. Hernandez-Soto, R.J. Noll, M. Fico, L. Gao, Z. Ouyang, R.G. Cooks, Anal. Chem. 80 (2008) 734–741.

[34] M. Yang, T.-Y. Kim, H.-C. Hwang, S.-K. Yi, D.-H. Kim, J. Am. Soc. Mass Spectrom. 19 (10) (2008) 1442–1448.

[35] P.H. Dawson, J.W. Hedman, N.R. Whetten, Anal. Chem. 42 (12) (1970) 103a–108a.

[36] E.R. Badman, R. Graham Cooks, J. Mass Spectrom. 35 (6) (2000) 659–671.

[37] N.S. Karellas, Q. Chen, J. Environ. Prot. 4 (2013) 99–105.

[38] J.M. Wells, E.R. Badman, R.G. Cooks, Anal. Chem. 70 (3) (1998) 438–444.

[39] Z. Ouyang, G. Wu, Y. Song, H. Li, W.R. Plass, R.G. Cooks, Anal. Chem. 76 (16) (2004) 4595–4605.

[40] L.S. Riter, E.C. Meurer, I. Cotte-Rodriguez, M.N. Eberlin, R.G. Cooks, Analyst 128 (2003) 1119–1122.

[41] J.N. Smith, A.D. Keil, R.J. Noll, R.G. Cooks, Analyst 136 (2011) 120–127.

[42] L. Gao, Q. Song, G.E. Patterson, R.G. Cooks, Z. Ouyang, Anal. Chem. 78 (2006) 5994–6002.

[43] G. Huang, L. Gao, J. Duncan, J.D. Harper, N.L. Sanders, Z. Ouyang, R.G. Cooks, J. Am. Soc. Mass Spectrom. 21 (2010) 132–135.

[44] K.-P. Hinz, R. Kaufmann, B. Spengler, Anal. Chem. 66 (1994) 2071–2076.

[45] T. Nordmeyer, K.A. Prather, Anal. Chem. 66 (1994) 3540–3542.

[46] P.F. DeCarlo, J.R. Kimmel, A. Trimborn, M.J. Northway, J.T. Jayne, A.C. Aiken, M. Gonin, K. Fuhrer, T. Horvath, K.S. Docherty, D.R. Worsnop, J.L. Jimenez, Anal. Chem. 78 (2006) 8281–8289.

[47] C. Mohr, P.F. DeCarlo, M.F. Heringa, R. Chirico, J.G. Slowik, R. Richter, C. Reche, A. Alastuey, X. Querol, R. Seco, J. Peñuelas, J.L. Jiménez, M. Crippa, R. Zimmermann, U. Baltensperger, A.S.H. Prévôt, Atmos. Chem. Phys. 12 (2012) 1649–1665.

[48] M. Crippa, P.F. DeCarlo, J.G. Slowik, C. Mohr, M.F. Heringa, R. Chirico, L. Poulain, F. Freutel, J. Sciare, J. Cozic, C.F. Di Marco, M. Elsasser, J.B. Nicolas, N. Marchand, E. Abidi, A. Wiedensohler, F. Drewnick, J. Schneider, S. Borrmann, E. Nemitz, R. Zimmermann, J.-L. Jaffrezo, A.S.H. Prévôt, U. Baltensperger, Atmos. Chem. Phys. 13 (2013) 961–981.

[49] X. Shu, B. Yang, J. Meng, Y. Wang, J. Shu, Spectrosc. Lett. 46 (2013) 227–234.

[50] J.D. Allan, J.L. Jimenez, P.I. Williams, M.R. Alfarra, K.N. Bower, J.T. Jayne, H. Coe, D.R. Worsnop, J. Geophys. Res. 108 (2003) 4090. http://dx.doi.org/10.1029/2002JD002358, 10 pages.

[51] M. Yang, P.T.A. Reilly, K.B. Boraas, W.B. Whitten, J.M. Ramsey, Rapid Commun. Mass Spectrom. 10 (1996) 347–351.

[52] D.M. Murphy, D.J. Cziczo, K.D. Froyd, P.K. Hudson, B.M. Matthew, A.M. Middlebrook, R. E. Peltier, A. Sullivan, D.S. Thomson, R.J. Weber, J. Geophys. Res. 111 (2006) D23S32. http://dx.doi.org/10.1029/2006JD007340, 15 pages.

[53] P.J. Silva, D.-Y. Liu, C.A. Noble, K.A. Prather, Environ. Sci. Technol. 33 (18) (1999) 3068–3076.

[54] N.L. Ng, M.R. Canagaratna, Q. Zhang, J.L. Jimenez, J. Tian, I.M. Ulbrich, J.H. Kroll, K.S. Docherty, P.S. Chhabra, R. Bahreini, S.M. Murphy, J.H. Seinfeld, L. Hildebrandt, N.M. Donahue, P.F. DeCarlo, V.A. Lanz, A.S.H. Prévôt, E. Dinar, Y. Rudich, D.R. Worsnop, Atmos. Chem. Phys. 10 (2010) 4625–4641.

[55] A.G. Rincón, A.I. Calvo, M. Dietzel, M. Kalberer, Environ. Chem. 9 (2012) 298–319.

[56] E. Gelhausen, K.-P. Hinz, A. Schmidt, B. Spengler, Atmos. Res. 102 (2011) 49–56.

[57] M.S. Reinard, K. Adou, J.M. Martini, M.V. Johnston, Atmos. Environ. 41 (2007) 9397–9409.

[58] L. Van Vaeck, H. Struyf, W. Van Roy, F. Adams, Mass Spectrom. Rev. 13 (1994) 209–232.

[59] B.R. Bzdek, M.R. Pennington, M.V. Johnston, J. Aerosol Sci. 52 (2012) 109–120.

[60] D.A. Barnett, B. Ells, R. Guevremont, R.W. Purves, J. Am. Soc. Mass Spectrom. 10 (12) (1999) 1279–1284.

[61] G.R. Asbury, H.H. Hill, J. Microcolumn Sep. 12 (3) (2000) 172–178.

[62] G.A. Eiceman, Trends Anal. Chem. 21 (4) (2002) 259–275.

[63] S. Armenta, M. Blanco, Anal. Chem. 84 (2012) 4560–4568.

[64] N.G. Adams, D. Smith, Int. J. Mass Spectrom. Ion Phys. 21 (1976) 349–359.

[65] T. Karl, T. Jobson, W.C. Kuster, E. Williams, J. Stutz, R. Shetter, S.R. Hall, P. Goldan, F. Fehsenfeld, W. Lindinger, J. Geophys. Res. 108 (2003) 4508. http://dx.doi.org/10.1029/2002JD003333, 15 pages.

[66] S. Hayward, C.N. Hewitt, J.H. Sartin, S.M. Owen, Environ. Sci. Technol. 36 (2002) 1554–1560.

[67] C. Warneke, S. Kato, J.A. de Gouw, P.D. Goldan, W.C. Kuster, M. Shao, E.R. Lovejoy, R. Fall, F.C. Fehsenfeld, Environ. Sci. Technol. 39 (2005) 5390–5397.

Chapter 11

Microscopic Single-Particle Analytical Methods for Aerosol Characterisation

Hong Geng,[1] Dhrubajyoti Gupta[2] and Chul-Un Ro[2,*]
[1]*Institute of Environmental Science, Shanxi University, Taiyuan, China;* [2]*Department of Chemistry, Inha University, Incheon, Korea*
**Corresponding author: E-mail: curo@inha.ac.kr*

Chapter Outline

1. Introduction **332**
2. Energy-Dispersive Electron Probe X-Ray Microanalysis **333**
2.1 Fundamentals 333
2.2 Applications 334
2.2.1 Characterisation of Urban Aerosols 336
2.2.2 Characterisation of Asian Dust Storm Particles 336
2.2.3 Characterisation of Antarctic and Arctic Aerosols 337
2.2.4 Characterisation of Atmospheric Aerosols in Subway Stations 337
3. Scanning Transmission X-Ray Microscopy **338**
3.1 Fundamentals 338
3.2 Applications 338
4. Molecular Vibrational Imaging Spectroscopies **339**
4.1 Fundamentals 339
4.2 Applications of ATR-FTIR Imaging for Characterisation of Individual Atmospheric Particles 340
4.3 Applications of RMS for Characterisation of Individual Atmospheric Particles 341
5. Miscellaneous Methods **342**
5.1 Time-of-Flight Secondary Ion Mass Spectrometry 342
5.2 Atomic Force Microscopy 342
5.3 Optical Microscopy 343
6. Application of Multiple Techniques for Characterisation of the Same Single Particles **344**
6.1 Combined Application of ED-EPMA and ATR-FTIR Imaging 345

Comprehensive Analytical Chemistry, Vol. 70. http://dx.doi.org/10.1016/bs.coac.2015.09.008

6.1.1 Standard Mineral Particles 345
6.1.2 Aged or Reacted Mineral Dust 348
6.1.3 Reacted Sea-Salt Aerosols Internally Mixed with Other Species 350
6.1.4 Silicate Particles 352
6.1.5 Antarctic Sea-Salt Aerosols 354
6.2 Combined Application of ED-EPMA and RMS 355
6.2.1 Chemical Mixing State of Individual Asian Dust Particles 355
6.2.2 Iron Speciation of Airborne Subway Particles 357
6.2.3 Internally Mixed Soot Particles 358
6.3 Combined Application of ED-EPMA, ATR-FTIR Imaging and RMS 358
7. Conclusions 363
References 364

1. INTRODUCTION

The physicochemical characterisation of atmospheric aerosols is of great interest because of their multiple roles in atmospheric processes and their diverse influences on environment. For example, atmospheric aerosols influence global climate, directly by scattering or absorbing solar radiation and indirectly by serving as cloud condensation nuclei. They also have a significant impact on human health and visibility. These roles and effects in turn depend on their morphology and compositions, and thus detailed information on physicochemical properties, size distribution and aerial behaviour of atmospheric aerosol particles is critical to accurately assess their impact on the Earth's climate as well as human health.

As atmospheric particles are chemically and morphologically heterogeneous, and the average composition and average aerodynamic diameter obtained by the traditional bulk analyses do not describe well the population of the particles, microanalytical single-particle analyses have been developed for studying atmospheric particles. Over the last 40 years, some advanced single-particle analytical techniques, mainly based either on mass spectroscopic or microscopic analytical approaches, have been used to provide detailed information on physicochemical properties of single particles formed in the atmosphere or in the laboratory [1–3]. Although mass spectroscopic methods have some advantages such as their possible real-time, online measurements and high sensitivity for non-refractory aerosols [4], microscopic methods can provide information on morphology and internal mixing state of individual aerosol particles, which is essential for atmospheric aerosol study. In addition, elemental and molecular compositions, surface compositions, functional groups and hygroscopic properties of individual particles can be investigated using various microscopic single-particle analytical techniques.

Herein, microscopic single-particle analytical techniques commonly used in atmospheric aerosol research are described, such as techniques based on scanning electron microscopy/energy-dispersive X-ray spectrometry (SEM/EDX), transmission electron microscopy/EDX (TEM/EDX), scanning transmission X-ray microscopy (STXM), Raman microspectrometry (RMS), attenuated total reflectance Fourier transform infrared spectroscopic (ATR-FTIR) imaging, atomic force microscopy (AFM) and time-of-flight secondary ion mass spectroscopy (TOF-SIMS). Except for a few, most techniques are relatively new, so that the aim of this chapter is to briefly introduce what these techniques can offer for physicochemical characterisation of individual atmospheric particles. In addition, the potential of combined use of various microscopic single-particle analytical techniques for more detailed aerosol characterisation is emphasised as any single analytical technique cannot provide multimodal information on physicochemical properties of aerosols.

2. ENERGY-DISPERSIVE ELECTRON PROBE X-RAY MICROANALYSIS

2.1 Fundamentals

Energy-dispersive electron probe X-ray microanalysis (ED-EPMA) appeared when the use of a solid-state Si(Li) EDX detector in conjunction with an electron microscope was described for the first time in 1968 by Fitzgerald and co-workers [5]. After a while, ED-EPMA based on SEM/EDX became one of the most common, non-destructive techniques which could provide information on morphology of (sub)micron individual particles from their secondary or back-scattered electron images, together with information on their chemical compositions from electron-induced, characteristic X-rays of chemical elements in the particles. When ED-EPMA measurements are performed in an automated, computer-controlled way, then a significant number of single aerosol particles can be probed in a reasonable analysis time [6]. After ultrathin window EDX detectors were commercially available in 1990s, the detection of low-Z elements such as C, N, O and F together with heavier elements ($Z \geq 11$) became feasible in ED-EPMA. However, accurate quantification of the low-Z elements had been a problem for a long time as the characteristic X-ray lines of the low-Z elements undergo extremely strong attenuation while propagating through the particle volume and the conventional correction procedures such as ZAF (atomic number (Z), X-ray absorption (A) and secondary fluorescence (F) correction) and $\varphi(\rho z)$ (the X-ray production and absorption as function of mass depth (ρz) in the specimen, where ρ is the sample density and z is the depth of electron penetration in sample) methods were not sufficient for the accurate estimation of the matrix effect for the low-Z elements. In 1999, a quantitative ED-EPMA technique based on a Monte Carlo simulation coupled with successive approximations, called low-Z particle EPMA, was developed [7]. In the quantification procedure based on the Monte Carlo calculation, electron trajectories are simulated, and generated characteristic and Bremsstrahlung X-rays are calculated for spherical, hemispherical and hexahedral particles sitting on a flat surface, so that the matrix effect of X-ray lines even from the low-*Z* elements can be accurately corrected and the concentrations of chemical elements of single particles can be determined [7,8]. The quantification procedure provided results accurate within 12% relative deviations between the calculated and nominal elemental concentrations when the method was applied to various types of standard particles such as NaCl, Al_2O_3, $CaSO_4 \cdot 2H_2O$, Fe_2O_3, $CaCO_3$ and KNO_3 [9]. The low-*Z* particle EPMA has since played important roles in characterisation of atmospheric aerosols as many environmentally relevant particles contain low-*Z* elements in the form of nitrates, sulphates, oxides or mixtures, including a carbon matrix, and quantitative information is critical for clear chemical speciation of individual particles [10–13].

Although the quantification procedure based on the Monte Carlo calculation was originally developed for SEM/EDX applications with a lower accelerating voltage (10–30 kV), it could also be an excellent tool for the quantitative single-particle analysis using TEM/EDX at a higher accelerating voltage (200 kV) [14]. In a Monte Carlo calculation newly modified for TEM/EDX applications, electron trajectories for a particle sitting on the thin formvar/carbon film (coated on TEM grids) are simulated and X-rays generated both from the particle and the film can be calculated. The accuracy of the quantification procedure, including the correction for C and O X-ray intensities from the formvar/carbon film, was evaluated by comparing calculated elemental atomic concentrations with nominal ones for 12 types of standard particles. The relative differences (denoted as Δ) between the calculated and nominal elemental atomic concentrations are within 18% with the smallest 4.4% (Δ_K in KCl) and the biggest 17.4% (Δ_C in $CaCO_3$) for electron beam refractory particles such as NaCl, KCl, SiO_2, Fe_2O_3, Na_2CO_3, $CaCO_3$, Na_2SO_4, K_2SO_4 and $CaSO_4$, which is sufficiently accurate for the reliable identification of chemical species in these particles (Figure 1). Intense and high-energy electron beams can cause oxygen depletion for nitrates (NO_3^-) or sulphates (SO_4^{2-}) through decomposition into ($NO_2 + O^-$) and ($NO_2^- + O$) or yielding sulphites (SO_3^{2-}), respectively [6]. Hence, for electron-beam-sensitive salt particles, such as $NaNO_3$, $Ca(NO_3)_2 \cdot 4H_2O$ and $(NH_4)_2SO_4$, the high electron energy and current employed in TEM/EDX measurements lead to large deviations (>30%) in quantification.

2.2 Applications

The low-*Z* particle EPMA, based either on SEM/EDX or TEM/EDX, has demonstrated its powerful applicability in the characterisation of environmental and geological single particles. By the application of the technique, the particle size distribution and quantitative chemical compositions of individual particles can be obtained without a complicated sample pretreatment process and many environmentally important atmospheric particles, e.g. sulphates, nitrates, ammonium and carbonaceous species, have been clearly identified [10–13]. Reaction processes of Asian dust and sea-salt particles

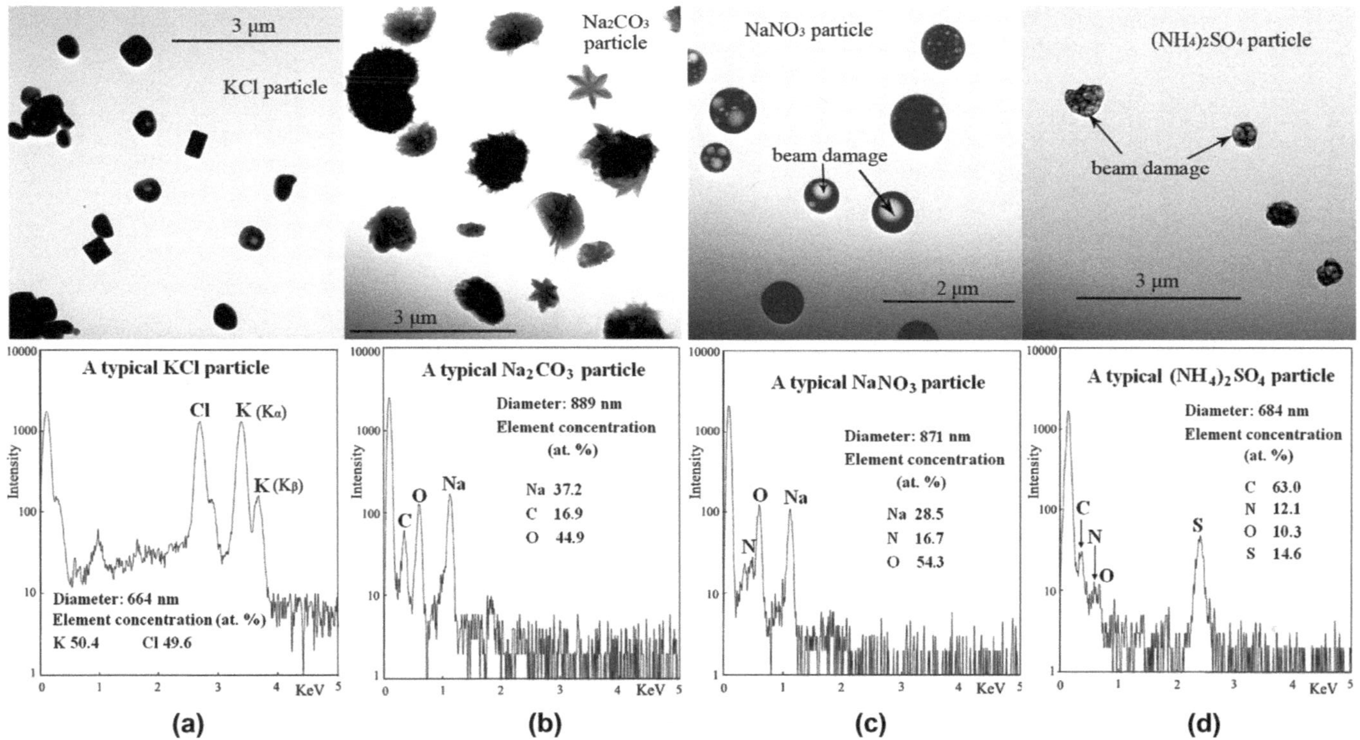

FIGURE 1 Transmission electron microscopy images and X-ray spectra of typical KCl, Na_2CO_3, $NaNO_3$ and $(NH_4)_2SO_4$ standard particles. *Reprinted with permission from Ref. [14], Figure 5, p. 8. Copyright (2010) John Wiley and Sons.*

with NO_x and/or SO_2 species could also be elucidated [15,16]. Up until now, the technique has been successfully applied to the characterisation of various types of ambient aerosols, including urban, indoor, subway, coastal and marine aerosols.

2.2.1 Characterisation of Urban Aerosols

Urban aerosols, which are often associated with emissions from vehicles, power plants, different burning processes and various other industries, have been of deep concern due to their adverse health and environmental effects. Low-Z particle EPMA was used to examine urban ambient particles collected in Incheon, Korea, when a typical haze episode occurred [11]. Overall 3600 individual particles in $PM_{2.5-10}$ and $PM_{1.0-2.5}$ fractions from 12 aerosol samples were analysed. The relative number abundances of various types of particles revealed that the haze occurrence was attributable to the increase of ambient organic carbon (OC) particles (especially smaller than 2.5 μm in diameter). TEM/EDX further showed that submicron airborne aerosols collected in Incheon during different seasons were abundant in carbonaceous particles and $(NH_4)_2SO_4/NH_4HSO_4$-containing particles mixed with OC and/or elemental carbon (EC), followed by Fe-rich, fly ash and transition or heavy metal-containing particles (the majority were Zn- and Pb-containing) [14].

2.2.2 Characterisation of Asian Dust Storm Particles

Based on the characterisation results for airborne particle samples collected in the marine boundary layer of the Bohai Sea and Yellow Sea using low-Z particle EPMA, it was suggested that Asian dust aerosols are important carriers of gaseous inorganic nitrogen species, especially NO_x and NH_3 [12]. By examining the chemical compositions and morphology of overall 4200 aerosol particles at stages 1–6 (with a 50% size cut-off of 16, 8, 4, 2, 1 and 0.5 μm in aerodynamic diameter, respectively) of a cascade May impactor, collected in Beijing, China and Incheon, Korea, during a same Asian dust storm (ADS) event, it was found that there were large differences in the chemical compositions between the ADS samples collected at the two sites [10]. When the ADS passed over the Yellow Sea, an accumulation of sea-salt components and secondary aerosols on individual ADS particles was observed. The significant increase of [Mg]/[Al] and [Mg]/[Si] ratios in Mg-containing aluminosilicates indicated that a significant evolution or ageing must have occurred on the ADS particles in the marine atmosphere during long-range transport from China to Korea. In addition, many (C, N, O, S)-rich droplet particles (most likely mixtures of $(NH_4)_2SO_4/NH_4NO_3$ and water-soluble organic matter) were encountered in Incheon samples, whereas none was observed in Beijing samples, implying that this type of particles was favourably formed in the marine atmospheric boundary layer.

2.2.3 Characterisation of Antarctic and Arctic Aerosols

Low-Z particle EPMA was used to characterise individual atmospheric particles collected at Korean scientific research stations located at King George Island in the Antarctic and Ny-Ålesund, Svalbard, in the Arctic [13,17]. For the Antarctic samples, two size modes of particles, i.e. $PM_{2.5-10}$ and $PM_{1.0-2.5}$, were analysed, and four types of particles were identified, with sulphur-containing sea-salt particles being the most abundant, in which internal mixture particles of sea – salts/$CH_3SO_3^-$/SO_4^{2-} were very frequently encountered. As nitrate-containing particles were not encountered, the oxidation of dimethyl sulphide (DMS) emitted from phytoplankton in the ocean is most likely to be responsible for the formation of the mixed sea – salts/$CH_3SO_3^-$/SO_4^{2-} particles [17]. On the other hand, in the summertime, in Arctic aerosol samples, most of the reacted sea-salt particles were nitrate-containing ones and many (C, N, O, S)-containing droplet particles were encountered, suggesting that the summertime Arctic atmosphere was disturbed by anthropogenic air pollutants such as NO_x (or HNO_3) [13].

2.2.4 Characterisation of Atmospheric Aerosols in Subway Stations

Low-Z particle EPMA was applied to characterise seasonal aerosol samples collected at underground subway stations in Seoul, Korea, and four major types of subway particles were significantly encountered: Fe-containing, soil-derived, carbonaceous and secondary nitrate and/or sulphate particles, in which Fe-containing particles were the most frequently encountered with relative number abundances in the range of 61–79% [18]. According to oxidation states, Fe-containing particles were further classified into three subtypes: iron metal, partially oxidised and fully oxidised species (denoted as 'Fe', 'Fe/FeO_x' and 'FeO_x', respectively). Iron metal particles were clearly distinguishable from oxidised iron particles because of both their high iron content and different secondary electron images (SEIs) from those of the iron oxide particles (in almost all cases, FeO_x-containing particles agglomerate by the coagulation of particles, possibly due to an attractive (magnetic) force between magnetic Fe_3O_4 particles). In mineralogy, iron oxide minerals are known to exist as FeO (wustite), Fe_2O_3 (haematite, maghemite) or Fe_3O_4 (magnetite). Frequently encountered FeO_x-containing particle types include FeO_x/C, FeO_x, FeO_x/Si/C, FeO_x/Ca/C, FeO_x/Mg/Si/C and FeO_x/Si/Ca/C [19].

TEM-EDX was performed to distinguish morphology and speciation of the metro particles collected in Shanghai, China [20]. The results showed that there were three subtypes of Fe-containing particles: free-floating Fe-rich particles, mineral-dust-hosted Fe-rich particles and other types of particles with minor Fe content. The free-floating Fe-rich particles were the most abundant species, mainly in the form of iron oxides with irregular shapes. Mineral-dust-hosted Fe-rich particles were also one of the typical particulate pollutants in the Shanghai metro system, especially in the above-ground line. They had smoother edges and lighter colours and were thought to be derived from the soil used for the construction of metro tunnels. Fly ash particles with minor Fe content were quite spherical, due to the molten nature of the material at high temperatures during steel production or other industrial processes.

3. SCANNING TRANSMISSION X-RAY MICROSCOPY

3.1 Fundamentals

STXM is a microscopic technique that provides chemical-species-sensitive images at a spatial resolution from several micrometres down to ~25 nm. It uses a monochromatic soft X-ray beam in the energy range of 100–2200 eV and can be operated under low or high vacuum, under atmospheric pressure or in a helium atmosphere [21]. Compared to TEM-based techniques, STXM induces less beam damage as it is usually coupled with X-ray absorption near-edge spectra. The energy of the X-ray beam is scanned across X-ray absorption edges, resulting in the generation of near-edge X-ray absorption fine structure (NEXAFS) spectra. The fine structure of NEXAFS spectra depends on elemental compositions and chemical bonding environments of the elements that are present in individual particles. Thus, information on organic and inorganic compositions of specimens at high spatial resolution can be obtained by STXM/NEXAFS, with a combination of high spatial resolution microscopy and NEXAFS spectroscopy [22]. However, STXM is rather slow (with a low data acquisition rate) as it is a scanning technique, and soft X-ray STXM needs thin samples (thickness of <300 nm) due to the high absorption cross-sections for K-edges of light elements such as C, N and O [23].

3.2 Applications

Using STXM imaging and NEXAFS, Fe metal was distinguished from other Fe-containing species (such as Fe_3O_4, Fe_2O_3, $Fe_2(SO_4)_3$, $FeSO_4 \cdot 6H_2O$ and dust) presented in particle samples collected at metro stations and the correlation between elemental components was investigated on a nanometre scale by using a contrast imaging technique [20]. Airborne organic aerosols of submicrometre size were differentiated into organic acid, organic-dust and organic-black carbon particles by examining their morphology and compositions of individual organic particles and probing the electronic structure of individual particles at a spatial resolution of approximately 30 nm [24,25]. When this method was applied to internally mixed submicron particles containing organic and inorganic moieties, it was demonstrated that not only was it effective but also simple and straightforward. Several algorithms were developed to exploit NEXAFS spectral features in the energy range from 278 to 320 eV (covering C–K and K–L edges) for a quantitative mapping of the spatial distribution of EC, OC, potassium and non-carbonaceous elements in

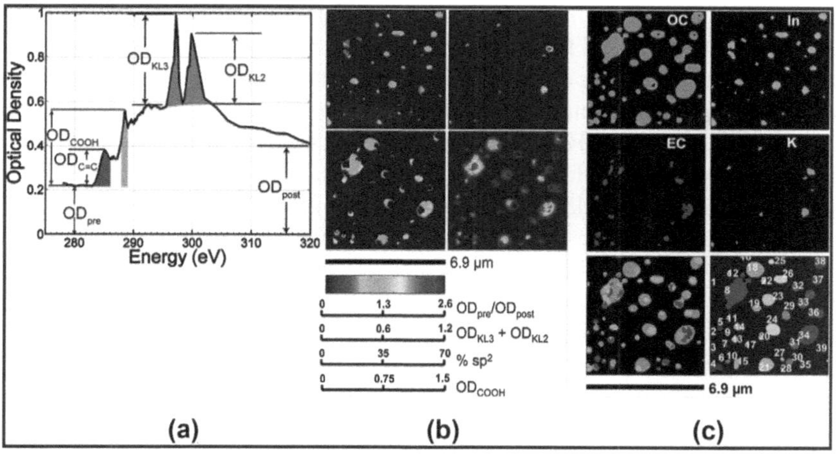

FIGURE 2 Characterisation of typical airborne particles collected in Mexico City during the MILAGRO 2006 campaign by scanning transmission X-ray microscopy/near-edge X-ray absorption fine structure (STXM/NEXAFS). (a) Characteristic carbon K-edge NEXAFS and spectral feature assignments: 'pre-edge' (OD_{pre}) occurs between 278 and 283 eV, where carbon–carbon double bonds ($OD_{C=C}$) are shown in red (dark grey in print versions), carboxylic acids (OD_{COOH}) in green (light grey in print versions) and potassium L_2 and L_3 edges (OD_{KL3}, OD_{KL2}) in orange (grey in print versions). OD_{post} is the post absorption edge at 320 eV. (b) STXM particle maps indicating different spectral components: OD_{pre}/OD_{post} represents total carbon absorption, $OD_{KL3} + OD_{KL2}$ is a potassium map (orange (grey in print versions) peaks), % sp^2 shows carbon–carbon double bonds (red (dark grey in print versions) peak) and OD_{COOH} is a map of carboxylic acid groups (green (light grey in print versions) peak). (c) Binary component maps after applying thresholds for organic carbon (OC), inorganic (In) constituents, elemental carbon (EC or soot) and potassium (K) containing salts. The lower left panel shows the total mixed particle compositions with images overlaid in the following order: (1) OC, (2) In, (3) K and (4) EC. The lower right panel shows particle identification indices. From the combined map, particle mixing states are identified. For example, particle #12 is identified as OC, particle #34 as 'In + K + EC + OC', particle #38 as 'In + OC' and particle #37 as 'In + OC'. *Adapted with permission from Ref. [3]. Copyright (2010) American Chemical Society.*

particles of mixed composition, as shown in Figure 2 [3]. Besides its imaging capabilities at high spatial resolution, the main advantage of STXM/NEXAFS over bulk methods such as X-ray diffraction, FTIR or nuclear magnetic resonance spectroscopy is to efficiently obtain structural information on very small quantities of sample. In brief, STXM/NEXAFS is a promising method for the submicrometre-scale characterisation of atmospheric aerosols, with the potential to improve our understanding of their hygroscopic behaviour, physical properties and formation and transformation mechanisms [26].

4. MOLECULAR VIBRATIONAL IMAGING SPECTROSCOPIES

4.1 Fundamentals

Studies of molecular characterisation and/or hygroscopic properties of size-segregated aerosol particles are often performed using vibrational spectroscopic techniques such as ATR-FTIR spectroscopy and RMS. Although both RMS and ATR-FTIR are powerful for functional group analysis and molecular speciation of organic and inorganic compounds including hydrated species under ambient conditions, the vibrational signals generated from them have different fundamentals [27]. ATR-FTIR is based on the attenuation of the evanescent wave generated by totally reflected IR radiation on an internal reflectance element (IRE) crystal, while RMS provides information on molecular vibrations based on the difference in wavelength between incident and scattered UV/VIS/NIR radiations (Raman scattering). According to selection rules, it is necessary for the molecule to have a permanent electric dipole for IR spectroscopy, whereas it is the polarisability of the molecule which is important for Raman spectroscopy. The fact that RMS and ATR-FTIR imaging provide spectra with a typical spectral range between 100–4000 cm^{-1} and 680–4000 cm^{-1}, respectively implies that RMS is useful to identify transition and heavy metal compounds in particle samples of which their characteristic vibrations mostly appear in far IR region (<680 cm^{-1}). Furthermore, RMS is more powerful for the investigation of the mixing state of individual particles than ATR-FTIR imaging as RMS has better lateral resolution due to the incident radiation (~1 μm for RMS and ~3.1 μm at 1726 cm^{-1} for ATR-FTIR imaging). In addition, Raman spectra can reflect the real nature of individual particles more, and their mostly sharp Raman peaks are very useful for unambiguous molecular speciation even for water-containing samples because H_2O molecules exhibit strong absorption in ATR-FTIR which may overlap the other IR bands. However, the laser beam employed in RMS can induce damage on many types of particles and the interference by fluorescence encountered in RMS often needs to be minimised, which are not the problems in ATR-FTIR measurements. As RMS images are acquired by point-by-point scanning mode, RMS mapping to obtain the spatial, chemical heterogeneity for individual particles takes a longer time than ATR-FTIR imaging does. ATR-FTIR imaging provides an ATR-FTIR spectrum at each pixel in the image field, obtaining information on the molecular species for all individual particles present on the image field within several hours. Therefore, the differences in the spectra from RMS and ATR-FTIR owing to different signal generation mechanisms (scattering vs absorption of energy) and different selection rules would make the two fingerprint techniques rather complementary. In recent years, a few applications of low-Z particle EPMA combined either with ATR-FTIR imaging, RMS or both were reported to be quite useful for the characterisation of the same individual atmospheric particles [28–33].

4.2 Applications of ATR-FTIR Imaging for Characterisation of Individual Atmospheric Particles

FTIR is a powerful technique for functional group analysis and molecular speciation of organic and inorganic compounds. Thus far, in research pertaining to atmospheric aerosols, many studies regarding the characterisation and quantification of functional groups in size-segregated aerosol samples have been performed using either transmission, reflectance or ATR-FTIR spectroscopic techniques [34,35]. The FTIR spectroscopic imaging technique can provide further information on the spatial distribution of chemical constituents within a sample, as the obtained IR image contains full spectral information at each pixel of the image corresponding to a unique spatial location on the sample. In atmospheric aerosol research, however, just a few studies on ambient aerosols or smog chamber particles using FTIR imaging have been reported [36,37]. In these studies, aerosols collected on a cascade impactor were analysed, and information on the average chemical species for agglomerated aerosol particles was obtained. However, no real application of FTIR imaging techniques alone has been made in the analysis of individual airborne particles, mainly because of the rather large spatial resolution available in conventional reflectance or transmission FTIR imaging techniques. Recently, ATR-FTIR imaging instruments with a spatial resolution of ~3.1 μm became commercially available although the spatial resolution remains inadequate for (sub)micron single-particle analysis. Therefore, the ATR-FTIR imaging approach should be combined with low-*Z* particle EPMA to investigate physiochemical properties of individual airborne particles, which will be described later.

4.3 Applications of RMS for Characterisation of Individual Atmospheric Particles

RMS is a well-established technique for the investigation of chemical characteristics of atmospheric aerosols which can offer molecular level compositional information on bulk materials as well as single particles and is suitable for identifying components within a heterogeneous sample with a spatial resolution of ~1 μm. RMS is useful because of its various analytical capabilities including identification of different polymorphs, determination of different oxidation states and crystallographic orientation, evaluation of temperature and stress effects on molecular structure, etc. Available literature on the application of RMS for the analysis of single atmospheric particles is increasing [31,33].

As a typical example of RMS application, volcanic ash particles were examined for characterisation of their mineral compositions with a spatial resolution down to several microns and for the determination of crystalline (ordered and disordered) or glassy (amorphous) structures [38]. The example not only demonstrates the potential of RMS in the identification and chemical characterisation of single volcanic ash particles but also provides the basis for more detailed RMS studies of airborne particles. RMS can be even more powerful when it is applied in combination with low-*Z* particle EPMA to the same individual particles, which will be described later.

5. MISCELLANEOUS METHODS

5.1 Time-of-Flight Secondary Ion Mass Spectrometry

TOF-SIMS is a powerful tool for the examination of organic and inorganic species on the surface of individual particles owing to its high sensitivity and capability for chemical speciation [39]. The surface moiety of individual particles can be investigated at a very small portion of area in the near-surface region of a particle. Even volatile and semivolatile organic compounds including polycyclic aromatic hydrocarbons and alkanes can be detected at the surface of the particles when a cryogenic stage is used to cool the samples analysed in a SIMS instrument [40]. TOF-SIMS uses a primary ion beam to probe the surface of a particle. The primary ion impact initiates and induces collision cascades so that neutrals, electrons and positive and negative secondary ions are emitted. The secondary ions escape just from the sample surface and may be analysed in the form of a spectrum (chemical composition), an image (distribution of chemical species) or a depth profile. It allows the acquisition of mass-resolved secondary ion images with high lateral resolution simultaneously for various ion species. Advantages of the technique are the wide mass range starting from H ($m/z = 1$) and the possibility of detecting elemental ions together with fragment, adduct and molecular ions which permits molecular compositional analysis. In addition to elemental compositional analysis ($Z = 1-92$), it is also feasible to generate secondary ion images of cluster ions, necessary for direct compound identification [41].

TOF-SIMS is one of the powerful candidates for characterisation of atmospheric aerosol particles. It requires samples with a very small amount of aerosol particles compared with some conventional techniques, e.g. inductively coupled plasma–mass spectrometry and gas chromatography–mass spectrometry, so that sampling methods to collect a very small amount of aerosols in a short sampling period can be well suited for TOF-SIMS measurement [40]. By the combined use of ED-EPMA (including field emission-SEM observation) and TOF-SIMS, Tomiyasu et al. found that a lot of diesel exhaust particles were adsorbed on the surface of iron particles and most of the adsorbed particles were carbonaceous ones containing a small amount of silicon in aerosol particles collected at locations near a traffic route and a railroad [42]. Liu et al. reported that dry $CH_3SO_3Na/NaCl$ particles from the sea are composed of a NaCl core surrounded by a CH_3SO_3Na shell [43].

5.2 Atomic Force Microscopy

AFM is a relatively young scanning probe microscopy technique. It allows insulators to be imaged at a nanoscale resolution and has been proven to be a convenient tool for the microscopic investigation of surface properties [44]. AFM characterisation involves structural morphology, grain size and distribution, etc. AFM topographic images provide not only the surface equivalent diameter but also the shape and physical state (e.g. liquid vs solid) of individual particles as most AFM modes can work in ambient air or even a liquid environment [27]. It is currently one of the foremost tools for imaging, measuring and manipulating matter (including atmospheric aerosols) at the nanoscale. On the other hand, AFM needs a flat surface for scanning, so it is somewhat difficult to set up the scan if static electricity is present on the surface of the sample and it takes a considerable time for the imaging of a large sample area.

AFM was used to investigate changes of morphology and surface properties of aggregated sootlike test aerosols as a function of relative humidity (RH), showing a significant alteration of morphology of hygroscopic particles depending on RH [45]. Three typical AFM micrographs for atmospheric particles collected simultaneously upwind and downwind of an industrial area in France showed that particles were mainly homogeneous single droplets with a size of 10—0.5 μm, which spread along the collecting substrate surface upon impaction and then formed hemispherical particles due to the hydrophilic nature of a silicon substrate [27].

5.3 Optical Microscopy

The past decade has witnessed an enormous growth in the extensive applications of optical microscopy for micron and submicron level investigations in microelectronics, nanophysics, biotechnology, pharmaceutical research, mineralogy, microbiology and the atmospheric environment. With the help of digital imaging and video, it can also be used to image very thin optical sections. In order to overcome the diffraction limit of visible light, other microscopic techniques have been designed such as AFM, SEM, TEM and X-ray microscopy. Although the use of electrons and X-rays provides much higher spatial resolution, optical microscopy, a robust and practical technique, plays an irreplaceably important role in observation or identification of aerosol particles. Now confocal optical systems are in operation in many aerosol research instruments, e.g. ATR-FTIR, RMS, AFM, etc., and provide optical images necessary for their measurements. For instance, the morphology of atmospheric particles can be investigated under an *in situ* condition by the combined use of optical and AFM imaging. After impaction, the optical images of several spots of particles provide a suitable 2D overview of the silhouette of particles deposited on silica plates [27]. Figure 3 shows several optical images of an image field of wet-dispersed $NaNO_3$ particles deposited on a TEM grid, which were obtained at different RHs during humidifying and dehydration processes [46]. All the wet-dispersed $NaNO_3$ particles are round in shape and undergo continuous growth or shrinkage in size during hydration and dehydration. The observation indicates that nebulised $NaNO_3$ particles, which represent completely reacted sea-salt aerosols, exist as amorphous particles. The hygroscopic measurement system employing an optical microscope is a good example of its use in aerosol research.

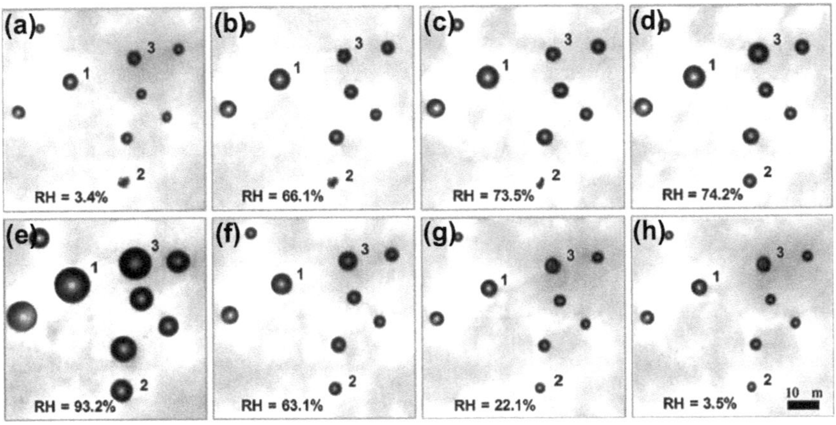

FIGURE 3 Optical images of wet-dispersed $NaNO_3$ aerosols obtained at (a)–(e) relative humidity (RH) = 3.4%, 66.1%, 73.5%, 74.2% and 93.2% during the humidifying process and at (f)–(h) RH = 63.1%, 22.1% and 3.5% during the dehydration process. *Reprinted with permission from Ref. [46]. Copyright (2012) Elsevier.*

6. APPLICATION OF MULTIPLE TECHNIQUES FOR CHARACTERISATION OF THE SAME SINGLE PARTICLES

Microscopic single-particle analytical techniques have been proven to be useful for physicochemical characterisation of individual airborne aerosol particles, as has been briefly described in this chapter. Overall, the various microscopic methods can provide information on morphology, internal mixing state, elemental and molecular compositions, surface compositions, functional groups and/or hygroscopic property of individual particles. However, these multimodal physicochemical characteristics of individual particles cannot be simultaneously investigated using any single technique as it is just capable of examining one or two physicochemical properties of individual particles. Recently, an integrated approach to obtain multimodal physicochemical information from the same individual particles has emerged, where different single-particle analytical techniques are applied in combination to probe the same individual particles of (sub)micrometre size [47]. In this approach, first, a single-particle analytical technique, e.g. ED-EPMA, is applied to obtain morphological and elemental compositional information on individual particles. And then, another technique, e.g. ATR-FTIR imaging or RMS, is applied to obtain information on molecular species for the same individual particles. In order to obtain multimodal information on physicochemical properties from the same individual particles, it is necessary to correctly relocate the same particles among different microscopic single-particle analytical instruments, which is feasible because of the inherent microscopic capability of the analytical instruments. In addition, each technique needs to be non-destructive, so that the same particles can be examined by multiple techniques. Although this integrated analytical approach is quite new and there may still be many aspects to be resolved, the following sections briefly describe what has appeared in the recent literature.

6.1 Combined Application of ED-EPMA and ATR-FTIR Imaging

Recently, the potential of the combined use of low-*Z* particle EPMA and ATR-FTIR imaging for the characterisation of the same individual aerosol particles has been demonstrated [28–30,48]. These single-particle analytical techniques provide complementary information on physicochemical characteristics of the same individual particles, i.e. low-*Z* particle EPMA on morphology and elemental concentrations and ATR-FTIR imaging on molecular species, crystal structures, functional groups and physical states. As the same individual particles can be probed by the two different single-particle analytical techniques, more detailed characteristics of aerosol particles on their different kinds of physicochemical properties can be obtained, which in turn provides insights into their sources, reactivity and fate. In this approach, the SEIs and X-ray spectral data of individual particles sitting on a collecting metallic substrate (e.g. Al or Ag foil) are first acquired by SEM/EDX. The particle sample is transferred into an ATR-FTIR imaging instrument, where the same particles are relocated by the help of an optical microscope in the ATR-FTIR imaging instrument. After the relocation, ATR-FTIR imaging measurements are performed to obtain ATR-FTIR spectra for the same individual particles.

6.1.1 Standard Mineral Particles

Low-*Z* particle EPMA and ATR-FTIR imaging techniques were successfully applied in combination for identification of various types of standard individual mineral particles [29]. Figure 4 shows SEIs from SEM and ATR-FTIR images of the same 29 individual National Institute of Standards and Technology (NIST) SRM 70a particles on Ag foil. The ATR-FTIR image of Figure 4(c) was obtained by the application of principal component analysis (PCA) after the first differentiation of original ATR-FTIR spectra at all of the pixels in the image. Figure 4(d) is a pseudo-colour ATR-FTIR image displayed for transmittance signal at 1000 cm^{-1}, where pink, red, yellow, green and sky-blue colours are used to display the transmission signal intensity of each pixel in the images, i.e. pink denotes the strongest and purple indicates the weakest. The SEIs clearly show morphologies and locations of the 29 particles before and after ATR-FTIR imaging measurement.

It was reported that the NIST SRM 70a feldspar material was obtained from the Kingman (Arizona) feldspar mine, and the material is a strained maximum microcline (one type of K-feldspar, $KAlSi_3O_8$) with about 10–15% of albite ($NaAlSi_3O_8$, Na-feldspar). According to the certificate of analysis for SRM 70a issued by the NIST, the sum of the major constituents, such as SiO_2 (67.1%), Al_2O_3 (17.9%), K_2O (11.8%) and Na_2O (2.6%), is 99.4%, where the

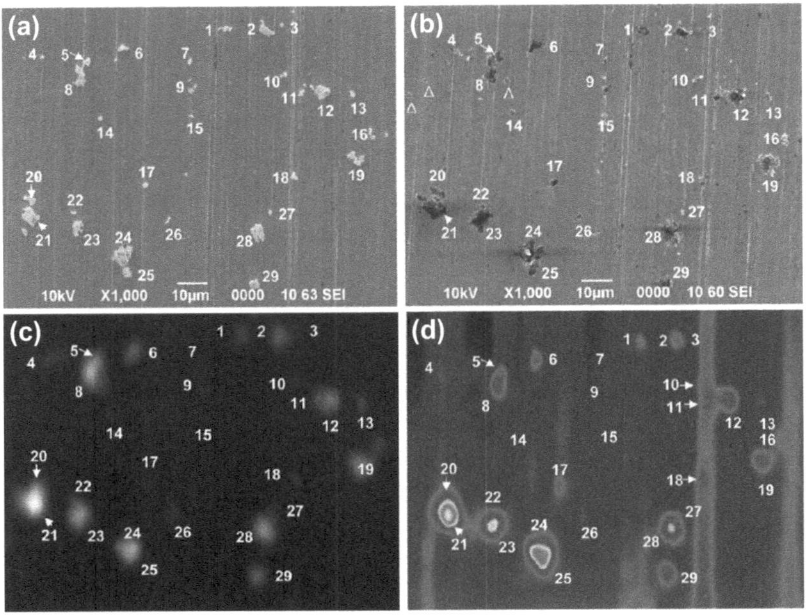

FIGURE 4 Secondary electron images (SEIs) from scanning electron microscopy (a) before and (b) after attenuated total reflectance Fourier transform infrared spectroscopic (ATR-FTIR) measurement and ATR-FTIR images (c) by a principal component analysis and (d) for transmittance signal at 1000 cm^{-1} of the same 29 K-feldspar SRM mineral particles on Ag foil. The same particles in the four images are denoted by matching particle numbers. *Reprinted with permission from Ref. [29]. Copyright (2010) American Chemical Society.*

certified values are given as the assumed oxide on an equivalent weight basis. On the other hand, based on X-ray and ATR-FTIR spectral data obtained from the same individual particles, four different types of feldspar minerals were identified: K-feldspar, Na-feldspar, (Na, K)-feldspar and (K, Fe)-feldspar, as shown in Figure 5. For a typical K-feldspar particle (Figure 5(a)), the atomic concentration ratios of [K]/[Al] and [Si]/[Al] are 0.9 and 3.1, respectively, which matches well to the chemical composition of K-feldspar ($KAlSi_3O_8$). For a Na-feldspar ($NaAlSi_3O_8$) particle (Figure 5(b)), the atomic concentration ratios of [Na]/[Al] and [Si]/[Al] are 0.9 and 2.9, respectively. Similarly, (Na, K)-feldspar (Figure 5(c)) and (K, Fe)-feldspar (Figure 5(d)) can be clearly distinguished from K- and Na-feldspars. However, for real atmospheric aerosol samples, mineral particles are complicated in composition and thus the sole use of ED-EPMA data is not sufficient for unambiguous identification of mineral particles.

In Figure 6, ATR-FTIR spectra of the four single particles, conventional transmission FTIR spectra of K- and Na-feldspar bulk samples (from the literature) and an ATR-FTIR spectrum obtained from bulk NIST SRM 70a powder are shown. The spectra of particles #5 and #2, which were identified

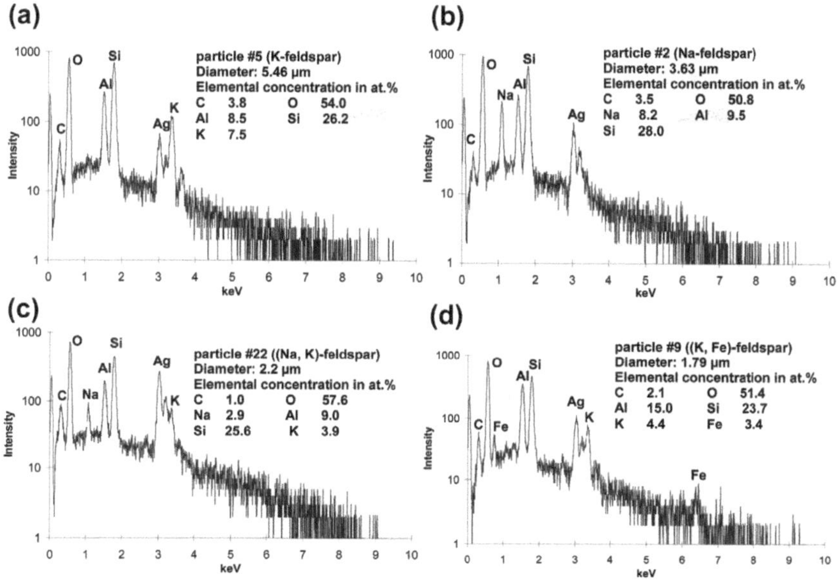

FIGURE 5 Typical X-ray spectra of four different feldspars observed in K-feldspar SRM mineral sample. *Reprinted with permission from Ref. [29]. Copyright (2010) American Chemical Society.*

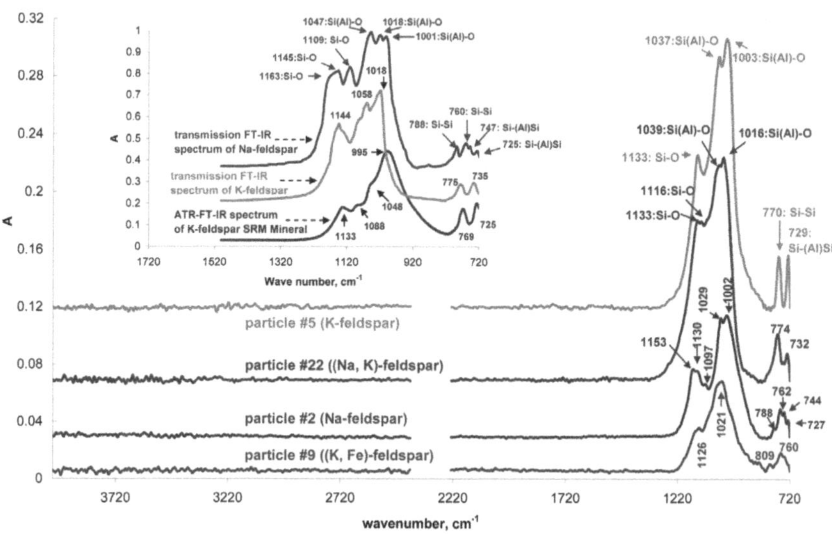

FIGURE 6 Typical attenuated total reflectance Fourier transform infrared (ATR-FTIR) spectra of four different feldspars observed in K-feldspar SRM mineral sample. Inset shows conventional transmission FTIR spectra of K- and Na-feldspar bulk samples from the literature and an ATR-FTIR spectrum obtained from bulk K-feldspar SRM mineral powder. *Reprinted with permission from Ref. [29]. Copyright (2010) American Chemical Society.*

by ED-EPMA as K- and Na-feldspar particles, respectively, look very similar to the K- and Na-feldspar spectra from the literature. The ATR-FTIR spectra for particles #22 and #9, which were identified by ED-EPMA as (Na, K)- and (K, Fe)-feldspar particles, respectively, were obtained for the first time.

6.1.2 Aged or Reacted Mineral Dust

By the combined application of low-*Z* particle EPMA and ATR-FTIR imaging, an Asian dust sample was examined to obtain detailed physicochemical characteristics of aged Asian dust particles, which experienced extensive chemical modification during long-range transport [32]. The application of this combined methodology to the characterisation of the aged Asian dust particles clearly showed that their internal mixing states are highly complicated. Figure 7 presents the ATR-FTIR and X-ray spectral data and SEIs of three typical Ca-containing particles in the Asian dust sample. The ATR-FTIR spectrum, X-ray spectrum and elemental atomic concentrations and SEIs of particle #89 are presented in Figure 7(a) and (d) and the insets, respectively. Based on low-*Z* particle EPMA, for particle #89, major elements were O (58% in atomic fraction), C (23%), Ca (12%), Si (2%), Al (1%), S (1%), Mg (1%) and Fe (1%), suggesting that it is a mixture of $CaCO_3$ and carbonaceous species with minor content of aluminosilicate and sulphate. Its bright and angular SEI obtained before ATR-FTIR imaging indicates its soil origin. The presence of the $CaCO_3$ moiety in particle #89 is confirmed by the strong IR peaks from CO_3^{2-} of calcite at 872 and 1404 cm^{-1} in its ATR-FTIR spectrum, and its ATR-FTIR spectral data indicate that the aluminosilicate moiety is montmorillonite as its IR peaks are at 790 and 1022 cm^{-1} [29]. The combined use of X-ray and ATR-FTIR shows that particle #89 is a mixture of calcite, EC and montmorillonite, in which the majority is calcite.

The ATR-FTIR spectrum, X-ray spectrum with elemental atomic concentrations and SEIs of particle #62 are presented in Figure 7(b) and (e) and the insets, respectively. The major elements of O (56%), C (21%), Ca (12%), N (9%), S (2%) and Cl (1%) suggest that particle #62 is a mixture of $CaCO_3$, nitrate and carbonaceous species with minor sulphate and chloride. Amorphous calcium carbonate (ACC) can be distinguished clearly from the crystalline $CaCO_3$ forms because ACC shows two split peaks of the asymmetric stretching vibration due to the deficient symmetric environment of the CO_3^{2-} ion. Hence, particle #62 in Figure 7(b), which shows two CO_3^{2-} peaks at 1420 and 1484 cm^{-1}, contains ACC. In addition, small and bright crystals inside dark particle #62 mainly contain Ca, C and O elements in their X-ray spectra, indicating that the small bright crystals are $CaCO_3$. The other three ACC particles encountered in this sample, except for particle #62, contain a small amount of silicate species, strongly suggesting their soil origin. Some carbonaceous species are organic because the IR peaks of the aliphatic CH group are present at 2865 and 2929 cm^{-1}. In addition, the IR peak at 1630 cm^{-1} is assigned to carboxylic ions (COO^-), most likely from humic substances

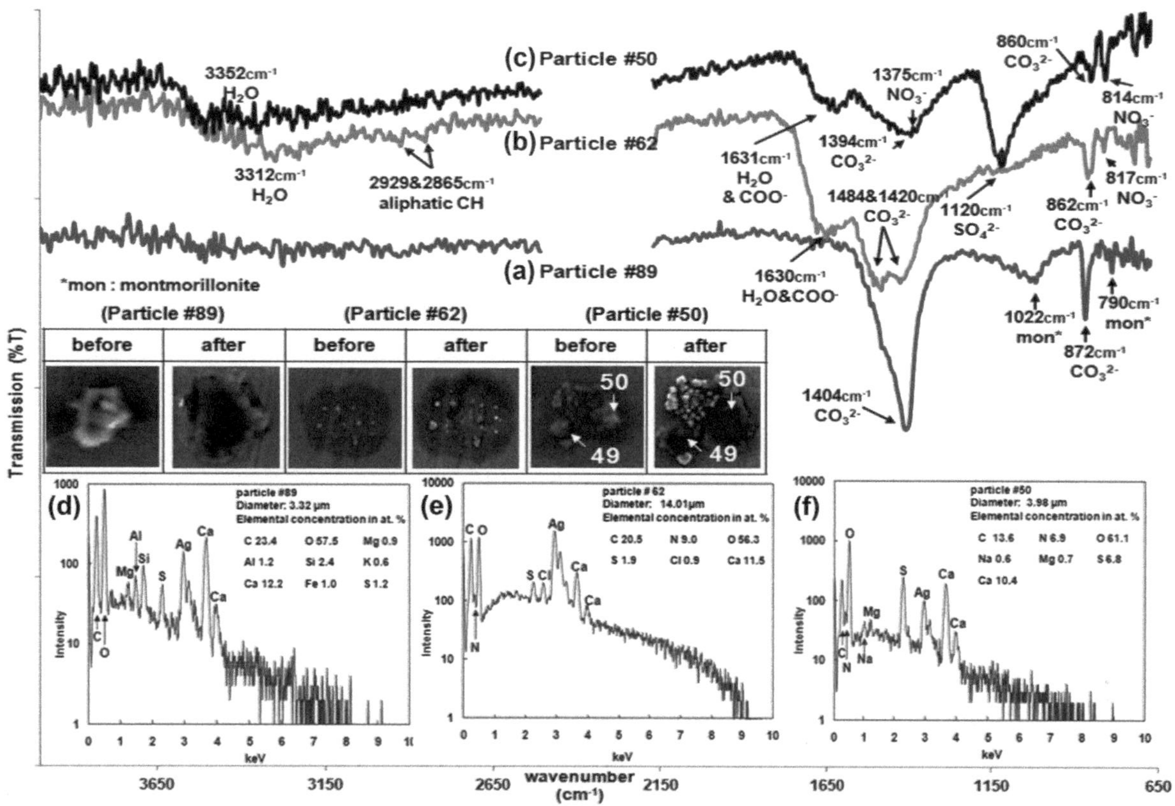

FIGURE 7 Attenuated total reflectance Fourier transform infrared (ATR-FTIR) and X-ray spectra of typical Ca-containing particles and secondary electron images of particles obtained before and after the ATR-FTIR imaging. The data for the 2200–2530 cm^{-1} region where atmospheric CO_2 peaks are present were deleted for clarity. *Reprinted with permission from Ref. [32]. Copyright (2013) Copernicus Publications.*

and/or organic compounds, such as oxygenated carboxylic or carbonyl compounds. The IR peak at 1630–1640 cm^{-1} can be from a water moiety and the small peaks at 817 and 1120 cm^{-1}, respectively, indicate that N and S elements detected by ED-EPMA are in the forms of nitrate and sulphate, respectively. The combined use of X-ray and ATR-FTIR spectral data shows that particle #62 is a mixture of ACC, organic, nitrate, sulphate and water.

The ATR-FTIR spectrum, X-ray spectrum and elemental atomic concentrations and SEIs of particle #50 are presented in Figure 7(c) and (f) and the insets, respectively. Its major elements of O (61%), C (14%), Ca (10%), N (7%) and S (7%) suggest that it is a mixture of $CaCO_3$, $Ca(NO_3)_2$, $CaSO_4$ and carbonaceous species. Its dark and circular SEI indicates that it was collected as a water droplet, which is supported by the presence of IR absorption peaks at 1631 and ~3352 cm^{-1} from the water moiety. Some bright particles inside the dark region would be crystalline $CaCO_3$. The presence of a strong IR peak at 1120 cm^{-1} and sharp peaks at 814 and 860 cm^{-1} confirms the presence of SO_4^{2-}, NO_3^- and CO_3^{2-} moieties, respectively. Organic species should be present due to the characteristic IR peak of carboxylic ions at 1631 cm^{-1}. These are typical examples of Ca-containing particles which experienced chemical modification through heterogeneous reactions of $CaCO_3$ particles with SO_x/H_2SO_4 and/or NO_x/HNO_3 during long-range transport. In short, these examples well demonstrate how powerful the combined application of low-*Z* particle EPMA and ATR-FTIR imaging is for detailed characterisation of the same individual particles.

6.1.3 Reacted Sea-Salt Aerosols Internally Mixed with Other Species

In the Asian dust sample, different types of $NaNO_3$-containing particles, which are reaction products between NaCl and NO_x/HNO_3, were frequently encountered although ED-EPMA just produced similar X-ray spectral data among the different $NaNO_3$-containing particles [32]. As a typical example of crystallised $NaNO_3$-containing particles, the ATR-FTIR spectrum, X-ray spectrum and elemental atomic concentrations and SEIs of particle #79 are presented in Figure 8(a) and (d) and the insets, respectively. The major elements are O (54%), Na (24%), N (22%) and Ca (1%), and its bright and cubic SEI indicates that it is a crystalline solid, which is supported by the absence of IR peaks of water. In addition, a sharp IR peak at 1786 cm^{-1} and the IR peaks at 833, 1064 and 1352 cm^{-1} are for the crystalline $NaNO_3$ [30].

For $NaNO_3$ particles internally mixed with other chemical species, a typical example of particle #33 is shown in Figure 8(b) and (e) and the insets, respectively. Its major elements are O (58%), Na (23%) and N (20%), and it has bright and cubic SEI. The presence of an IR peak of crystalline $NaNO_3$ species at 1788 cm^{-1} supports that it is a crystalline solid. The $NaNO_3$ moiety is major due to its strong IR peaks at 833 and 1346 cm^{-1}. Minor Na_2SO_4 and

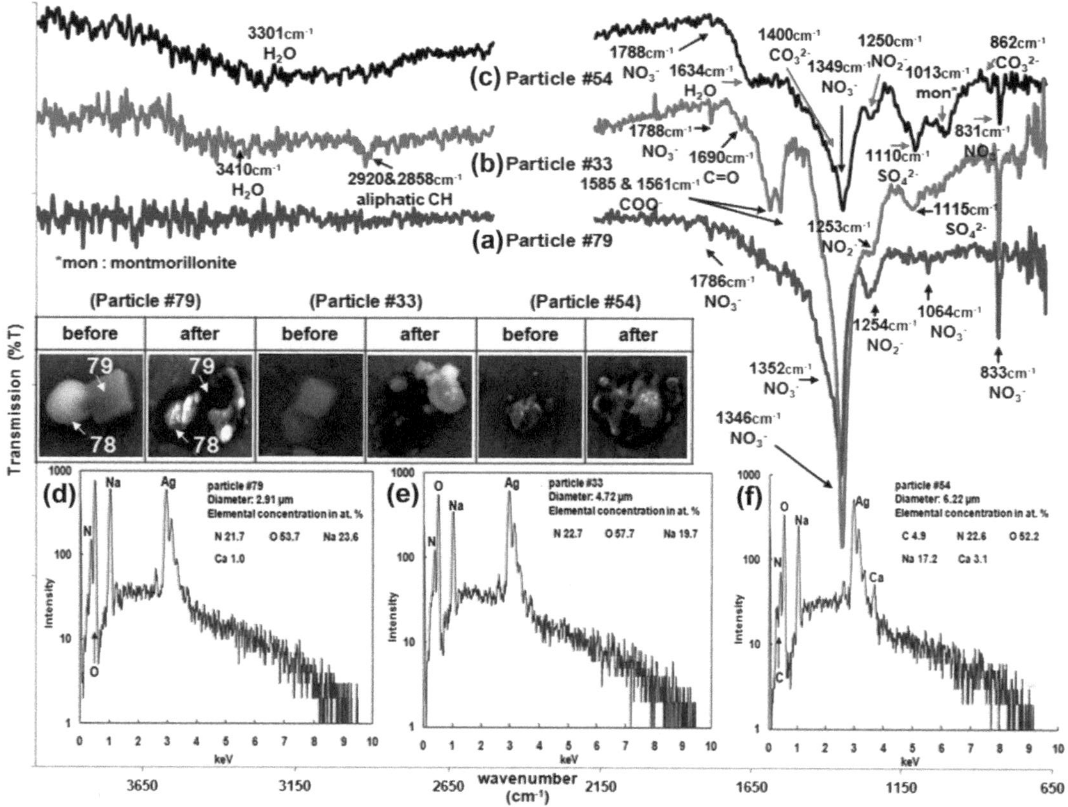

FIGURE 8 Attenuated total reflectance Fourier transform infrared (ATR-FTIR) and X-ray spectra of typical $NaNO_3$-containing particles and secondary electron images of particles obtained before and after ATR-FTIR imaging. The data for the 2200–2530 cm^{-1} region where atmospheric CO_2 peaks are present were deleted for clarity. *Reprinted with permission from Ref. [32]. Copyright (2013) Copernicus Publications.*

water moieties based on the IR peaks at 1115 and ~3410 cm^{-1}, respectively, as well as organics based on the IR peaks at ~2920, 1585 and 1561 cm^{-1}, are also present.

A $NaNO_3$-containing particle #54 internally mixed with carbonate and silicate minerals is presented in Figure 8(c) and (f) and the insets, respectively. The faint, dark circle surrounding bright crystalline particles, as observed in its SEI obtained before ATR-FTIR imaging, indicates that it was collected as a water droplet containing $NaNO_3$ species, which is confirmed by the presence of IR peaks of water at 1634 and ~3301 cm^{-1} and of crystalline $NaNO_3$ at 831, 1349 and 1788 cm^{-1}.

6.1.4 Silicate Particles

In the Asian dust sample, many different types of silicate particles were encountered [32]. For a typical silicate particle without water (particle #24 in Figure 9), its ATR-FTIR spectrum, X-ray spectrum and elemental atomic concentrations and SEIs are shown in Figure 9(a) and (d) and the insets, respectively. Its major elements are O (60%), Si (15%), Al (14%), C (5%), K (3%) and Na (1%), suggesting that it is an aluminosilicate particle with carbonaceous species. Based on its ATR-FTIR spectral data, it was identified as muscovite internally mixed with minor carbonate and organic species; i.e. IR peaks at 753, 821, 926, 1006, 1160 and 3630 cm^{-1} are for muscovite, 1400 cm^{-1} for carbonate and ~1620 cm^{-1} for carboxylic ion. The IR peak at 3630 cm^{-1} is assigned to crystal OH in muscovite, and no water is present in this particle.

As an example of swelling silicate particles mixed with nitrate, organic and water, the ATR-FTIR spectrum, X-ray spectrum and elemental atomic concentrations and SEIs of particle #40 are shown in Figure 9(b) and (e) and the insets, respectively. Its elemental concentrations are O (55%), Si (10%), C (12%), N (7%), Al (6%), Ca (4%), Mg (4%) and Fe (2%). Based on its ATR-FTIR spectral data, its major species is vermiculite supported by its IR peaks at 753, 833, 910, 1006 and 1158 cm^{-1}, and minor components are carbonate (866 and 1404 cm^{-1}), nitrate (824 and 1367 cm^{-1}), water (1639 and 3305 cm^{-1}) and organics (1618 cm^{-1} (COO^-) and 1705 cm^{-1} (C=O)).

ATR-FTIR spectrum, X-ray spectrum and elemental atomic concentrations and SEIs of particle #45, an example of non-swelling silicate particles mixed with nitrate, organic and water, are shown in Figure 9(c) and (f) and the insets, respectively. Its major elements are O (58%), Si (20%), Al (7%), K (5%), C (6%) and N (2%), and ATR-FTIR spectral data reveal that its major species is K-feldspar ($KAlSi_3O_8$) supported by its IR peaks at 993, 1086 and 1109 cm^{-1} and the minor species to be carbonate (866 and 1410 cm^{-1}), water (1639 and 3314 cm^{-1}) and organic (1623 cm^{-1} (COO^-)). By the combined use of low-Z particle EPMA and ATR-FTIR imaging, the mineralogy, reactivity and internal mixing state of silicate particles could be clearly elucidated.

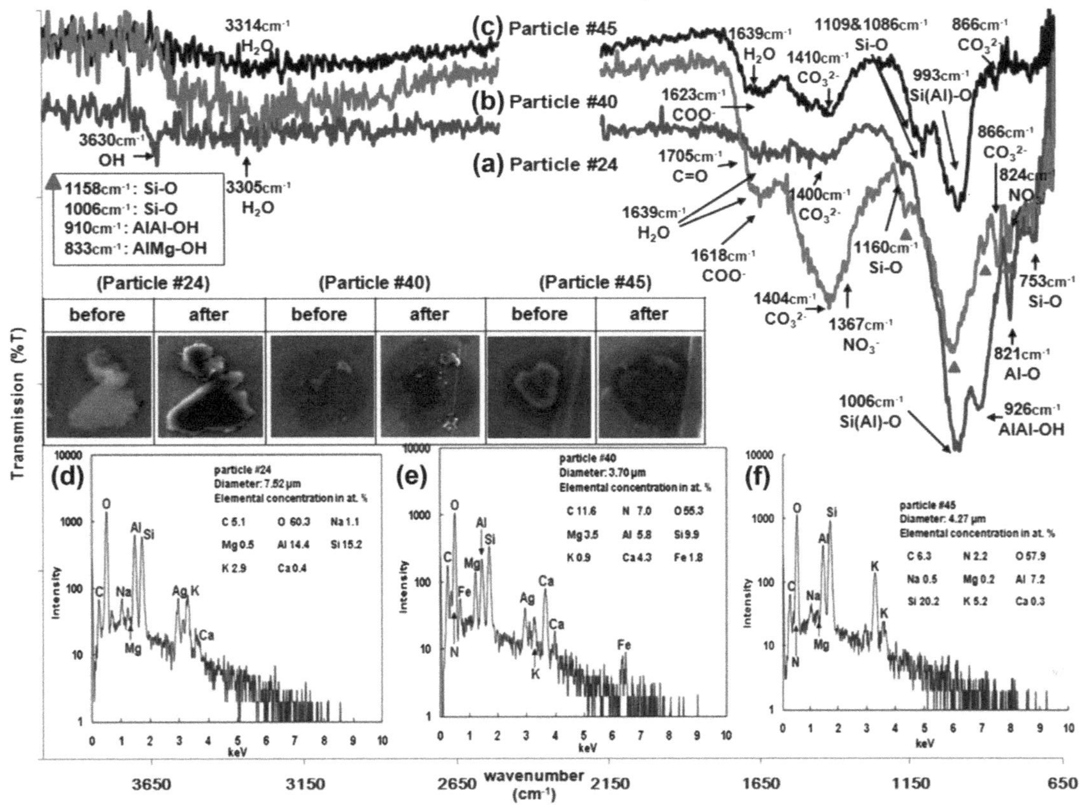

FIGURE 9 Attenuated total reflectance Fourier transform infrared (ATR-FTIR) and X-ray spectra of typical silicate particles and secondary electron images of particles obtained before and after ATR-FTIR imaging. The data for the 2200–2530 cm^{-1} region where atmospheric CO_2 peaks are present were deleted for clarity. *Reprinted with permission from Ref. [32]. Copyright (2013) Copernicus Publications.*

6.1.5 Antarctic Sea-Salt Aerosols

Low-*Z* particle EPMA and ATR-FTIR imaging techniques were applied in combination to characterise aerosols collected at a Korean scientific research station located at King George Island in the Antarctic [17]. Many sulphur-containing sea-salt particles were identified. Typical X-ray and ATR-FTIR spectra of sulphur-containing particles are shown in Figure 10. The particles were categorised either as sea-salt sulphate ($ss - SO_4^{2-}$) with a sulphur content similar to that of the seawater (particle #2, Figure 10) or non-sea-salt sulphate ($nss - SO_4^{2-}$) with a higher sulphur than the seawater (particle #57, Figure 10).

Each spectrum shows common IR absorption peaks at 3325, 1640, 1364, 1251 and 1123 cm^{-1}. However, the IR spectrum of particle #2 is somewhat different from that of particle #57. That is, IR peaks at 1476, 1427, 1311, 1251, 1123 and 1640 cm^{-1} are stronger in the spectrum of particle #2, whereas IR peaks at 2922 and 2853 cm^{-1} are stronger for particle #57. Overall, the peak height of SO_4^{2-} IR peak at ~1123 cm^{-1} on average in genuine sea-salt particles (0.012 ± 0.002) is lower than that in $nss - SO_4^{2-}$-containing sea-salt particles (0.031 ± 0.012). Na_2SO_4, CH_3SO_3Na, DMS, dimethyl sulphoxide (DMSO) and methanesulphonic acid (MSA) are reported to be intermediate or final chemicals in the formation of non-sea-salt sulphates from phytoplankton

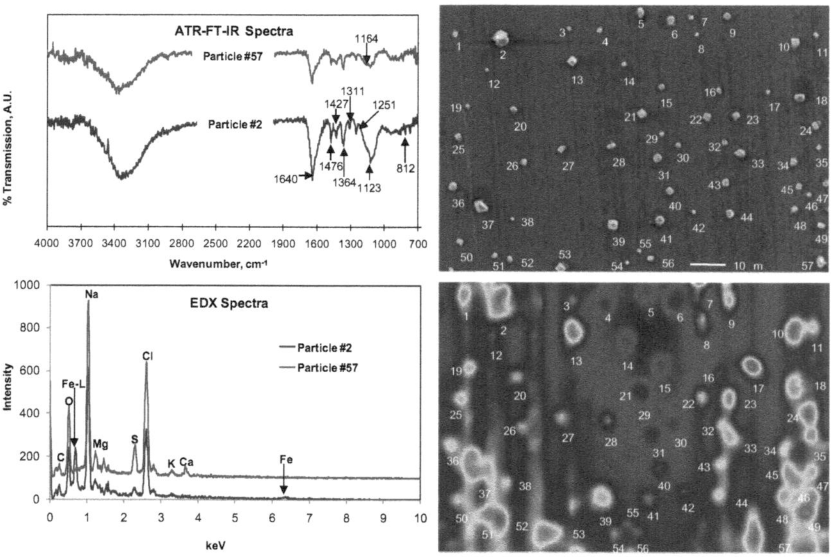

FIGURE 10 Exemplar attenuated total reflectance Fourier transform infrared (ATR-FTIR) (upper left panel) and X-ray spectra (lower left panel) of two sea-salt particles. Secondary electron image (upper right panel) and ATR-FTIR image (lower right panel) of typical aerosol particles collected from the Antarctic. EDX, energy-dispersive X-ray spectrometry. *Reprinted with permission from Ref. [17]. Copyright (2011) American Chemical Society.*

activities. Although the SEIs of the particles show that they look crystalline, it is clear that a significant amount of water exists in the particles according to ATR-FTIR analysis. The IR peaks at 2922 and 2853 cm^{-1} are from the DMS molecule, which is contained in particle #57. The peak at 1640 cm^{-1} is too broad to be only from water molecule; it seems that another IR peak (1587 cm^{-1} from humic-like natural organic matter, NOM [17]) is convoluted with the peak at 1640 cm^{-1}. IR peaks at 1476, 1364, 1251 and 812 cm^{-1} seem to be from biogenic species and/or humic or humic-like substances in the marine environment. The peaks at 1427 and 1311 cm^{-1} are either from DMS or DMSO molecules. A very broad, strong peak in the 1190–1040 cm^{-1} region seems to have merged with 2 IR peaks arising from SO_3^{2-} of CH_3SO_3Na at ~1046 and ~1170 cm^{-1}, and also with that at ~1091 cm^{-1} from SO_4^{2-} of Na_2SO_4. The IR peaks of MSA at 1174 and 1047 cm^{-1} may also have merged into the broad IR peak in the 1190–1040 cm^{-1} region. The sharp IR peak at 1123 cm^{-1} in the ATR-FTIR spectra of particle #2 indicates that it likely only contains ss – SO_4^{2-}. The sulphur species in the sulphur-containing particles seem to be complex mixtures of SO_4^{2-}, $CH_3SO_3^-$, DMS, DMSO, MSA and NOM-like species.

6.2 Combined Application of ED-EPMA and RMS

ED-EPMA provides morphological and elemental compositional information on individual atmospheric aerosol particles, while RMS provides information on their microstructure, chemical and physical mixing states and molecular species. The combination of ED-EPMA and RMS can provide unambiguous chemical and structural information on heterogeneous environmental particles with a spatial resolution of <1 μm [31,33]. The combination may be achieved in a parallel interface [33,49] or as standalone techniques [31,50]. Similar to the low-Z particle EPMA and ATR-FTIR combined technique, well-dispersed aerosol particles on Al, Ag or Si-wafer substrates can be relocated for sequential measurements with visual pattern recognition for particles on image fields. Carbon-coated TEM grid substrates can be even better for the relocation of particles, where the Cu-grid provides reference coordinates for image fields. A few applications of the combined technique for characterisation of airborne aerosol particles have been recently reported.

6.2.1 *Chemical Mixing State of Individual Asian Dust Particles*

Airborne particles collected during an ADS event were investigated by the combined use of low-Z particle EPMA and RMS on the same individual particles, where internally mixed and aged Asian dust aerosol particles were effectively characterised [31]. Typical examples of particles internally mixed with different minerals of soil origin are shown in Figure 11. The elemental composition of particle F (Figure 11(a) and (b)), which was assigned to the

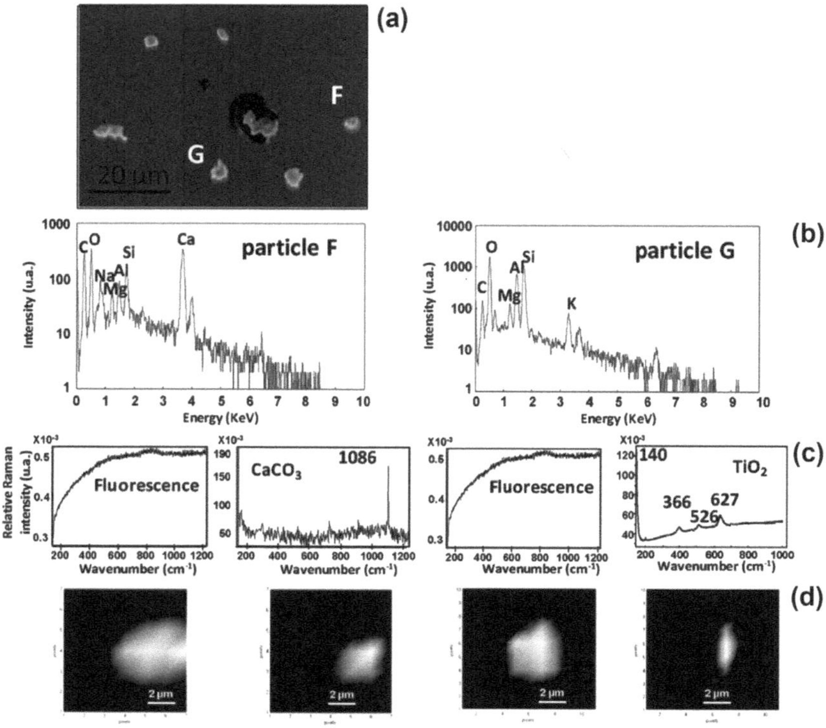

FIGURE 11 (a) Secondary electron image, (b) X-ray spectra, (c) resolved Raman spectra and (d) corresponding Raman images (after smoothing procedure) of clay mineral particles internally mixed with calcite (particle F) and anatase (particle G). *Reprinted with permission from Ref. [31]. Copyright (2012) American Chemical Society.*

(Ca,Mg)CO_3-AlSi particle type by low-*Z* particle EPMA, where '(Ca,Mg) CO_3-AlSi' is for the mixture of $CaCO_3$, $MgCO_3$ and aluminosilicates, is 24.8% C, 37.9% O, 1.4% Mg, 2.7% Al, 6.5% Si and 26.1% Ca. The Raman spectra and images (Figure 11(c) and (d)) clearly indicate that particle F is an intimate mixture of two solids, such as calcite and a clay mineral within an individual particle, rather than two agglomerated particles. Particle G in Figure 11(a) was assigned to the AlSi(Mg,K)-C type, where 'AlSi(Mg,K)-C' is for aluminosilicates with minor Mg and K contents mixed with carbonaceous species, on the basis of its elemental composition (Figure 11(b)), i.e. 10.6% C, 53.0% O, 1.7% Mg, 11.4% Al, 20.0% S and 1.8% K. Raman imaging analysis determined particle G to be an internal mixture of a clay mineral and TiO_2 (anatase) (Figure 11(c) and (d)). The clay minerals were detected using both techniques. Ti was not detected by ED-EPMA as its content was below the detection limit, but a small amount of TiO_2 (<1 wt%), which can act as a photocatalyst, was detected by RMS because of its high Raman cross-section.

6.2.2 Iron Speciation of Airborne Subway Particles

Airborne subway particles collected in underground tunnels of Seoul, Korea, were analysed by the combined use of low-*Z* particle EPMA and RMS [50]. The majority of collected particles were found to be internal mixtures of magnetite (toxic), haematite and iron metal generated mainly from mechanical wear and friction processes at rail—wheel—brake interfaces. Because iron-containing particles have different toxicity and magnetic properties depending on their oxidation states, it is important to determine the iron species of underground subway particles in the context of both indoor public health and control measures.

The SEIs, visible optical images, X-ray spectra, and Raman spectra of two exemplar subway particles are shown in Figure 12. Figure 12(a) shows a typical iron metal particle mixed with magnetite and EC (assigned as $Fe/Fe_3O_4/C(Si)$). The SEI shows that this angular and sliced particle is rather large (~5.5 μm in its equivalent diameter, which is defined as the diameter of a circular particle with the same area as the particle), and an aggregate of some small particles. Its sliced morphology with dark contrast indicates that its major chemical species is iron metal. Indeed, the atomic concentration of Fe element obtained from low-*Z* particle EPMA is 67.4% compared to an oxygen content of 21.3% (i.e. O/Fe = 0.27). The optical image obtained by the optical microscope in the Raman microspectrometer ensured that the same particle was analysed by both low-*Z* particle EPMA and RMS. The Raman spectrum clearly indicates that iron oxide species mixed with iron metal is magnetite with characteristic Raman peaks at 663, 528 and 300 cm^{-1}, as shown in Figure 12(a). The carbonaceous species with C = 9.7% from X-ray analysis is EC or soot based on its characteristic Raman peaks at 1325 and 1598 cm^{-1}.

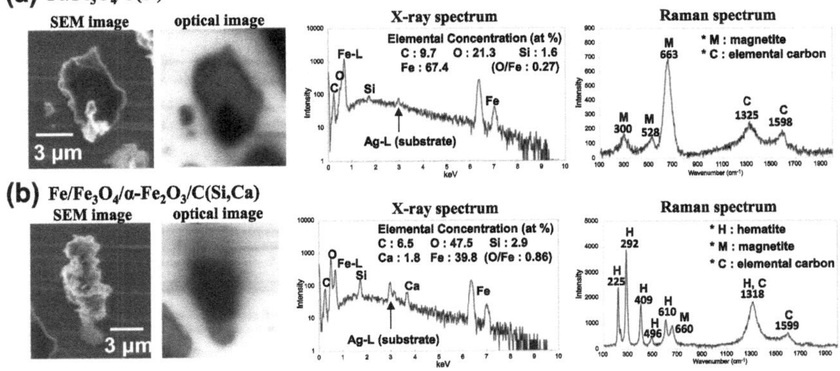

FIGURE 12 Two exemplar subway particles with their secondary electron images, optical images, X-ray spectra and Raman spectra. SEM, scanning electron microscopy. *Reprinted with permission from Ref. [50]. Copyright (2013) American Chemical Society.*

Figure 12(b) shows a typical particle mixed with magnetite, haematite and iron metal (assigned as $Fe/Fe_3O_4/\alpha\text{-}Fe_2O_3/C(Si,Ca)$). The SEI shows that this angular particle is ~6.4 μm in its equivalent diameter, and the optical image also confirms that X-ray and Raman data could be obtained for this same particle. The Raman spectrum shows that this particle is composed of magnetite and haematite with Raman peaks at 662 cm^{-1} and at 610, 496, 409, 292 and 225 cm^{-1} [51], respectively.

6.2.3 *Internally Mixed Soot Particles*

The crystallite mineral needles were found to be deeply embedded in a soot matrix, thereby demonstrating the intimate contact between soot and minerals using SEM and RMS [52]. Figure 13 presents secondary and backscattered electron images and Raman spectra of a typical internally mixed soot sample. The SEI (Figure 13(a)) illustrates the porous morphology of the soot matrix. On the other hand, the image of the backscattered electrons (Figure 13(b)) demonstrates the material is composed of $CaSO_4$ crystallites (shown in white) deeply embedded in soot matrix (shown in dark grey), i.e. an internal mixture characterised by an intimate contact between soot and minerals is identified. If the soot is externally (physically) mixed with $CaSO_4$, separated particles of soot and minerals would have been observed. Raman spectra reveal the soot-characteristic first-order graphitic (G) and disordered (D) peaks at ~1600 and 1335 cm^{-1}, respectively (Figure 13(c)). In this study, it was reported that neither the salt species nor the salt content in soot influences the structure of the graphitic soot layers, demonstrating the effectiveness of these complementary techniques.

6.3 Combined Application of ED-EPMA, ATR-FTIR Imaging and RMS

Previous sections showed that the characterisation of individual particles including chemical speciation and mixing state analysis could be performed in more detail using ED-EPMA and either ATR-FTIR imaging or RMS in combination than using each single-particle technique alone [17,29–33,49,50,52]. As RMS and FTIR are also two complementary vibrational spectroscopic techniques, it would be more powerful for the characterisation of heterogeneous airborne particle ensembles if spectral data using the three single-particle analytical techniques could be obtained for the same individual particles. The same individual particles deposited on Al or Ag foils could be relocated among the three standalone instruments, resulting in the successful acquisition of quality spectral data of the same individual particles from the three techniques [47].

Figure 14(a–d) shows an SEI obtained before ATR-FTIR imaging measurements, an ATR-FTIR PCA image, a visible light optical image obtained from a Raman instrument after ATR-FTIR imaging and an SEI obtained after the RMS measurement, respectively, of the same 51 airborne individual

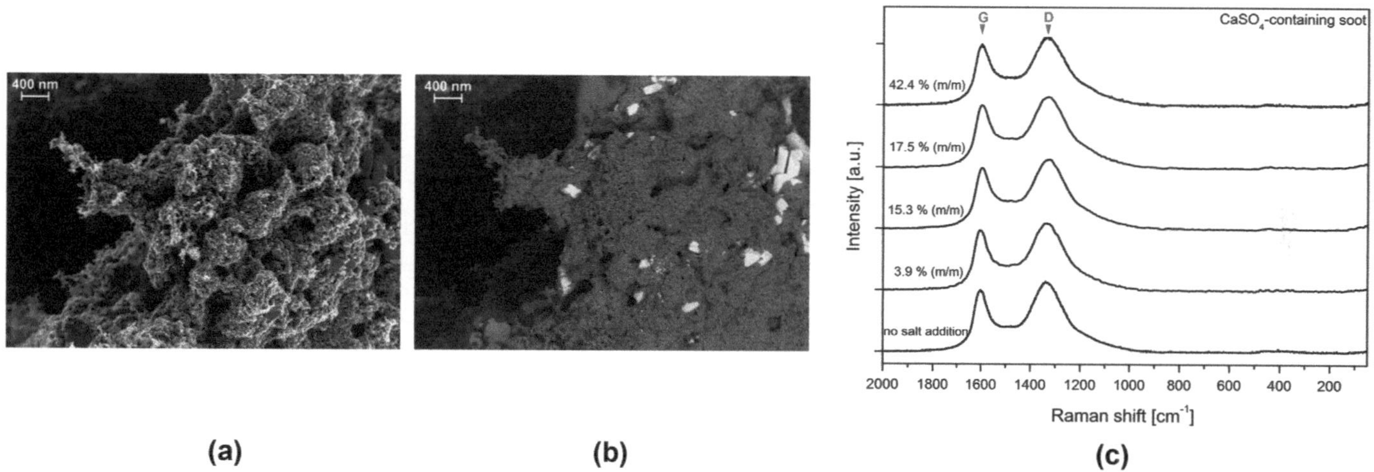

FIGURE 13 (a) Secondary electron image and (b) backscattered electron image of internally mixed soot sample containing 42 wt% $CaSO_4$ and (c) Raman spectra of soot with different $CaSO_4$ content in wt%. Soot-characteristic G and D peaks are marked in grey. *Reprinted with permission from Ref. [52]. Copyright (2014) Elsevier.*

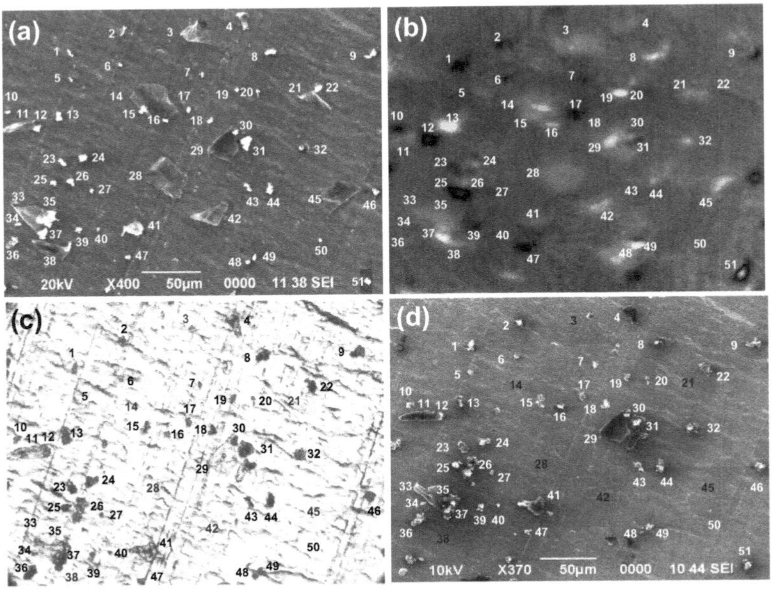

FIGURE 14 (a) Secondary electron image (SEI) from scanning electron microscopy before attenuated total reflectance Fourier transform infrared spectroscopic (ATR-FTIR) measurement, (b) ATR-FTIR principal component analysis image, (c) optical image from a Raman spectrometer and (d) SEI after Raman microspectrometry measurement of the same 51 airborne indoor particles on Al foil. *Reproduced from Ref. [47] with permission from The Royal Society of Chemistry.*

particles of an indoor aerosol sample collected at an underground shopping area in Seoul, Korea [47]. The same patterns of the particle location among the images ensured that the same micrometre-sized particles were observed in the three different analytical instruments. First, the morphological and X-ray spectral data of all 51 particles in the image field were obtained by SEM/EDX. Second, the sample of $\sim 1 \times 1\ cm^2$ in size was moved to the ATR-FTIR imaging instrument, and the ATR-FTIR spectra for all the pixels on the same image field were obtained after relocating the same particles examined by SEM/EDX. ATR-FTIR imaging was performed on the image field of $350 \times 270\ \mu m^2$ in size. The sample was then transferred offline again into a Raman spectrometer where the Raman mapping data for all the individual particles on the same image field were obtained. After the RMS measurements, an SEI of the same image field was obtained by SEM to check what modification had occurred during the ATR-FTIR measurements due to IRE crystal contact with the sample and during the Raman measurements due to laser beam damage.

Figure 15 shows SEIs and optical images from SEM and Raman spectroscopy, respectively, of three exemplar particles encountered frequently in this indoor sample, together with their X-ray, Raman and ATR-FTIR spectra.

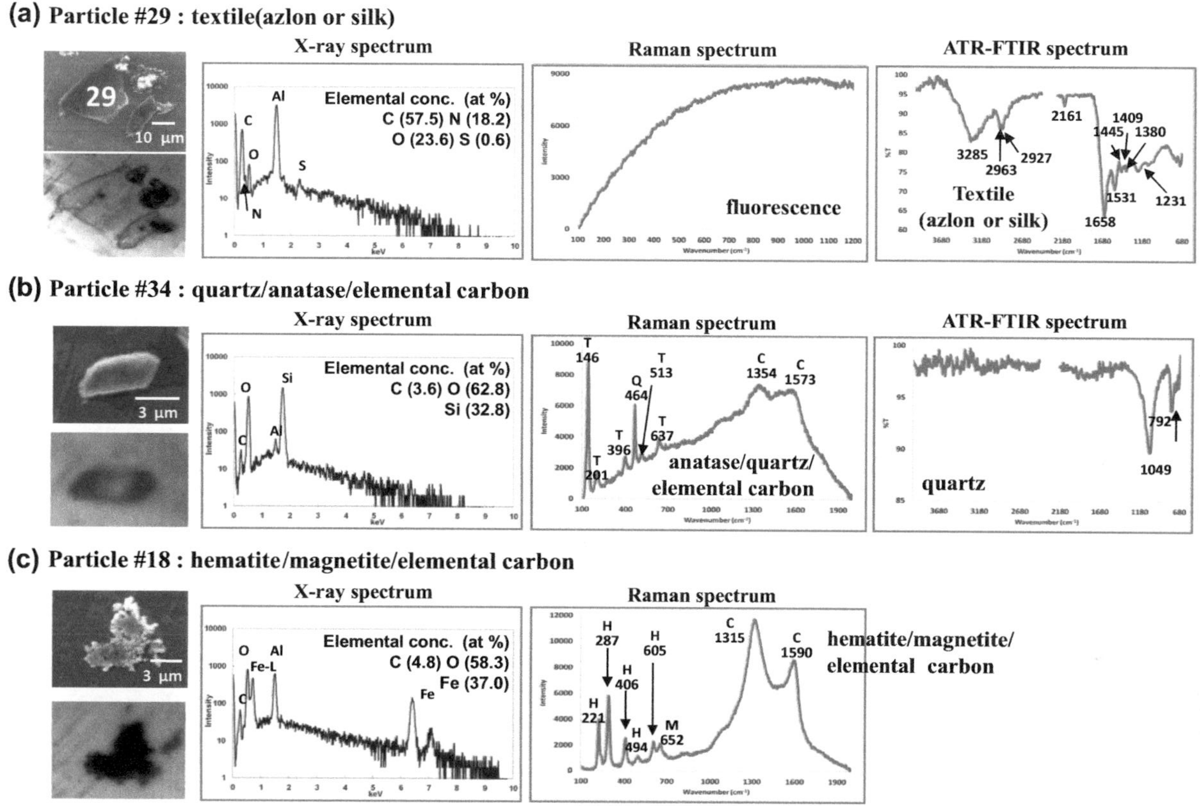

FIGURE 15 Three exemplar indoor particles with their morphological images and X-ray, Raman and attenuated total reflectance Fourier transform infrared (ATR-FTIR) spectra. *Reproduced from Ref. [47] with permission from The Royal Society of Chemistry.*

Particle #29 appeared peculiar and flat on its SEI and transparent on its optical image. Based on its X-ray spectral data, particle #29 is a type of organic particle containing C, N and O with a small amount of S. Raman spectroscopy did not provide useful information on its molecular species because only a fluorescence signal, which is probably from some organic compounds, was observed. On the other hand, ATR-FTIR analysis clearly identified it as a textile particle, either azlon or silk, which was achieved by comparing its spectrum with that from a PerkinElmer FTIR spectral library.

Particle #34 appears angular both on the SEI and optical image, which is typical for mineral particles. X-ray analysis indicated that it is a SiO_2 mineral with minor carbonaceous species (atomic concentrations of Si, O and C are 32.8%, 62.8% and 3.6%, respectively). Airborne SiO_2 mineral particles can normally exist either as quartz or cristobalite, and RMS and ATR-FTIR analyses identified particle #34 as quartz based on the Raman and IR peaks. Regarding the carbonaceous species present in its X-ray spectrum, the ATR-FTIR spectrum does not show any related peak, probably because it is IR-inactive EC. Indeed, the Raman spectrum of particle #34 shows strong Raman bands at 1354 and 1573 cm^{-1}, which are characteristic of EC, such as graphite and soot with a large Raman cross-section. In addition, a fluorescence signal is present, indicating the presence of OC (probably humic or humic-like substances in very small amounts that are not detectable by ATR-FTIR analysis). The combined application of the three single-particle analytical techniques indicates that particle #34 is a quartz particle mixed with a minor amount of anatase and EC along with very minor amounts of OC.

Particle #18 appears agglomerated with many nano-sized particles, which is typical for iron-containing underground subway particles. X-ray analysis indicated that it is an iron oxide particle with minor carbonaceous species (atomic concentrations of Fe, O and C are 37.0%, 58.3% and 4.8%, respectively). Since X-ray analysis cannot provide clear chemical speciation of the iron oxides in particle #18 and ATR-FTIR is blind to the analysis of iron oxides due to the absence of characteristic IR peaks in the 680–4000 cm^{-1} spectral range, RMS is helpful for detecting the presence of both haematite and magnetite based on the haematite Raman bands at 225, 291, 410, 496 and 610 cm^{-1} and a magnetite band at 658 cm^{-1}. In addition, the carbonaceous species present in the X-ray spectrum is only EC, which was suggested by the presence of Raman bands at 1319 and 1594 cm^{-1}, no fluorescence in the Raman spectrum and no peak in the ATR-FTIR spectrum. The combined application of the three single-particle analytical techniques indicates that particle #18 is a haematite (α-Fe_2O_3) and magnetite (Fe_3O_4) mixture with a small amount of EC (notated as haematite/magnetite/EC in Figure 15(c)).

$CaCO_3$-containing particles are also abundant in this sample. When silicates are mixed with $CaCO_3$, RMS shows fluorescence and/or EC, which are similarly observed for the silicate particles. The combined use of low-Z particle EPMA and ATR-FTIR analyses was found to be useful for the chemical speciation of these $CaCO_3$-containing particles, along with additional information from RMS on carbonaceous species. As described for particle #18 above, ATR-FTIR cannot detect iron-containing particles. On the other hand, RMS is powerful because it covers the spectral range of 100–670 cm^{-1}, where the Raman bands of iron oxides are observed.

7. CONCLUSIONS

In this chapter, the applicability of various microscopic single-particle and surface analytical techniques to obtain a complete understanding of multimodal physicochemical properties of atmospheric aerosols is described. The full description of multimodal physicochemical properties of individual particles, such as mixing state, chemical heterogeneity (including elemental and molecular compositions), physical state (liquid vs. solid), crystallinity (amorphous or crystalline solids) and morphology (size and shape), is essential to better understand their sources, evolution and fate (multiphase chemical reactions/transformation, agglomeration, coalescence, etc.) and to assess their impacts on climate, ambient air quality and human health.

The ED-EPMA technique based on SEM/EDX or TEM/EDX effectively provides morphology (size and shape) and elemental composition of (sub)micron aerosol particles, which form the fundamental basis for single-particle aerosol characterisation. The low-*Z* particle EPMA technique is capable of the quantitative determination of low-*Z* elements, such as C, N and O, in microscopic volume by utilising Monte Carlo calculations in an iterative approach. The ability of analysing low-*Z* elements in a microscopic volume enlarges the applicability of this technique for the analysis of atmospheric aerosol particles. The other microspectrometric techniques discussed in this chapter provide critical but complementary physicochemical characteristics of aerosol particles. The STXM/NEXAFS technique, which requires an access to a synchrotron facility, can provide high-resolution (~25 nm) chemical maps of some inorganic and organic species within individual particles. The vibrational spectroscopic techniques like ATR-FTIR imaging and RMS are powerful for functional group analysis and molecular speciation of organic and inorganic compounds including hydrated species under ambient conditions. They have advantages for identification of different polymorphs, determination of different oxidation states and crystallographic orientation, evaluation of temperature and stress effects on molecular structure, etc. TOF-SIMS is a powerful tool for the examination of the inorganic and organic moiety including volatile and semivolatile organic compounds over a very small surface region (tens of nanometres) of a particle. AFM topographic images provide not only the surface equivalent diameter but also the shape and the physical state (e.g. liquid vs solid) of impacted individual particles. Confocal optical microscopy is combined with RMS, AFM, ATR-FTIR, etc. and has also been demonstrated to be useful in observing the hygroscopic behaviour of laboratory generated or ambient aerosols. Overall, the various microscopic methods can provide information on morphology, internal mixing state, elemental and molecular compositions, surface compositions, functional groups and/or hygroscopic property of individual particles. However, these multimodal physicochemical characteristics for the same individual particles cannot be simultaneously investigated using any single technique alone. The combination of these techniques can provide more detailed information on physicochemical properties of the same individual particles, as demonstrated in the applications for various types of aerosols, such as aged mineral dust, reacted sea-salt particles, subway particles, carbonaceous particles like soot, etc. Though powerful and promising, this combined approach employing different techniques for the characterisation of the same individual aerosol particles still faces a few challenges, regarding (1) the use of suitable universal substrates for collecting aerosol particle samples and for analysis by all the complementary techniques like RMS, SEM/EDX, ATR-FTIR, AFM, optical microscopy, etc.; (2) different lateral resolutions of the complementary techniques; (3) possible damage of samples due to electron, ion, laser or photon beams and (4) the maintenance of character, structure and physical state of samples during transfer between the stand-alone instruments for analysis, to name a few.

REFERENCES

[1] S. Ghorai, A.V. Tivanski, Anal. Chem. 82 (2010) 9289–9298.

[2] A. Laskin, in: R. Signorell, J. Reid (Eds.), Fundamental and Application in Aerosols Spectroscopy, Taylor and Francis Books, Inc, Boca Raton, FL, 2010, pp. 243–272.

[3] R.C. Moffet, T. Henn, A. Laskin, M.K. Gilles, Anal. Chem. 82 (2010) 7906–7914.

[4] G. Zhang, B. Han, X. Bi, S. Dai, W. Huang, D. Chen, X. Wang, G. Sheng, J. Fu, Z. Zhou, Atmos. Res. 153 (2015) 286–295.

[5] R. Fitzgerald, K. Keil, K.F.J. Heinrich, Science 159 (1968) 528–530.

[6] A. Laskin, J.P. Cowin, M.J. Iedema, J. Electron Spectrosc. Relat. Phenom. 150 (2006) 260–274.

[7] C.-U. Ro, J. Osán, R. Van Grieken, Anal. Chem. 71 (1999) 1521–1528.

[8] C.-U. Ro, J. Osán, I. Szalóki, J. de Hoog, A. Worobiec, R. Van Grieken, Anal. Chem. 75 (2003) 851–859.

[9] C.-U. Ro, J. Osán, I. Szalóki, K.-Y. Oh, H. Kim, R.V. Grieken, Environ. Sci. Technol. 34 (2000) 3023–3030.

[10] H. Geng, H. Hwang, X. Liu, S. Dong, C.-U. Ro, Atmos. Chem. Phys. 14 (2014) 3307–3323.

[11] H. Geng, J. Ryu, S. Maskey, H. Jung, C.-U. Ro, Atmos. Chem. Phys. 11 (2011) 1327–1337.

[12] H. Geng, Y.-M. Park, H.-J. Hwang, S. Kang, C.-U. Ro, Atmos. Chem. Phys. 9 (2009) 6933–6947.

[13] H. Geng, J. Ryu, H. Jung, H. Chung, K. Ahn, C.-U. Ro, Environ. Sci. Technol. 44 (2010) 2348–2353.

[14] H. Geng, S. Kang, H.-J. Jung, M. Choël, H. Kim, C.-U. Ro, J. Geophys. Res. Atmos. 115 (2010) D15306.

[15] H. Hwang, C.-U. Ro, Atmos. Environ. 40 (2006) 3869–3880.

[16] C.-U. Ro, K.-Y. Oh, H. Kim, Y.P. Kim, C.B. Lee, K.-H. Kim, J. Osan, J. de Hoog, A. Worobiec, R. Van Grieken, Environ. Sci. Technol. 35 (2001) 4487–4494.

[17] S. Maskey, H. Geng, Y. Song, H. Hwang, Y. Yoon, K. Ahn, C.-U. Ro, Environ. Sci. Technol. 45 (2011) 6275–6282.

[18] S. Kang, H. Hwang, Y. Park, H. Kim, C.-U. Ro, Environ. Sci. Technol. 42 (2008) 9051–9057.

[19] H. Jung, B. Kim, M. Malek, Y. Koo, J. Jung, Y. Son, J. Kim, H. Kim, C.-U. Ro, J. Hazard. Mater. 213–214 (2012) 331–340.

[20] L. Guo, Y. Hu, Q. Hu, J. Lin, C. Li, J. Chen, L. Li, H. Fu, Sci. Total Environ. 496 (2014) 443–452.

[21] H. Bluhm, K. Andersson, T. Araki, K. Benzerara, G.E. Brown, J.J. Dynes, S. Ghosal, M.K. Gilles, H.C. Hansen, J.C. Hemminger, A.P. Hitchcock, G. Ketteler, A.L.D. Kilcoyne, E. Kneedler, J.R. Lawrence, G.G. Leppard, J. Majzlam, B.S. Mun, S.C.B. Myneni, A. Nilsson, H. Ogasawara, D.F. Ogletree, K. Pecher, M. Salmeron, D.K. Shuh, B. Tonner, T. Tyliszczak, T. Warwick, T.H. Yoon, J. Electron Spectrosc. Relat. Phenom. 150 (2006) 86–104.

[22] M. Obst, J.J. Dynes, J.R. Lawrence, G.D.W. Swerhone, K. Benzerara, C. Karunakaran, K. Kaznatcheev, T. Tyliszczak, A.P. Hitchcock, Geochim. Cosmochim. Acta 73 (2009) 4180–4198.

[23] S. Behrens, A. Kappler, M. Obst, Environ. Microbiol. 14 (2012) 2851–2869.

[24] S. Takahama, S. Liu, L.M. Russell, J. Geophys. Res. Atmos. 115 (2010) D01202.

[25] S. Takahama, A. Johnson, J. Guzman Morales, L.M. Russell, R. Duran, G. Rodriguez, J. Zheng, R. Zhang, D. Toom-Sauntry, W.R. Leaitch, Atmos. Environ. 70 (2013) 500–512.

[26] J. Cosmidis, K. Benzerara, N. Nassif, T. Tyliszczak, F. Bourdelle, Acta Biomater. 12 (2015) 260–269.

[27] S. Sobanska, G. Falgayrac, J. Rimetz-Planchon, E. Perdrix, C. Brémard, J. Barbillat, Microchem. J. 114 (2014) 89–98.

[28] J. Ryu, C.-U. Ro, Anal. Chem. 81 (2009) 6695–6707.

[29] H.-J. Jung, M.A. Malek, J. Ryu, B. Kim, Y.-C. Song, H. Kim, C.-U. Ro, Anal. Chem. 82 (2010) 6193–6202.

[30] Y.-C. Song, J. Ryu, M.A. Malek, H.-J. Jung, C.-U. Ro, Anal. Chem. 82 (2010) 7987–7998.

[31] S. Sobanska, H. Hwang, M. Choël, H.-J. Jung, H.-J. Eom, H. Kim, J. Barbillat, C.-U. Ro, Anal. Chem. 84 (2012) 3145–3154.

[32] Y.C. Song, H.J. Eom, H.J. Jung, M.A. Malek, H.K. Kim, H. Geng, C.U. Ro, Atmos. Chem. Phys. 13 (2013) 3463–3480.

[33] A. Worobiec, S. Potgieter-Vermaak, A. Brooker, L. Darchuk, E. Stefaniak, R. Van Grieken, Microchem. J. 94 (2010) 65–72.

[34] C. Coury, A.M. Dillner, Atmos. Environ. 43 (2009) 940–948.

[35] S.F. Maria, L.M. Russell, Environ. Sci. Technol. 39 (2005) 4793–4800.

[36] E.J. Palen, D.T. Allen, S.N. Pandis, S. Paulson, J.H. Seinfeld, R.C. Flagan, Atmos. Environ. 27A (1993) 1471–1477.

[37] E.J. Palen, D.T. Allen, S.N. Pandis, S. Paulson, J.H. Seinfeld, Atmos. Environ. 26A (1992) 1239–1251.

[38] N. Ivleva, S. Huckele, B. Weinzierl, R. Niessner, C. Haisch, T. Baumann, Anal. Bioanal. Chem. 405 (2013) 9071–9084.

[39] P. Lazzeri, G. Clauser, E. Iacob, A. Lui, G. Tonidandel, M. Anderle, Appl. Surf. Sci. 203–204 (2003) 767–771.

[40] R.E. Peterson, B.J. Tyler, Atmos. Environ. 36 (2002) 6041–6049.

[41] A.P. Nair, B.J. Tyler, R.E. Peterson, Appl. Surf. Sci. 231–232 (2004) 538–542.

[42] B. Tomiyasu, K. Suzuki, T. Gotoh, M. Owari, Y. Nihei, Appl. Surf. Sci. 231–232 (2004) 515–519.
[43] Y. Liu, B. Minofar, Y. Desyaterik, E. Dames, Z. Zhu, J.P. Cain, R.J. Hopkins, M.K. Gilles, H. Wang, P. Jungwirth, A. Laskin, Phys. Chem. Chem. Phys. 13 (2011) 11846–11857.
[44] G. Binnig, C.F. Quate, C. Gerber, Phys. Rev. Lett. 56 (1986) 930–933.
[45] P. Gwaze, H.J. Annegarn, J. Huth, G. Helas, Atmos. Res. 86 (2007) 93–104.
[46] H. Kim, M.-J. Lee, H.-J. Jung, H.-J. Eom, S. Maskey, K.-H. Ahn, C.-U. Ro, Atmos. Environ. 60 (2012) 68–75.
[47] H.-J. Jung, H.-J. Eom, H.-W. Kang, M. Moreau, S. Sobansk, C.-U. Ro, Analyst 139 (2014) 3949–3960.
[48] M.A. Malek, B. Kim, H.-J. Jung, Y.-C. Song, C.-U. Ro, Anal. Chem. 83 (2011) 7970–7977.
[49] N. Goienaga, A. Sarmiento, M. Olivares, J.A. Carrero, L.A. Fernández, J.M. Madariaga, Anal. Chem. 85 (2013) 7173–7181.
[50] H.-J. Eom, H.-J. Jung, S. Sobanska, S.-G. Chung, Y.-S. Son, J.-C. Kim, Y. Sunwoo, C.-U. Ro, Anal. Chem. 85 (2013) 10424–10431.
[51] I. Chourpa, L. Douziech-Eyrolles, L. Ngaboni-Okassa, J.-F. Fouquenet, S. Cohen-Jonathan, M. Souce, H. Marchais, P. Dubois, Analyst 130 (2005) 1395–1403.
[52] H. Bladt, N.P. Ivleva, R. Niessner, J. Aerosol Sci. 70 (2014) 26–35.

Section IV

Concluding Comments

Chapter 12

Concluding Comments and Future Outlook

Patricia B.C. Forbes
Department of Chemistry, Faculty of Natural and Agricultural Sciences, University of Pretoria, Pretoria, South Africa
E-mail: patricia.forbes@up.ac.za

Chapter Outline

References 371

This book provides a comprehensive reference for those embarking on ambient air pollution monitoring studies. The context of ambient air pollution monitoring and important quality considerations, which should be borne in mind when undertaking monitoring campaigns, were discussed in Chapter 1, whilst Section II of the book detailed different sampling and sample preparation methodologies which are employed for a range of different ambient air pollutants. The sampling options available range from those which are cost-effective and simple, such as passive samplers and biomonitors (covered in Chapters 2 and 3), to those which require more extensive capital outlay and maintenance, as in the continuous real-time monitors discussed in Chapter 6. Further, more traditional approaches to sampling air, as in the case of canister, bag and sorbent sampling (Chapter 4), were outlined; as were more recent sampling options, such as solid-phase microextraction and the employment of needle trap devices (Chapter 7), as well as denuder-based technologies for simultaneous sampling of gas- and particle-phase analytes (Chapter 5).

It is clearly evident that a wide range of options for sampling ambient air are available, and the best option for a particular sampling campaign depends on a number of factors, including target analyte(s); expected concentration of the analyte(s); the phase of the analyte(s) (gas or particle); sampling location; sampling interval; resources available; technical expertise; expected meteorological conditions and the required outcomes of the monitoring campaign. There are also a number of possible sample preparation techniques which can be employed once the ambient air samples have been taken, which have been

Comprehensive Analytical Chemistry, Vol. 70. http://dx.doi.org/10.1016/bs.coac.2015.10.002

described in the relevant chapters in Section II and are chosen largely based on the target analyte(s); means of sample collection (sample matrix); expected analyte concentration and resources available (including instrumental resources).

Section III of the book focuses on a number of the various means of analysing samples containing ambient air pollutants. These include those techniques applicable to quantify and characterise inorganic analytes (Chapter 8) as well as those which are organic in nature (Chapter 9), whilst mass spectrometry discussed in Chapter 10 and microscopy-based techniques (Chapter 11) apply to both classes of compounds.

It is expected that the fundamental principles as well as the examples of applications and practical advice presented in this book will assist readers in determining the most appropriate sampling; sample preparation and analytical methodologies to employ for their application.

There are numerous challenges in the monitoring of ambient air pollutants, including the low concentrations of target analytes typically present; access to sampling locations; as well as portability and cost of sampling and analytical equipment. Technologies are constantly being developed and existing methods are being enhanced to address these issues and to allow for an expanding range of compounds to be analysed. An example is the determination of endocrine-disrupting chemicals (such as phthalates, oestradiol and bisphenol A) in airborne particles, as reviewed in Ref. [1]. It is not possible to cover every technique available in this book, therefore, a number of monitoring methods which show promise and find application in ambient air pollutant monitoring have not been described in the preceding chapters, such as the laser-induced fluorescence detection of polycyclic aromatic hydrocarbons [2,3]; and the use of Light Detection and Ranging (LIDAR) and remote sensing for atmospheric measurements [4,5], thus, the reader is referred to the referenced texts for more information on these topics.

Due to the heterogeneity of particles present in ambient air, single-particle analyses have the potential to assist with the elucidation of the sources of aerosol particles, as well as their chemical activity and environmental and biological effects. Aerosol optical tweezers [6], for example, allow for the monitoring of changes in particle size. Analytical methods with high temporal resolution and portability are also immensely useful in understanding atmospheric chemistry due to the dynamic nature of the troposphere in terms of ambient gaseous and particulate air pollutant composition and concentrations. Indeed, real-time monitoring of aerosols has become accessible to the general public with the advent of smartphone applications, which use the camera and sensors of the phone, as well as its computing and communication capabilities in conjunction with, for example, a passive spectropolarimetric add-on [7]. Information on the amount of aerosol, the particle size distribution, as well as the chemical composition thereof can be obtained in this manner. Such developments have the potential to enable ambient air pollutant monitoring with high temporal and geographic resolution, which would further enhance our understanding of the complex chemistry of our ambient atmosphere and in assessing the potential impacts of ambient air pollutants.

REFERENCES

[1] N. Salgueiro-González, M.J. López de Alda, S. Muniategui-Lorenzo, D. Prada-Rodríguez, D. Barceló, Trends Anal. Chem. 66 (2015) 45–52.

[2] P.B.C. Forbes, E.R. Rohwer, Environ. Pollut. 157 (2009) 2529–2535.

[3] V.V. Gridin, T. Inoue, T. Ogawa, I. Schechter, Instrum. Sci. Technol. 28 (2) (2000) 131–141.

[4] D.E. Heard (Ed.), Analytical Techniques for Atmospheric Measurement, Blackwell Publishing, UK, 2006.

[5] M. Marć, M. Tobiszewski, B. Zabiegała, M. de la Guardia, J. Namieśnik, Anal. Chim. Acta 853 (2015) 116–126.

[6] M.D. Summers, D.R. Burnham, D. McGloin, Opt. Express 16 (11) (2008) 7739–7747.

[7] iSPEX-EU. www.ispex-eu.org (accessed 29.09.15).

Index

Note: Page numbers followed by "f" indicate figures and "t" indicate tables.

A

ACC. *See* Amorphous calcium carbonate (ACC)
Accelerated solvent extraction (ASE), 85–87. *See also* Microwave-assisted extraction (MAE)
Accuracy, 6
Acrocarpous mosses, 65
Activated-carbon-based parallel-plate denuder, 174
Active samplers, 16–17, 17t, 18f
ADS particles. *See* Asian dust storm particles (ADS particles)
Adsorbant-based denuders, 173–174
AED. *See* Aerodynamic equivalent diameter (AED)
Aerodynamic diameter (d_a), 242
Aerodynamic equivalent diameter (AED), 144–145
Aerosol MS (AMS), 321–323
Aerosol optical tweezers, 370–371
Aerosol TOFMS (ATOFMS), 322–323
Aerosols chemical profile, 244
  chemical composition, 244–245
  chemical markers of emission sources, 245–254
AFM. *See* Atomic force microscopy (AFM)
AFS. *See* Atomic fluorescence spectrometry (AFS)
Aged mineral dust, 348–350
Air flow considerations, 165–166
Air pollutants, 3, 369–371
  chromatography applications in, 295–302
  chromatography detectors, 289–295
  cycling, 5
  emission of acid-forming air pollutants, 5
  human health effects, 4
  monitoring, 4
  stationary phases in chromatographic analysis, 283
    analyte-specific stationary phases, 287–288
    developments in LC stationary phases, 288–289
    in GC, 284
    orthogonality for GC X GC separations, 288
    polar *vs.* nonpolar stationary phases, 285–286
Air pollution, 3
  potential effects, 4–5
  quality control importance in environmental analyses, 5–7
  statistical tools for environmental monitoring, 7–8
Air sampling, 109–111
  air samples, 110, 222–223
  applications, 232t
  passive, 14–16
  selection and characteristics of SPME sorbent coatings, 212–214
  sorbent coatings for, 212–214
Airborne particulate matter (APM), 240. *See also* Atmospheric aerosol; Particulate matter (PM)
Airborne subway particles iron speciation, 357–358
Altef®, 116
Ambient air, 5
Ambient air pollution, 369–370
American Society for Testing and Materials (ASTM), 126
Ammonia sampler (NH_3 sampler), 15
Amorphous calcium carbonate (ACC), 348–350
AMS. *See* Aerosol MS (AMS)
Analyte extraction methods, 85–98
  MAE, 87–88
  QuEChERS, 93–95
  sample cleanup, 95–98
  Soxhlet extraction, 85–87, 86t
  USAE, 88–93
Analyte-specific stationary phases, 287–288

Analytical techniques, 98–100, 99t
inorganic analytes, 100
organic analytes, 98–99
Analytical techniques, 254
IC for anions and cations, 261–263
spectroscopic techniques for elemental analysis, 255–261
Annular denuders, 170–171
Annular type denuders, 162, 163f
Antarctic and Arctic aerosols characterisation, 337
Antimony (Sb), 247–248
APCI. *See* Atmospheric pressure chemical ionisation (APCI)
APM. *See* Airborne particulate matter (APM)
ASE. *See* Accelerated solvent extraction (ASE)
Asian dust storm particles (ADS particles), 336
ASTM. *See* American Society for Testing and Materials (ASTM)
Atmospheric aerosol, 240, 332. *See also* Airborne particulate matter (APM); Particulate matter (PM)
analytical techniques, 254–263
atmospheric particles physical aspects, 241–244
in subway stations, 337–338
Atmospheric chemistry, 240
Atmospheric particles, 240, 332
physical aspects, 241–242
morphology, 243–244
size distribution, 242–243
Atmospheric pressure chemical ionisation (APCI), 314
Atmospheric samples, 289, 296
Atmospheric science, 3–4
ATOFMS. *See* Aerosol TOFMS (ATOFMS)
Atomic absorption spectrometry, 256–258
Atomic fluorescence spectrometry (AFS), 241, 248, 258
Atomic force microscopy (AFM), 332–333, 342–343
ATR-FTIR spectroscopy. *See* Attenuated total reflectance Fourier transform infrared spectroscopy (ATR-FTIR spectroscopy)
Attenuated total reflectance Fourier transform infrared spectroscopy (ATR-FTIR spectroscopy), 332–333, 339–340
applications, 340–341
Automated continuous air monitoring, 184
data handling, quality and processing, 200–204
measurement motives and consequences thereof, 184–199
air quality *vs.* scientific research interest, 184–190
continuously operating instruments and measurement resolutions, 187t–189t
data logging requirements, 197–199
measurement accreditation, 193–195
monitoring devices selection, 190–193
site selection, 195–197
Automated continuous atmospheric measurements, 184

B

Badge sampler, 18
Bags, whole-air sampling with, 112–120
Barbula unguiculata (*B. unguiculata*), 66
Base peak, 294
BC. *See* Black carbon (BC)
Benzene, toluene, ethyl benzene and xylene (BTEX), 20, 213, 317
Benzene, toluene and xylene (BTX), 20
BFRs. *See* Brominated flame-retardants (BFRs)
Bias, 41
Bio-Beads S-X3, 97–98
Biogenic VOC (BVOC), 325
Biogenic volatile organic compounds (BVOCs), 300–301
Bioindicator studies, 54
Biomonitoring studies, 54–55
Biomonitors, 54
for air pollution monitoring, 55–56
analyte extraction methods, 85–98
analytical techniques, 98–100
bioindicator studies, 54
biomonitoring studies, 54–55
conifer needles, 67–76
environmental aspects, 56
lichens, 56–65
mosses, 65–67
other plants, 78
Pseudevernia furfuracea lichen false colour imaging, 54f
sampling considerations, 78–85
sampling of biomonitors for inorganic analytes, 84–85
sampling of biomonitors for organic analytes, 83–84
tree bark, 76–78
Black carbon (BC), 197
Blowoff effects, 160–161, 161f

Blunt-tip NT, 223f–224f
Boltzmann constant, 324
Borosilicate glass, 136
Breakthrough volumes (BVs), 138
"Breakthrough" phenomenon, 138
Brominated flame-retardants (BFRs), 74
BTEX. *See* Benzene, toluene, ethyl benzene and xylene (BTEX)
BTX. *See* Benzene, toluene and xylene (BTX)
BVOC. *See* Biogenic VOC (BVOC)
BVOCs. *See* Biogenic volatile organic compounds (BVOCs)
BVs. *See* Breakthrough volumes (BVs)

C

$CaCO_3$-containing particles, 362–363
Calibration blank, 42
Canister
  cleaning, 123–124
  types, 123
  whole-air sampling with, 120–130, 121f
Capacity factor, 269
Carboxen (Car), 213, 225
Carboxylic ions (COO^-), 348–350
Carrier gas, 279–280
Cartridge sampler, 18
Catch sample. *See* Grab sample
Cerium, 251
Certified reference materials (CRMs), 5–6, 255
Chemical composition, 244–245
Chemical ionisation (CI), 314
Chemical markers of emission sources, 245
  geological elements, 245–246
  marine aerosols, 253
  secondary aerosols, 253–254
  TREs, 246–253
Chemical vapour generation system (CVG system), 258
Chemometrics, 7–8
Chlorine, 253
Chromatography, 268
  applications in air pollutants analysis, 295–296
    PAHs and derivatives, 299–300
    POPs, 296–298
    VOCs, 300–302
  capacity factor, 269
  detectors
    ECD, 291–292
    FID, 290
    MSD, 294–295
    NPD, 292–293
    PID, 293–294
    requirements of detectors, 289–290
  efficiency, 269
    separation, 269–271
  optimizing resolution and separation, 272
  peak capacity, 271
  relative retention, 268–269
  resolution, 271–272
  selectivity, 269
  two-dimensional gas chromatography, 303f
  types
    GC, 272–274, 273f
    GC X GC, 275–277, 278f
    HPLC, 274–275
CI. *See* Chemical ionisation (CI)
Coarse particles, 244–245
Coated solid sorbents, 140–141
Coiled denuders, 162–163, 164f
Cold vapour atomic absorption (CV-AAS), 256
Commercially available diffusion passive samplers, 16f, 20f
Comparability, 41
Composite cylindrical denuders, 162, 174
Composite sample. *See* Time-integrated sample
Conifer needles, 67
  inorganic air pollutants, 74–76, 75t
  organic analytes, 73–74
  using pine needles as biomonitors, 77t
Conventional filter/adsorbent-based sampling systems, disadvantages of, 159–162, 160f
COO^-. *See* Carboxylic ions (COO^-)
Criteria pollutants, 185
CRMs. *See* Certified reference materials (CRMs)
Crustose lichen, 57–59
Cryogenic modulators, 276–277
CV-AAS. *See* Cold vapour atomic absorption (CV-AAS)
CVG system. *See* Chemical vapour generation system (CVG system)
Cylindrical denuders, 162, 163f, 168–170

D

DART. *See* Direct analysis in real-time (DART)
Data cleaning, 203–204
Data handling, quality and processing, 200–204

Data handling, quality and processing (*Continued*)
data cleaning, 203–204
data quality, 201–203
working with large data sets, 200–201
Data logging requirements, 197–199
Data processing, 200, 204
Data quality, 201–203
maintaining electronic diary, 202–203
regular checks and data transfer importance, 201–202
Data quality objective (DQO), 41
DCP-OES. *See* Direct current plasma–optical emission spectrometry (DCP-OES)
DDT spraying program, 73–74
Dead time, 269
DEBITS programme. *See* Deposition of Biogeochemically Important Trace Species programme (DEBITS programme)
Denudation, 156, 157f
Denudation theory, 165
air flow considerations, 165–166
gas phase collection efficiency, 168–171
particle transmission efficiency, 166–168
Denuder difference method, 171
Denuder-based sampling, 156–159
Denuders, 156
analyte extraction methods and analytical techniques, 179
conventional filter/adsorbent-based sampling systems, disadvantages of, 159–162
denuder-based sampling, 156–159
reasons for determining gas and particle-associated analyte concentrations, 159
sampling applications
inorganic analytes sampling, 171, 172t
organic analytes sampling, 173–175, 176t–177t
sampling considerations, 175–178
types, 162–165
Deposition of Biogeochemically Important Trace Species programme (DEBITS programme), 17–18, 44–46
Diazotising agent, 37
Diesel engine exhaust emissions, 4
Diffusive laws, interface model on, 218–219
Dimethyl sulphide (DMS), 337
Dimethyl sulphoxide (DMSO), 354–355
2,4-dinitrophenylhydrazine-coated cartridges (DNPH-coated cartridges), 130
Direct analysis in real-time (DART), 301
Direct current plasma–optical emission spectrometry (DCP-OES), 255–256
Direct thermal desorption-GC (DTD-GC), 283
Dirty reaction cell window, 193f
dispersive SPE (dSPE), 93–95
Divinylbenzene (DVB), 213
DMS. *See* Dimethyl sulphide (DMS)
DMSO. *See* Dimethyl sulphoxide (DMSO)
DNPH-coated cartridges. *See* 2,4-dinitrophenylhydrazine -coated cartridges (DNPH-coated cartridges)
DQO. *See* Data quality objective (DQO)
Drift gas, 312
Drift time, 324
dry weight (dw), 73–74
DSASE. *See* Dynamic sonication-assisted solvent extraction (DSASE)
dSPE. *See* dispersive SPE (dSPE)
DTD-GC. *See* Direct thermal desorption-GC (DTD-GC)
DVB. *See* Divinylbenzene (DVB)
DVB/Car/PDMS devices, 214
dw. *See* dry weight (dw)
Dynamic sonication-assisted solvent extraction (DSASE), 93

E

EC. *See* Elemental carbon (EC)
ECD. *See* Electron capture detector (ECD)
ED-EPMA, ATR-FTIR imaging and RMS, combined application of, 358–363
ED-EPMA. *See* Energy-dispersive electron probe X-ray microanalysis (ED-EPMA)
ED-EPMA and ATR-FTIR imaging, combined application of, 345
aged mineral dust, 348–350
Antarctic sea-salt aerosols, 354–355
reacted sea-salt aerosols, 350–352
silicate particles, 352
standard mineral particles, 345–348
ED-EPMA and RMS, combined application of, 355
airborne subway particles iron speciation, 357–358
individual Asian dust particles chemical mixing state, 355–356
internally mixed soot particles, 358
EDS. *See* Energy-dispersive spectroscopy (EDS)
EDX. *See* Energy-dispersive X-ray spectrometry (EDX)

Effective length of denuder, 169
EI. *See* Electron impact (EI); Electron ionisation (EI)
Electro-thermal atomic absorption spectrometry (ETAAS), 241, 257–258
Electron capture detector (ECD), 275–276, 291–292
Electron impact (EI), 297, 313
Electron ionisation (EI), 294, 313
Electronic diary maintaining, 202–203
Electropolished canister, 123
Elemental analysis, spectroscopic techniques for, 255
  atomic absorption spectrometry, 256–258
  atomic fluorescence spectroscopy, 258
  NAA, 260–261
  particle-induced X-ray emission, 260–261
  plasma-based techniques, 259–260
  X-ray fluorescence, 260–261
Elemental carbon (EC), 336
Emerging microextraction technologies, 209–210
Energy-dispersive electron probe X-ray microanalysis (ED-EPMA), 333–334
  applications, 334–338
  KCl, Na_2CO_3, $NaNO_3$ and $(NH_4)_2SO_4$, 335f
  quantification procedure, 334
Energy-dispersive spectroscopy (EDS), 243
Energy-dispersive X-ray spectrometry (EDX), 332–333
Environmental analyses, quality control importance in, 5–7
Environmental monitoring, statistical tools for, 7–8
Environmental Protection Agency (EPA), 64, 185, 186t
Environmental samples, 5–6
EPA. *See* Environmental Protection Agency (EPA); U. S. Environmental Protection Agency (EPA)
Epidemiological studies, 240–241
Equilibrium-based microextraction methods, 211–212
ETAAS. *See* Electro-thermal atomic absorption spectrometry (ETAAS)
European Integrated project on Aerosol, Cloud, Climate, and Air Quality Interactions (EUCAARI), 184
European Union (EU), 120
External calibration of gaseous samples, 216–218
Extraction mode, 218–219

F

FAAS. *See* Flame atomic absorption spectrometry (FAAS)
FCC. *See* Fluid catalytic cracking (FCC)
Federal Reference Method (FRM), 147
Ferm-type sampler, 18–19, 19f
FI. *See* Flow injection (FI)
FI-CV-AAS. *See* Flow injection-cold vapour-AAS (FI-CV-AAS)
Fibre-based geometry, 214
Fick's first law of diffusion, 22, 228–229
Ficus benghalensis leaves, 78
FID. *See* Flame ionisation detector (FID)
Filter, 318
Flame atomic absorption spectrometry (FAAS), 241, 257
Flame ionisation detector (FID), 130, 290
FLD. *See* Fluorescence detector (FLD)
Flow injection (FI), 255
Flow injection-cold vapour-AAS (FI-CV-AAS), 253
Flow meter, 114f
Fluid catalytic cracking (FCC), 252–253
Fluorescence detector (FLD), 99
Fly ash, 243
Foliose lichen, 57–59, 58f
Fossil fuel combustion, 253–254
Fourier transform infrared spectroscopy (FTIR), 340–341
Fractionators, 144–147
FRM. *See* Federal Reference Method (FRM)
Fruticose lichen, 57–59
FTIR. *See* Fourier transform infrared spectroscopy (FTIR)
Fuel additives, 251

G

Gas chromatography (GC), 39–40, 40f, 93–95, 209–210, 268, 272–274, 273f
  liner for, 281f
Gas chromatography–mass spectrometry (GC-MS), 126, 299, 313, 319–320
Gas phase collection efficiency, 168–171
  annular denuders, 170–171
  cylindrical denuders, 168–170
Gas sampling bags, 112
Gas-tight syringe, 115f

Gas–liquid chromatography (GLC), 272–273
partition chromatography, 284
Gasoline, 252–253
GAW program. *See* Global Atmosphere Watch program (GAW program)
GC. *See* Gas chromatography (GC)
GC coupled to a high-resolution mass spectrometer detector (GC-HRMS), 297
GC X GC. *See* Two-dimensional gas chromatography (GC X GC)
GC X GC separations, orthogonality for, 288
GC-ECD method, 292
GC-HRMS. *See* GC coupled to a high-resolution mass spectrometer detector (GC-HRMS)
GC-MS. *See* Gas chromatography–mass spectrometry (GC-MS)
GCB. *See* Graphitised carbon black (GCB)
GCxGC-TOFMS. *See* Two dimensional GCetime-of-flight mass spectrometry (GCxGC-TOFMS)
Gel filtration, 284–285
Gel-permeation. *See* Gel filtration
Geological elements, 245–246
GFF. *See* Glass fibre filters (GFF)
GHGs. *See* Greenhouse gases (GHGs)
Glass fibre filters (GFF), 160
GLC. *See* Gas–liquid chromatography (GLC)
Global Atmosphere Watch program (GAW program), 17–18, 193
Gormley–Kennedy solution, 169
Governing equations, 22–25
Grab sample, 110
Gradient method, 262
Graphitised carbon black (GCB), 93–95
Greenhouse gases (GHGs), 183–184

H

HCB. *See* Hexachlorobenzene (HCB)
HCH. *See* Hexachlorocylohexane (HCH)
HCL. *See* Hollow cathode lamp (HCL)
Height equivalent theoretical plate (HETP), 269–271
HEN. *See* High-efficiency nebuliser (HEN)
HETP. *See* Height equivalent theoretical plate (HETP)
Hexachlorobenzene (HCB), 67
Hexachlorocylohexane (HCH), 64
HG. *See* Hydride generation (HG)
hi-vol air sampler methodology. *See* High volume air sampler methodology (hi-vol air sampler methodology)
High volume air sampler methodology (hi-vol air sampler methodology), 159–160
High volume sampling, 149–150, 150t
High-efficiency nebuliser (HEN), 256
High-performance liquid chromatography (HPLC), 98–99, 140–141, 274–275
High-performance liquid chromatography coupled to fluorescence detection (HPLC-FLD), 299
High-resolution continuous source atomic absorption spectrometry (HR-CS AAS), 257
High-resolution gas chromatography (HRGC), 320
High-resolution mass spectrometers (HRMS), 320
High-resolution TOFMS (HR-TOFMS), 319–320
High-resolution-ICP-MS (HR-ICP-MS), 250–251
Hold-up time, 269
Hollow cathode lamp (HCL), 257
HPLC. *See* High-performance liquid chromatography (HPLC)
HPLC-FLD. *See* High-performance liquid chromatography coupled to fluorescence detection (HPLC-FLD)
HR-CS AAS. *See* High-resolution continuous source atomic absorption spectrometry (HR-CS AAS)
HR-ICP-MS. *See* High-resolution-ICP-MS (HR-ICP-MS)
HR-TOFMS. *See* High-resolution TOFMS (HR-TOFMS)
HRGC. *See* High-resolution gas chromatography (HRGC)
HRMS. *See* High-resolution mass spectrometers (HRMS)
Hydride generation (HG), 248
Hydrogen, 291
Hylocomium splendens (*H. splendens*), 66
Hypnum cupressiforme (*H. cupressiforme*), 66, 76–78
Hypogymnia physodes (*H. physodes*), 73

I

IC. *See* Ion chromatography (IC)
ICP-MS. *See* Inductively coupled plasma–mass spectrometry (ICP-MS)

ICP-OES. *See* Inductively coupled plasma–optical emission spectrometry (ICP-OES)
ID. *See* Inner diameter (ID)
Ideal gas law, 23–24
IGAC project. *See* International Global Atmospheric Chemistry project (IGAC project)
ILs. *See* Ionic liquids (ILs)
Impinger herein, 134
Impingers/bubblers, air sampling with, 133–136, 133f, 135t
  advantages/limitations, 136
  applications, 134–135
  methodology, 134
  preparation, 134
  storage, 135–136
IMS. *See* Ion mobility spectrometry (IMS)
In-field concentration, 132–143
  impingers/bubblers, air sampling with, 133–136
  solid sorbents, air sampling with, 136–143
In-needle capillary adsorption trap (INCAT), 221f
INAA. *See* Instrumental neutron activation analysis (INAA)
INCAT. *See* In-needle capillary adsorption trap (INCAT)
Individual Asian dust particles chemical mixing state, 355–356
Individual atmospheric particles
  ATR-FTIR imaging applications, 340–341
  RMS applications, 341
Inductively coupled plasma–mass spectrometry (ICP-MS), 59, 241
Inductively coupled plasma–optical emission spectrometry (ICP-OES), 59, 241
Inner diameter (ID), 282
Inorganic air pollutants, 74–76, 75t
Inorganic analytes, 100
  extraction
    MAE, 88
    USAE, 91–93, 92t
  sampling, 171, 172t
  sampling of biomonitors for, 84–85
Inorganic gases. *See also* Organic gases
  laboratory methods for sampling of, 35
    passive sampler chromatogram, 38f
    precautions and known interferences, 37–39
    sample analysis, 37
    sampler extraction and analytical preparation, 36–37
    sampler preparation, 35–36
  samplers for, 17–20
Inorganic ions, 262
Instrumental neutron activation analysis (INAA), 261
Integrated Organic Gas and Particle Sampler (IOGAPS), 173
Integrated Organic Vapor/Particle Sampler (IOVPS), 173
Interface model on diffusive laws, 218–219
Interferences, 6–7
Internal reflectance element (IRE), 339–340
Internally mixed soot particles, 358
International Global Atmospheric Chemistry project (IGAC project), 17–18
IOGAPS. *See* Integrated Organic Gas and Particle Sampler (IOGAPS)
Ion chromatography (IC), 36, 241, 262
  for anions and cations, 261–263
Ion mobility, 324
Ion mobility coefficient, 324
Ion mobility spectrometry (IMS), 312, 323–325. *See also* Mass spectrometry (MS)
Ion trap, 318
Ion-trap analyser, 295
Ionic liquids (ILs), 289
Ionisation methods, 313
  electron ionisation, 313
  laser desorption/ionisation, 315–316
  secondary ionisation methods, 314–315
IOVPS. *See* Integrated Organic Vapor/Particle Sampler (IOVPS)
IRE. *See* Internal reflectance element (IRE)
Isocratic method, 262

K

KI. *See* Potassium iodide (KI)
Kinetic calibration techniques, 219–220

L

Laser desorption/ionisation, 315–316
Laser microprobe mass spectrometry, 323
Laser-induced fluorescence, 370
LC. *See* Liquid chromatography (LC)
LC stationary phases, developments in, 288–289
Lead (Pb), 248–249

Leptodon smithii (*L. smithii*), 66
Lichens (*Parmelia* spp.), 55–57
air pollutants accumulation, 57–58
as biomonitors of inorganic air pollutants, 58–59, 60t–61t
as biomonitors of organic air pollutants, 59–65, 62t–63t
lack of cuticle in, 57
Limit of detection (LOD), 7, 257, 297
Limit of quantitation (LOQ), 7
Linear uptake range, 16
Liner types, 280
for gas chromatography, 281f
for split inlet, 281
splitless liners, 281–282
Liquid chromatography (LC), 93–95, 269–270
Lock mass, 319–320
LOD. *See* Limit of detection (LOD)
LOQ. *See* Limit of quantitation (LOQ)
Lung sampler, 115–116

M

m/z. *See* Mass-to-charge ratios (m/z)
MAAP. *See* Multi-angle absorption photometer (MAAP)
MAE. *See* Microwave-assisted extraction (MAE)
MAHS-SPME. *See* Microwave-assisted headspace solid-phase microextraction (MAHS-SPME)
Manganese, 251–252
Mapping regional pollutant concentrations
DEBITS network, 44–46
exceedance of critical loads in South Africa, 43–44
Marine aerosols, 253
Mass analysers, 318–319
Mass resolution, 319–320
Mass spectrometer, 294
Mass spectrometer detector (MSD), 275–276, 294–295
Mass spectrometry (MS), 273, 311–312
components, 312f
instrumentation
AMS, 321–323
GC-MS, 319–320
mass analysers, 318–319
portable/miniature mass spectrometers, 320–321
PTR-MS, 325–326
ionisation methods, 313
electron ionisation, 313
laser desorption/ionisation, 315–316
secondary ionisation methods, 314–315
methods of sample, 316–317
Mass-to-charge ratios (m/z), 311–312
Mean AED, 145–147
Measurement accreditation, 193–195
Mechanical flow controllers, 124–126, 125f, 125t
Membrane introduction, 317
Membrane introduction mass spectrometry (MIMS), 317
Mercuric chloride solution, 14–15
Mesa Labs BGI Very Sharp Cut Cyclone (Mesa Labs BGI VSCC®), 145–147, 145f–146f
Methanesulphonic acid (MSA), 354–355
Method precision determinations, 6
Method validation, 6
Methods of sample, 316
MIMS, 317
TD, 316
Methylcyclopentadienyl Mn tricarbonyl, 251–252
Methylpolysiloxane phases, 287
Microscale purge and trap, 131–132
Microscopic single-particle analytical techniques, 332–333
AFM, 342–343
ED-EPMA, 333–338
molecular vibrational imaging spectroscopies, 339–341
optical microscopy, 343
STXM, 338–339
techniques for same single particles characterisation, 344–345
combined application of ED-EPMA, ATR-FTIR imaging and RMS, 358–363
combined application of ED-EPMA and ATR-FTIR imaging, 345–355
combined application of ED-EPMA and RMS, 355–358
TOF-SIMS, 342
Microwave-assisted extraction (MAE), 85–88. *See also* Ultrasound-assisted extraction (USAE)
inorganic analytes extraction, 88
microwave digestions in air pollution monitoring, 90t
organic analytes extraction, 87–88, 89t

Microwave-assisted headspace solid-phase microextraction (MAHS-SPME), 88
Microwave-induced plasma (MIP), 255–256
Midpolarity phases, 286
MIMS. *See* Membrane introduction mass spectrometry (MIMS)
Miniature mass spectrometers, 320–321
MIP. *See* Microwave-induced plasma (MIP)
Molecular vibrational imaging spectroscopies, 339–340
  ATR-FTIR imaging applications, 340–341
  RMS applications, 341
Monitoring devices selection, 190–193
Monodimensional GC, 276, 296
Monoelemental techniques, 255
Mosses, 55, 65
  as biomonitors of air pollutants, 65–67, 68t–72t
  bryophytes, 65
MS. *See* Mass spectrometry (MS)
MS/MS. *See* Tandem mass analysis (MS/MS)
MSA. *See* Methanesulphonic acid (MSA)
MSD. *See* Mass spectrometer detector (MSD)
Multi-angle absorption photometer (MAAP), 197–198
Multi-capillary denuder, 174
Multichannel silicone rubber trap denuder, 178f
Multichannel systems. *See* Composite cylindrical denuders
Multilayer foil bags, 116

N

NAA. *See* Neutron activation analysis (NAA)
NAAQS. *See* National Ambient Air Quality Standards (NAAQS)
NADP/AMoN. *See* National Atmospheric Deposition Program's Ammonia Monitoring Network (NADP/AMoN)
NADP/NTN. *See* NADP's National Trends Network (NADP/NTN)
NADP's National Trends Network (NADP/NTN), 46–48
$NaNO_2$. *See* Sodium nitrite ($NaNO_2$)
NaOH. *See* Sodium hydroxide (NaOH)
Naphthylethylenediamine dihydrochloride (NEDA), 37
National Ambient Air Quality Standards (NAAQS), 185
National Atmospheric Deposition Program's Ammonia Monitoring Network (NADP/AMoN), 17–18, 46–49
National Institute of Safety and Health (NIOSH), 39–40
National Institute of Standards and Technology (NIST), 345
Near-edge X-ray absorption fine structure spectra (NEXAFS spectra), 338
NEDA. *See* Naphthylethylenediamine dihydrochloride (NEDA)
Needle trap (NT), 223f
  and matching narrow neck liner, 226f
Needle trap devices (NTD), 210, 221–222, 221f
  active sampling, 227–228
  air sampling applications, 232t
  application of NT sampling, 231–233
  extended tip needle trap, 224f, 227f
  geometry and extraction mechanism, 222–227
  interception, impaction and diffusion onto solid sorbent particles, 230f
  narrow neck liner design, 224f
  particulate matter characterisation, 229–231
  passive TWA-based sampling, 228–229
Needle trap sampling
  application, 231–233
  particulate matter characterisation using, 229–231
Negative ion chemical ionisation (NICI), 297
Neutral loss scan (NLS), 297–298
Neutron activation analysis (NAA), 241, 260–261
Neutron activation analysis, 261
NEXAFS spectra. *See* Near-edge X-ray absorption fine structure spectra (NEXAFS spectra)
NH_3 sampler. *See* Ammonia sampler (NH_3 sampler)
NICI. *See* Negative ion chemical ionisation (NICI)
Nickel, 291
NIOSH. *See* National Institute of Safety and Health (NIOSH)
NIST. *See* National Institute of Standards and Technology (NIST)
Nitrated PAHs (nitro-PAHs), 292
Nitric acid (HNO_3), 191
nitro-PAHs. *See* Nitrated PAHs (nitro-PAHs)
Nitrogen, 291
Nitrogen dioxide (NO_2), 15
Nitrogen phosphorus detector (NPD), 275–276, 292–293

NLS. *See* Neutral loss scan (NLS)
NO/NO_x measurement, 193–195, 193f
Non-chromatographic-based denuder systems, 171
Nonpolar stationary phases, 285–286
Normal phase HPLC (NP-HPLC), 274
NPD. *See* Nitrogen phosphorus detector (NPD)
NT. *See* Needle trap (NT)
NTD. *See* Needle trap devices (NTD)

O

OC. *See* Organic carbon (OC)
Occupational Safety and Health Administration (OSHA), 126
OCPs. *See* Organochlorine pesticides (OCPs)
Octanol-air partition coefficient, 158
On-site sampling, SPME calibration for, 214–216. *See also* Air sampling
  external calibration of gaseous samples, 216–218
  interface model on diffusive laws, 218–219
  kinetic calibration techniques, 219–220
Optical microscopy, 343
Organic analytes, 73–74, 98–99
  extraction
    MAE, 87–88
    USAE, 93, 94t
  sampling, 173–175, 176t–177t
    of biomonitors for, 83–84
Organic carbon (OC), 144, 156, 336
Organic gases. *See also* Inorganic gases
  laboratory methods for sampling
    precautions and known interferences, 41
    sampler deployment, 39
    sampler desorption and analysis, 39–41
    sampler preparation, 39
  samplers for, 20–22
Organochlorine pesticides (OCPs), 73, 297
OSHA. *See* Occupational Safety and Health Administration (OSHA)
Ozone (O_3), 14–15

P

PAH-columns, 287
PAHs. *See* Polycyclic aromatic hydrocarbons (PAHs)
Palmes Tube' diffusion passive sampler, 15
PAN. *See* Peroxyacetyl nitrate (PAN)
Parallel-plate denuders, 162–163, 164f
part-per-billion by volume (ppbv), 110–111
part-per-trillion by volume (pptv), 110–111
Particle diffusion coefficients, 166–168
Particle phase, 156, 159–160
Particle transmission efficiency, 166–168
Particle-induced X-ray emission, 260–261
Particulate matter (PM), 111, 319. *See also* Airborne particulate matter (APM)
Partition coefficient, 158
parts per trillion (ppt), 209–210
Passive air sampling, 14–16
Passive diffusion of air pollutants, 14
Passive samplers, 14–17, 17t, 18f
  for ambient air quality monitoring, 15
  holders and shields, 32f
  methodology, 33–34
    laboratory methods for sampling of inorganic gases, 35–39
    laboratory methods for sampling of organic gases, 39–41
    network operations, 34–35
  network design, 25
    active and passive measurements comparison, 33f
    collocation with reference measurements, 30–32
    concentration profile of pollutant, 32f
    field siting criteria, 28–30, 31f
    interlaboratory comparisons, 34f
    sampling frequency and duration, 28, 29t
    selection and quantity of sampling locations, 27, 27t
  pollutant gas, coating solution, analytical reproducibility and detection limits, 19t
Passive sampling devices, 15–16
  case studies, 43–49
  commercially available diffusion passive samplers, 16f, 20f
  governing equations, 22–25
  linear uptake range, 16
  physical characteristics, 16
  physical constants and parameters for gases, 22t
  samplers
    for inorganic gases, 17–20
    for organic gases, 20–22
  uptake curve for passive diffusion sampler, 17f
Paul trap, 318
PBDEs. *See* Polybrominated diphenyl ethers (PBDEs)

PCA. *See* Principal component analysis (PCA)
PCBs. *See* Polychlorinated biphenyls (PCBs)
PCDD/Fs. *See* Polychlorinated dibenzo dioxins and Furans (PCDD/Fs)
PCDDs. *See* Polychlorinated dibenzo-*p*-dioxins (PCDDs)
PCNs. *See* Polychlorinated naphthalenes (PCNs)
PDMS. *See* Polydimethylsiloxane (PDMS)
Peak capacity, 271, 276
Peclet number (P_e), 166
PEG. *See* Polyethylene glycol (PEG)
Permeation process, 317
Peroxyacetyl nitrate (PAN), 37–39, 191, 291–292
Peroxypropionyl nitrate (PPN), 291–292
Persistent organic pollutants (POPs), 59, 292, 296–298, 320
PGEs. *See* Platinum group elements (PGEs)
Photoionisation detector (PID), 293–294
PID. *See* Photoionisation detector (PID)
Pine needles, 73–74, 75t
Piston pump, 114f
PIXE. *See* Proton-induced X-ray emission (PIXE)
Plasma-based techniques, 249, 259–260
Platinum group elements (PGEs), 247, 249–251
Pleurocarpous mosses, 65
Pleurozium schreberi (*P. schreberi*), 66
PM. *See* Particulate matter (PM)
Polar stationary phases, 285–286
Polybrominated diphenyl ethers (PBDEs), 174, 297
Polychlorinated biphenyls (PCBs), 55, 111, 157–158, 297
Polychlorinated dibenzo dioxins and Furans (PCDD/Fs), 55, 297
Polychlorinated dibenzo-*p*-dioxins (PCDDs), 209–210
Polychlorinated naphthalenes (PCNs), 96–97
Polycyclic aromatic hydrocarbons (PAHs), 54–55, 111, 156, 209–210, 285, 299, 315–316
  and derivatives, 299–300
Polydimethylsiloxane (PDMS), 140, 174, 212–213
Polyethylene glycol (PEG), 210
Polypropylene valve, 113f
Polytetrafluoroethylene (PTFE), 136–137
Polytrichum formosum (*P. formosum*), 66–67
Polyurethane foam (PUF), 159–160
  cartridges, 139, 139f
polyvinylidene fluoride. *See* Tecaflon® (polyvinylidene fluoride)
POPs. *See* Persistent organic pollutants (POPs)
Portable mass spectrometers, 320–321
Portable standard gas generating vials, 216–218
Potassium carbonate (K_2CO_3), 19–20
Potassium hydroxide (KOH), 15
Potassium iodide (KI), 19–20
ppbv. *See* part-per-billion by volume (ppbv)
PPN. *See* Peroxypropionyl nitrate (PPN)
ppt. *See* parts per trillion (ppt)
pptv. *See* part-per-trillion by volume (pptv)
Precision, 41
Preconcentrator, 130–131, 131f
Primary–secondary amine (PSA), 93–95
Principal component analysis (PCA), 7–8, 345
Programmable temperature vaporizer (PTV), 283
Proton transfer reaction (PTR), 314–315
Proton-induced X-ray emission (PIXE), 241
Proton-transfer reaction mass spectrometry (PTR-MS), 301, 325–326
PSA. *See* Primary–secondary amine (PSA)
PTFE. *See* Polytetrafluoroethylene (PTFE)
PTR. *See* Proton transfer reaction (PTR)
PTR-MS. *See* Proton-transfer reaction mass spectrometry (PTR-MS)
PTV. *See* Programmable temperature vaporizer (PTV)
PUF. *See* Polyurethane foam (PUF)
Purge valve time, 280
Pyrex® glass tubes, 133

Q

Q-ICP-MS. *See* Quadrupole-ICP-MS (Q-ICP-MS)
QA. *See* Quality assurance (QA)
QC. *See* Quality control (QC)
QCL. *See* Quantum cascade laser (QCL)
qMS. *See* quadrupole mass spectrometry (qMS)
Quadrupole analyser, 294
Quadrupole mass spectrometers, 322
quadrupole mass spectrometry (qMS), 292–293
Quadrupole-ICP-MS (Q-ICP-MS), 255–256

Quality assurance (QA), 25
comparison with reference methods, 42
data validation and reporting, 42–43
field, 42
laboratory, 41–42
procedures, 41
Quality control (QC), 25
comparison with reference methods, 42
data validation and reporting, 42–43
field, 42
importance in environmental analyses, 5–7
laboratory, 41–42
procedures, 41
Quantum cascade laser (QCL), 191
Quartz wool, 222–223
QuEChERS, 93–95
Quercus ilex L., 78

R

Radial sampler, 18
Radio frequency (RF), 293
Radiochemical neutron activation analysis (RNAA), 261
Raman microspectrometry (RMS), 332–333
applications, 341
Rare-earth elements (REEs), 244
Reacted mineral dust, 348–350
Real-time aerosol analysis, 322
Receptor model analysis (RM analysis), 245
REEs. *See* Rare-earth elements (REEs)
Relative humidity (RH), 119, 343
Relative retention, 268–269
Resolution, 271–272
Retention factor. *See* Capacity factor
Reverse phase high performance liquid chromatography (RP-HPLC), 274
Reverse phase liquid chromatography (RP-LC), 286
Reverse phase liquid–solid adsorption chromatography, 286
Reynolds number (Re), 218–219
RF. *See* Radio frequency (RF)
RH. *See* Relative humidity (RH)
Rigid box, 115f
RM analysis. *See* Receptor model analysis (RM analysis)
RMS. *See* Raman microspectrometry (RMS)
RNAA. *See* Radiochemical neutron activation analysis (RNAA)
Robinio pseudoacacia L., 78
RP-HPLC. *See* Reverse phase high performance liquid chromatography (RP-HPLC)
RP-LC. *See* Reverse phase liquid chromatography (RP-LC)

S

Safe sampling volumes (SSVs), 138
Sample
cleanup, 95–98
introduction, 277
liner types, 280–282
split *vs.* splitless liquid injections, 279–280
TD injections, 282–283
vapourising split/splitless injector, 279f
preparation, 369–370
Samplers
commercially available sorbents characteristics, 21t
for inorganic gases, 17–20
for organic gases, 20–22
Sampling, 222–223
timer, 129
volume, 227–228
Sandwich fibre, 214
Saran® (polyvinylidene chloride), 116
SAX. *See* Strong anion exchange (SAX)
Scanning electron microscopy (SEM), 248, 332–333
Scanning transmission X-ray microscopy (STXM), 332–333, 338
applications, 338–339
Schmidt number (Sc), 218–219
SDME. *See* Single drop microextraction (SDME)
Sea salt aerosol (SSA), 253
Sea-spray aerosol, 253
Secondary aerosols, 253–254
Secondary electron images (SEIs), 337, 346f
Secondary ionisation methods, 314–315
Secondary organic aerosol (SOA), 254
Sector field-inductively coupled plasma–mass spectrometry (SF-ICP-MS), 250
SEIs. *See* Secondary electron images (SEIs)
Selective ion monitoring (SIM), 275–276, 294
Selectivity, 6–7, 269
Self-reversal-electro-thermal atomic absorption spectrometry (SR-ETAAS), 258
SEM. *See* Scanning electron microscopy (SEM)

Semi-volatile organic compounds (SVOCs), 57–58, 111, 157–158, 209–210, 253–254, 312–313
Sensitivity, 41
Separation efficiency, 269–271
SF-ICP-MS. *See* Sector field-inductively coupled plasma–mass spectrometry (SF-ICP-MS)
Shape-to-charge ratios (Ω), 312
Shinwa needle trap, 222–223, 223f
Silica column chromatography, 97
Silicon-lined canister (SL canister), 123
Silicone-based denuders, 170
SIM. *See* Selective ion monitoring (SIM); Single ion monitoring (SIM)
Single drop microextraction (SDME), 221f
Single ion monitoring (SIM), 299
Single particle soot photometer (SP2), 197–198
Single sample extraction, 142–143
Single-scatter albedo (SSA), 200, 201f
Site selection, 195–197
Size distribution, 242–243
Size exclusion chromatography, 97–98
SL canister. *See* Silicon-lined canister (SL canister)
Smartphone applications, 370–371
SMCs. *See* Synthetic musk fragrances (SMCs)
SOA. *See* Secondary organic aerosol (SOA)
Sodium hydroxide (NaOH), 19–20
Sodium nitrite ($NaNO_2$), 19–20
Solid phase dynamic extraction (SPDE), 221–222, 221f
Solid phase microextraction (SPME), 209–210. *See also* Needle trap devices (NTD)
  analyte-dependant selection, 215t
  calibration for on-site sampling, 214–220
  configuration of commercially available, 216f
  devices and techniques, 210
  equilibrium-based microextraction methods, 211–212
  fibre geometry and operation, 214
  fibre in needle, 221f
  from in-vial standard gas generating system, 217f
  interface model, 218f
  sorbent coatings for air sampling, 212–214
Solid sorbents, 137f, 141t
  advantages/limitations, 142–143
  air sampling with, 136–143
  applications, 140
  coated solid sorbents, 140–141
  methodology, 137–140
  preparation, 140
  storage, 141–142
Solid-phase extraction (SPE), 93, 96f
SOPs. *See* Standard operating procedures (SOPs)
Sorbent tubes. *See* Solid sorbents
Sorbent-based denuders, 174–175
Sorption-based denuder, 168
Soxhlet extraction, 85–87, 86t
SP2. *See* Single particle soot photometer (SP2)
SPDE. *See* Solid phase dynamic extraction (SPDE)
SPE. *See* Solid-phase extraction (SPE)
Split injection, 280
Split inlet, liner for, 281
Split liquid injections, 279–280
Split ratio, 280
Splitless injection, 279–280
Splitless liners, 281–282
Splitless liquid injections, 279–280
SPME. *See* Solid phase microextraction (SPME)
Squamulous lichen, 57–59
SR-ETAAS. *See* Self-reversal-electro-thermal atomic absorption spectrometry (SR-ETAAS)
SR-XAS. *See* Synchrotron radiation X-ray absorption (SR-XAS)
SRMs. *See* Standard reference materials (SRMs)
SSA. *See* Sea salt aerosol (SSA); Single-scatter albedo (SSA)
SSVs. *See* Safe sampling volumes (SSVs)
Stainless-steel canisters, 126
Standard mineral particles, 345–348
Standard operating procedures (SOPs), 193
Standard reference materials (SRMs), 287–288
Stationary phases, 283
  in air pollutants chromatographic analysis, 283
  analyte-specific stationary phases, 287–288
  developments in LC stationary phases, 288–289
  in GC, 284
  orthogonality for GC X GC separations, 288
  polar *vs.* nonpolar stationary phases, 285–286

Statistics, 7–8
Strong anion exchange (SAX), 97
STXM. *See* Scanning transmission X-ray microscopy (STXM)
Sulphur dioxide (SO_2), 14–15
SUMMA® like canister, 123
SUMMA®, 123
SVOC/PM, 143–149
  applications, 147
  fractionators, 144–147
  methodology, 144
  preparation, 144
  sampling methods and application suitability, 148t
  storage, 149
  SVOCs/artifacts, 148–149
SVOCs. *See* Semi-volatile organic compounds (SVOCs)
SVOCs/artifacts, 148–149
Synchrotron radiation X-ray absorption (SR-XAS), 248
Synthetic musk fragrances (SMCs), 74

T

Tandem mass analysis (MS/MS), 322
TD. *See* Thermal desorption (TD)
TDS. *See* Total dissolved solids (TDS)
TE-1000PUF high volume sampler, 147f
Tecaflon® (polyvinylidene fluoride), 116
Tedlar® bags, 112f, 116
Teflon, 116
TEM/EDX. *See* Transmission electron microscopy/EDX (TEM/EDX)
Thermal desorption (TD), 119, 316
  injections, 282–283
3D trap, 318
TIC. *See* Total ion chromatogram (TIC)
Time-integrated sample, 110
Time-of-flight analysers (TOF analysers), 295, 315–316, 318–319
Time-of-flight mass spectrometer (TOF-MS), 39–40
Time-of-flight secondary ion mass spectroscopy (TOF-SIMS), 332–333, 342
Time-weighted average sampling (TWA sampling), 110, 219
TO. *See* Toxic Organic (TO)
TOF analysers. *See* Time-of-flight analysers (TOF analysers)
TOF-MS. *See* Time-of-flight mass spectrometer (TOF-MS)
TOF-SIMS. *See* Time-of-flight secondary ion mass spectroscopy (TOF-SIMS)
Tortula intermedia (*T. intermedia*), 66
Tortula muralis (*T. muralis*), 66
Total dissolved solids (TDS), 100
Total ion chromatogram (TIC), 294
Total reflection X-ray fluorescence (TXRF), 260
Total suspended particulate (TSP), 144
Toxic Organic (TO), 120–121
Traffic-related elements (TREs), 246
  antimony, 247–248
  brake pad composites, 251
  lead, 248–249
  PGEs, 249–251
  zinc, 252
Transmission efficiency. *See* Transport efficiency
Transmission electron microscopy/EDX (TEM/EDX), 332–333
Transport efficiency, 167
Tree bark, 76
  as biomonitor of air pollutants, 76–78
  examples of biomonitoring studies of inorganic and organic analytes, 79t–81t
TREs. *See* Traffic-related elements (TREs)
Tritium, 291
TSP. *See* Total suspended particulate (TSP)
Tube sampler, 18
TWA sampling. *See* Time-weighted average sampling (TWA sampling)
Two dimensional GCetime-of-flight mass spectrometry (GCxGC-TOFMS), 99
Two-dimensional gas chromatography
  with quadrupole mass spectrometry, 303f
Two-dimensional gas chromatography (GC X GC), 275–277, 278f
TXRF. *See* Total reflection X-ray fluorescence (TXRF)

U

U.S. Environmental Protection Agency (EPA), 20–22, 120–121, 159–160, 285
UCM. *See* Unresolved complex mixture (UCM)
Ultrahigh purity (UHP), 123–124
Ultrasound-assisted extraction (USAE), 88–91. *See also* Microwave-assisted extraction (MAE)
  inorganic analytes extraction, 91–93, 92t
  organic analytes, 93, 94t

Uninterrupted power supplies (UPSs), 204
United States (US), 120, 185
University Research Glassware Corporation systems (URG Corporation systems), 162–163
Unresolved complex mixture (UCM), 290
UPSs. *See* Uninterrupted power supplies (UPSs)
Uptake rates, 24, 26t
Urban aerosols characterisation, 336

V

Vacuum ultraviolet photoionisation (VUV photoionisation), 322
Van Deemter equation, 269–270, 270f
Vapour generation (VG), 256
Velocity theory, 269–270
Very volatile organic compounds (VVOCs), 111
VG. *See* Vapour generation (VG)
VOCs/SVOC in-field concentration, 132–143
Volatile organic compounds (VOCs), 111, 126, 209–210, 253–254, 291–292, 300–302, 312
Volatility, 111
VUV photoionisation. *See* Vacuum ultraviolet photoionisation (VUV photoionisation)
VVOCs. *See* Very volatile organic compounds (VVOCs)

W

Whole-air sampling, 112–132
  with bags, 112–120
    advantages/limitations, 119–120
    applications, 116, 118t
    methodology, 113–116
    preparation, 116
    storage, 119
    types, 116, 117t
  with canisters, 120–130, 121f
    advantages/limitations, 128–130
    applications, 126
    methodology, 120–122
    preparation, 123–130
    sample preconcentration, 130–132
    storage, 126–128
Wind erosion, 245
World Health Organization (WHO), 111
World Meteorology Organization (WMO), 17–18, 193

X

X-ray diffraction analysis (XRD), 251
X-ray fluorescence (XRF), 241, 260–261
XAD-2® sorbent, 139
XAD-4-based denuder systems, 173
XAD-4-coated commercial annular denuders, 171

Z

Zinc, 252

Kolibri 360 GmbH
Ferdinand-Jühlke-Straße 7
99095 Erfurt, Deutschland
produktsicherheit@kolibri360.de